KALAMAZOO VALLEY
COMMUNITY COLLEGE

Presented By

T. Hollowell

Fundamentals of Oral Histology and Embryology

Fundamentals of Oral Histology and Embryology

D. Vincent Provenza, B.S., M.S., Ph.D., F.A.C.D.

Professor and Chairman, Department of Anatomy
Assistant Dean for Biological Sciences
Baltimore College of Dental Surgery
Dental School
University of Maryland at Baltimore

Second Edition

LEA & FEBIGER *Philadelphia*

1988

Lea & Febiger
600 Washington Square
Philadelphia, PA 19106–4198
U.S.A.
(215)922-1330

Library of Congress Cataloging-in-Publication Data

Provenza, D. Vincent (Dominic Vincent), 1918–
Fundamentals of oral histology and embryology.

    Includes bibliographies and index.
    1. Mouth. 2. Teeth. 3. Histology. 4. Embryology,
Human. I. Title. [DNLM: 1. Mouth—anatomy & histology.
2. Mouth–embryology. 3. Odontogenesis. 4. Tooth—
anatomy & histology. WU 101 P969f]
QM306.P76   1987   611′.31   87–2961
ISBN 0-8121-1081-1

Print No. 4 3 2 1

To my beloved wife, Lucia,
and especially to S.S.
and His spouse, M.S.

Preface

Traditionally, textbook revisions have been motivated by changes or modifications of concepts ensuing from new knowledge. The scope and content of such revisions, particularly those of books in the anatomic sciences, must meet the requirement of dental health care students and practitioners for the scientific information necessary to increase their clinical competence and evaluate research findings and trends. Moreover, clinicians are increasingly asking that students be provided stronger backgrounds in basic science before they embark on their clinical experiences. Thus, the tendency to limit subject matter to core information is giving way to broadening and intensifying the information taken from selected disciplines to complement the biomedical research and clinical advances evolving to accommodate dental care needs. Currently, health science educators are increasingly embracing the philosophy that the integration of basic science and clinical knowledge, wherever possible, allows increased retention of the subject matter by the student. They further believe that integration potentiates mastery of scientific principles, increases competence in problem-solving, and provides a more rational base for judgments, all of which are essential in clinical practice.

Students are often confronted with the use of older dental terms. The problem presented by terminology differences has been addressed in this revision by introducing current terms together with older ones so that a smoother transition may be effected.

The concensus among educators is that knowledge and skill-acquisition are related to repetitive experience. Thus, terms and concepts should be approached, expanded, and intensified in recurrent patterns. In this revision I have made liberal use of this principle of learning (repetitive association).

To this end, I have sought, in this edition of FUNDAMENTALS OF ORAL HISTOLOGY AND EMBRYOLOGY, to increase selectively the knowledge base of the student while concomitantly, with the aid of my clinical colleagues, *increasing* the relevance of the discipline.

Chapter 1, *"The Cell,"* presents a review of cell functions, morphology, cell cycles, and mitosis, especially as they relate to organelles and inclusions. The second chapter deals with the basic tissues and briefly with histogenesis. Chapter 3 involves embryology, especially of the face, oral, and paraoral structures, among which are the checks, nose, palate, sinuses, tongue, tonsils, and salivary glands, including their soft- and hard-tissue components. Chapter 4 concerns odontogenesis (formation of enamel, dentin, and dental pulp). It also includes the formation of the tooth's attachment apparatus (cementum, periodontal ligament, alveolar process, and attachment epithelium), the chronology of dentition, and the eruption and replacement of primary teeth. Chapters 5, 6, and 7 describe the definitive dental tissues (enamel, dentin, and dental pulp). The definitive tissues comprising the attachment apparatus are described in Chapter 8. Chapter 9 presents a clinically focused discussion of the lip, oral mucosa, salivary glands, tongue, tonsils, nasal passage, and sinuses. Chapter 10 is concerned with the temporomandibular joint.

The chapters concerned with cytology and basic tissues have been retained to provide instant accessibility of information and thus relieve students of oral histology of the necessity of consulting other sources.

Many of the figures in the text have been published previously in *Oral Histology: Inheritance and Development* editions 1 and 2 and *Fundamentals of Oral Histology* edition 1.

With the expansion of information presented in this edition, I have sought to maintain a simplicity of writing style in order to facilitate comprehension and accommodate diverse learning needs.

Baltimore, Maryland   D. Vincent Provenza

Acknowledgments

Without the help and encouragement of my colleagues and friends, publication of this book, the second edition of FUNDAMENTALS OF ORAL HISTOLOGY AND EMBRYOLOGY, would have been extremely difficult. Accordingly, I wish to acknowledge the contributions of my esteemed colleagues and friends.

Chapter 2, Basic Tissues, was written by:

Werner Seibel, B.A., M.A., Ph.D.
Associate Professor, Department of Anatomy
Baltimore College of Dental Surgery
Dental School
University of Maryland at Baltimore

For the sections entitled "Clinical Implications" in Chapters 7 and 10, I wish to acknowledge:

Chapter 7. Dental Pulp

Thomas Dumsha, D.D.S.
Associate Professor, Department of Endodontics
Baltimore College of Dental Surgery,
Dental School,
University of Maryland at Baltimore

Chapter 10. Temporomandibular Joint

Daniel M. Laskin, D.D.S., M.S.,
Professor and Chairman, Department of Oral and Maxillofacial Surgery

Director, Temporomandibular Joint and Facial Pain Research Center,
Medical College of Virginia

For reviews of selected chapters, I am indebted to:

Chapter 3. Development of Orofacial Structures

Malcolm C. Johnston, D.D.S., Ph.D,
Professor, Department of Anatomy,
School of Medicine and Dentistry,
University of North Carolina

Chapter 4. Development of Dental and Paradental Structures
Chapter 5. Enamel

John D. Termine, Ph.D.,
Chief, Bone Research Branch,
National Institute of Dental Research,
National Institutes of Health

Chapter 8. Attachment Apparatus

> Lawrence S. Freilich, D.D.S., Ph.D,
> Assistant Professor, Departments of Anatomy and Periodontics,
> Georgetown University,
> School of Medicine and Dentistry

> John J. Bergquist, D.D.S.,
> Professor and Chairman, Department of Periodontics,
> Baltimore College of Dental Surgery,
> Dental School,
> University of Maryland at Baltimore

Chapter 10. Temporomandibular Joint

> B. Antonia Balciunas, D.D.S., M.S.D.,
> Assistant Professor, Department of Oral Diagnosis

> Melvyn A. Steinberg, D.D.S.,
> Assistant Professor, Department of Fixed Restorative Dentistry,
> Baltimore College of Dental Surgery,
> Dental School, University of Maryland at Baltimore

I am most grateful to Craig Aebli for many of the new drawings, to Dennis Driscoll and Leslie McCleary for their technical assistance, and to Edna Seitz for preparation of the manuscript.

I am especially thankful for the cooperation and support provided by Dean Errol L. Reese, Baltimore College of Dental Surgery, Dental School, University of Maryland at Baltimore.

I acknowledge a great debt to the many authors and publishers permitting the reproduction of the illustrative material in the original or modified forms, and I extend my sincere thanks to my editors and publisher, Lea & Febiger, for their patience, assistance, and kind cooperation.

Finally, I am most indebted to my beloved wife, Dr. Lucia Serio Provenza, for her encouragement, patience, cooperation, and especially her editorial expertise.

D.V.P.

Contents

CHAPTER 1

The Cell

The branch of biology dealing with the structure and interrelationships of the various parts of the body is known as *anatomy*. It is divided into two parts: *macroscopic (gross anatomy)* and *microscopic*. The latter, which requires microscopic magnification, is subdivided into several disciplines: *embryology* (developmental anatomy), *histology* (tissues), and *cytology* (cell ultrastructure and fine structure). The biologic science concerned with the transmission or receipt of hereditary traits is known as *genetics*. The study of body form and structure is called *morphology*. Disease-induced changes in the structure and function of body components are studied in the discipline of *pathology*.

Cells possess certain functional (physiologic) characteristics, traditionally known as *life functions*. These include irritability, conductivity, contractility, absorption-assimilation, secretion, respiration (oxidation), and reproduction. Irritability is the ability of cells to react to stimuli, which may be chemical, electrical, or physical. It may be associated with the properties of conductivity and/or contractility. For example, a stimulus may be transmitted over the cell surface to arrive at and stimulate other cell parts. In some cells the response is one of contraction.

Furthermore, all cells require nourishment for energy and growth. Food for these activities enters the cells by the process known as absorption. The material that is then made part of the cell is said to be assimilated. Assimilated food may be used to form various internal cell structures, or it may be used to produce new products to be secreted by the cell. Food material not assimilated by the cell may be oxidized for energy using the oxygen taken into the cell by the process of respiration. The by-products of oxidation are wastes, and are excreted. Absorbed food, as appropriate, may be used for

growth, for the formation of other cells (reproduction), or for both.

Cells are the fundamental units of life; there are over 50 billion in the adult body. Of these, about 50 million die and are replaced every second. Cells share many cytologic features (see discussions titled Organelles; Inclusion). Structural and functional dissimilarities occur among cells, which may be of sufficient magnitude to categorize them into special groups, called *tissues*. Cells of a given tissue must be structurally similar and must serve in one or more similar/complementary functions. There are four basic tissue groups: epithelial, connective, muscle, and nervous. Epithelial cells form surface coverings or internal linings for the body; connective tissue elements support, bind, and pack body structures; muscles are contracting structures; and nerve cells are excitatory or stimulating. Basic tissues assume specialized forms. For example, fat, blood, bone, cartilage, and certain dental tissue (cementum, dentin, and pulp) are specialized subgroups of connective tissue. Different tissues and/or their subgroups are organized into organs. Finally, different organs that are functionally contributory may be arranged into *systems*. The system with which the mouth and its associated organs—for example, teeth, salivary glands, tongue, tonsils—are functionally associated is known as the digestive system. Clinically, the functional organization composed of the muscles of mastication, nervous components, and other organs related to the oral cavity is referred to by some oral histologists as the "stomatognathic" system (Gr. *stoma*, mouth; *gnaths*, jaw). The science of *cytology* is concerned with the study of the cell, *histology*, with tissues, and *organology* with organs. This chapter is concerned with these three subdisciplines of the anatomic sciences, especially as they relate to the stomatognathic system.

Cells vary greatly in shape. Generally, cells that are free and isolated are spheroid; others may be columnar, cuboidal, discoidal, pyramidal, astral, fusiform, or spindle-shaped. For the most part, cell shape or morphology is a product of cell function. Furthermore, the shape of cells may be altered by environmental conditions, such as crowding. For example, solitary or loosely aggregated fat cells may be spheroid or oval but, when crowded, they may be flattened or irregular in shape. Similarly, red blood cells in an unrestricted environment are biconcave discs; however, when traveling through blood vessels smaller in bore than their diameter, they bend in half to facilitate their passage through these minute channels.

Cells vary not only in shape but in size. Most body cells are so small that they can be studied only when magnified. Two magnifying instruments are commonly used—the *light (optical) microscope* and the *electron microscope*. The optical microscope permits less detailed examination, because only magnifications up to 2000 diameters are possible. Some cell parts, such as the nucleus, can be seen with optical microscopy, but others cannot because they are so small that their resolution lies beyond the magnifying limits of the light microscope. To study the fine structure—*ultrastructure*—of cells, an electron microscope must be used (Fig. 1–1). This instrument achieves magnifications of 200,000 diameters, thereby permitting detailed study of the most minute structures, especially that of unit membranes (see below). Objects observed by optical microscopy are measured in microns (μ) or micrometers (μm)* ($1~\mu$m = 0.001 mm = 0.000039 inch). Human cells range in diameter from 10 to 25 μm). Structures observed electron microscopically are measured in angstroms (Å), preferably in nanometers (nm) ($1~nm = 10~Å = 0.001~\mu$m).

Tissue Processing

Tissue slices ("thick" sections) several micrometers thick may be studied using different types of microscopes, of which the most common is the optical microscope. The electron micro-

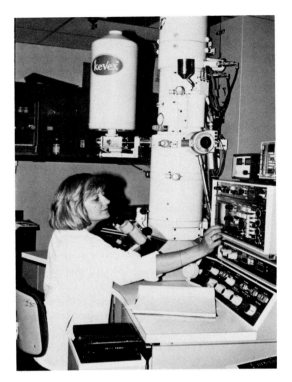

Fig. 1–1. Electron microscope and operator.

scope, through examination of "thin" sections, permits the study of the cell's fine structure, or ultrastructure. Except for light and electron microscopy, other magnifying systems are not particularly important to the student of oral histology and embryology. Furthermore, excluding phase microscopy, which permits direct examination of fresh tissue sections, all other systems require tissue processing to prepare histologic sections for light and electron microscopic inspection. Six steps are involved in tissue preparation for light and electron microcope examination: fixation, dehydration, embedding, sectioning, mounting, and staining. Histologic techniques require that tissue samples be fresh, small, carefully excised and handled, and fixed. These requirements are essential to reduce artifacts (artificial conditions). Living tissues may be stained by special dyes injected into the organs (tissue) to be studied. This process is known as supravital staining, and the dyes are known as vital stains. Fixatives coagulate protoplasm and attempt to prevent degeneration of the cytoplasmic components. Additionally, they may enhance the staining properties of the cell. Frequently used laboratory fixatives include for-

---

*1 μ is equivalent to 1 μm, but we will be using the more generally accepted μm in this text.

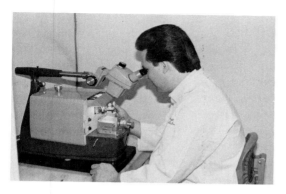

Fig. 1–2. Ultramicrotome with operator engaged in thin sectioning.

malin, glutaraldehyde, osmium tetroxide, and others composed of various combinations of alcohols, acids, bichloride of mercury, and potassium bichromate. Subsequent steps involve washing the tissue free of most of the fixative, by dehydration via an increasing graded alcohol series, and "clearing" the tissue of the dehydrating solution using benzene, cedar oil, chloroform, or xylol. The tissue segment is then infiltrated by liquid paraffin, celloidin, or a plastic, which on solidification can be sliced into sections of desired thickness. For most studies, section thicknesses may range from 3 to 12 μm. These are obtained using an instrument, the microtome, which cuts thin slices of tissue, using a mechanism that advances at precise settings. Preparatory to staining, serial or single sections may be floated in a warm water bath, collected onto egg albumin-coated glass slides, and dried on a warming table. Because most stains are dissolved in water, the paraffin in these histologic sections must be removed by immersing the slides in a decerating agent, such as toluol or xylol. Following deceration, the slides are immersed in an alcohol series of decreasing concentration and then stained. Preparation of tissues for electron microscopic study follows the same steps as those for optical microscopy. There are several differences, however. For example, the tissue segments must be much smaller and extremely fresh. Also, the tissues are usually double-fixed with formalin-glutaraldehyde and osmium tetroxide. The embedding medium is a plastic, usually Epon, Araldite, or Durcupan, and an ultramicrotome equipped with a diamond knife is used to obtain "thin" sections from 25 to 50 nm thick (Fig. 1–2). These are mounted

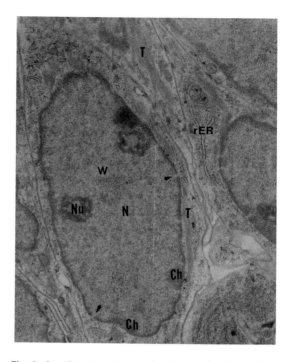

Fig. 1–3. Electron micrograph of segments of several adjacent cells showing the cytosol (cell matrix) containing rough-surfaced endoplasmic reticulum (rER), free ribosomes, and mitochondria. Note that the karyoplasm of the nucleus (N) contains uniformly distributed chromatin (W) suspended in its karyolymph. Also shown are the nucleolus (Nu), chromatin masses (Ch), nuclear pores *(arrowheads)*, and tonofilaments (T). (Uranyl acetate and lead citrate stain; ×5,000) (Provenza, D.V., and Sisca, R.F.: Electron micrograph study of human dental primordia. Arch. Oral Biol., *16*:121, 1971.)

on copper grids and stained with uranyl acetate, lead citrate or both.

General Considerations

The living material of which cells are composed has traditionally been referred to as *protoplasm*. It is composed of two basic substances: cytoplasm, constituting the entire milieu of the cell excluding the nucleus, which is comprised of karyoplasm (Fig. 1–3). Components of the cytoplasm include organelles and inclusions suspended in a matrix, the cytosol, also known as cytoplasmic matrix. Organelles are living components of the cell suspended in its cytosol (Fig. 1–3). They are also called organoids, or "tiny organs" and, just as the heart, brain, liver, and other such structures are necessary components of the body, so too are organelles important to cell function. These cytoplasmic structures are present in varying population densities according to age, health, functional state, and competency, and other factors involved in the cell's life activities. Among the organelles present in a cell are the nucleus, mitochondria, Golgi complex (apparatus), ribosomes, endoplasmic reticulum (rough and smooth), vesicles, lysosomes, centrosomes, microbodies, microtubules, and filaments (Fig. 1–4).

Inclusions are present in the cytoplasm either as transient or permanent substances; they include carbohydrates, proteins, pigments, crystalloids, lipids, gas, and other metabolites or foreign materials. Both organelles and inclusions form a colloidal suspension in the cytosol.

Some cytologic components cannot be observed because they may have been removed during tissue preparation for histologic slides. For example, lipids are present in membranes of organelles. Hence, when preparing tissues for histologic slides, the organization of the membrane system will be altered or destroyed if a fat solvent is used, thereby affecting resolution of the cell components. On the other hand, certain techniques do not require embedding in wax for sectioning, so fat-dissolving chemicals are not needed. In such instances, membrane organization remains unchanged and, if the tissue is spared other damaging factors (such as those given below), the membrane portions of the organelles remain intact.

Cells cannot grow indefinitely in size. Rather, definite limits exist that are believed to be governed by cell volume and surface area.

This chapter is concerned with the structural features of cells and their perpetuation through mitotic activity. Coats, basement membranes, and junctional complexes of cells will be discussed in Chapter 2.

ORGANELLES

Plasmalemma

Employing optical microscopy, the limiting cell (plasma) membrane, or *plasmalemma*, is not easily observed. With extremely high magnification (+100,000), obtained electron microscopically, the limiting membrane is seen as a single unit comprised of a light lipid central layer, from 3 to 3.5 nm thick, sandwiched between two dark protein layers, 2 to 4 nm thick. The single membrane comprised of the two pro-

tein laminae and the central lipid lamina constitute the unit membrane, which is trilaminar (trilamellar) (see Fig. 2–6). Incorporated into the unit membrane may be found protein particles. Although the concept of the protein-lipid complex for the unit membrane is the most widely accepted, other theories have been proposed that allow for differences in the structure and organization of the plasma membrane. The plasmalemma is semipermeable, and controls exchanges between the internal and external cell environments. For the most part, the cell membrane forms a continuous boundary for the cell and, although occasional local breaks occur that interrupt the continuity of the membrane, they are quickly repaired by membrane-limited (-bound) organelles, thus avoiding excessive loss of cytoplasm. Most organelles are membrane-bound; therefore their contents are structurally isolated from chemicals in the cytosol. The membranous organelles are discussed below. Their unit membranes are structurally similar to those of the plasmalemma.

Nucleus

This is the most prominent organelle in the cell (Figs. 1–3 and 1–4). The nucleus can be easily resolved with optical microscopy, even at lower magnifications. Except for the mature red blood cell, it is present in all cells. Some cells, such as osteoclast, skeletal muscle, and liver parenchyma, contain more than one nucleus. The shape of the nucleus generally conforms to that of the cell. Within cells of unrestrained space, nuclei are round. Nuclei of cells in which the internal volume is more restricted tend to reflect the cells' profile; in these the nuclei are oval to flat. Additionally, the shape of the nucleus may vary depending on the functional state of the cell. For example, in the relaxed smooth muscle cell the nucleus is oval; however, during contraction, the nucleus may be wrinkled because of the shortened condition of the fiber. Similarly, in some cuboidal or columnar secretory cells, as secretions rise and accumulate at the secretory surface the nucleus is pushed basalward, and may be compressed into a flattened or even a concave mass against the basal cell membrane. After the secretion has been released, the nucleus becomes round again and assumes a more central position in the cell.

Optical microscopic examination reveals that

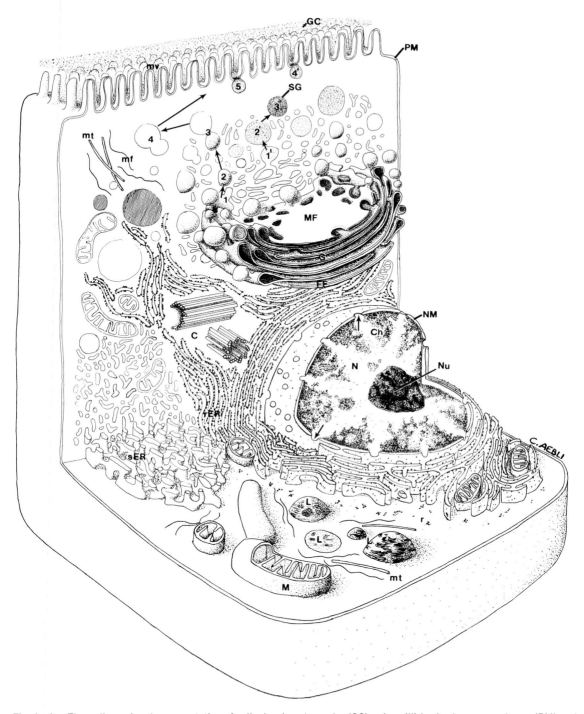

Fig. 1–4. Three dimensional representation of cell, showing glycocalyx (GC), microvilli (mv), plasma membrane (PM) and organelles, with secretion granules (SG), secretory path (1–5 and 1'–4'), Golgi system (G), Golgi maturing (trans) face (MF), Golgi-forming (cis) face (FF), centrioles/centrosome (C), nucleus (N), nucleolus (Nu), nuclear pores *(arrows),* nuclear envelope (NM), chromatin (Ch), mitochondria (M), ribosomes (r), smooth-surfaced endoplasmic reticulum (sER), rough-surfaced endoplasmic reticulum (rER), lysosome (L), microtubules (mt), and microfilaments (mf).

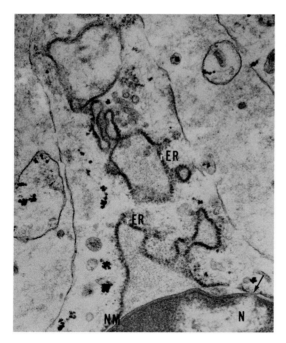

Fig. 1–5. Segment of nucleus (N) showing nuclear envelope (NM) and nuclear pore *(arrow)*. Note that the outer leaflet of the nuclear envelope is continuous with the membrane of the rough-surfaced endoplasmic reticulum (rER). (Uranyl acetate and lead citrate stain; ×11,500) (Sisca, R.F., Provenza, D.V., and Fischlschweiger, W.: Ultrastructural characteristics of the human enamel organ in an early stage of development. J. Baltimore Coll. Dent. Surg., 22:18, 1967.)

nuclei not engaged in cell division (mitosis) contain loosely organized chromatin of two varieties, heterochromatin and euchromatin. The latter is active metabolically, providing signals for protein synthesis. The former, however, is not believed to be involved in these activities. Chromatin is composed principally of deoxyribonucleic acid (DNA). The intense staining round body located off-center in the nucleus is known as the nucleolus (Figs. 1–3 and 1–4). It is composed mostly of ribonucleic acid (RNA), and is especially prominent in younger cells, cancer cells and those actively engaged in protein synthesis.

The limiting membrane of the nucleus, the *nuclear envelope*, is tough and elastic. Examined with the electron microscope, it is found to consist of two unit membranes, the *outer* and *inner membranes*, each 7.5 nm thick, with an intervening space, the perinuclear cistern, which is between 40 and 70 nm wide. Euchromatin adhering to the inner membrane is responsible for

patency of the nuclear envelope with optical microscopy. The membranes of the envelope are not continuous; rather, they may be interrupted at intervals to form round gaps, the nuclear pores (70 nm in diameter) (Figs. 1–3, 1–4, and 1–5). The pores are not open channels; rather they are spanned by a protein diaphragm, single layered and electron dense. It is entirely possible that the pores are sites at which materials are selectively exchanged between the nucleus and the cytoplasm. The outer membrane of the nuclear envelope becomes continuous with the membrane of the rough endoplasmic reticulum, which is involved in protein synthesis (Figs. 1–4 and 1–5). Additionally, the nuclear envelope may exhibit outpocketing, which increase the surface area of the nucleus. The unit membranes of the nuclear envelope are formed by the fusion of minute spherules called vesicles, which are derived from the endoplasmic reticulum. Chromatin and nucleoli are suspended in a nuclear matrix. The latter is fibrillar, forming a nucleoskeleton extending to the inner membrane, where it forms a *fibrous lamina*. A description of the nuclear material in dividing cells will be provided in the discussion of mitosis (below).

Mitochondria

These organelles accommodate the cell's energy needs by producing adenosine triphosphate (ATP), which results from cell respiration. The shape, size, and structural complexity of a mitochondrion vary from cell to cell. These features appear to be related to the cell's metabolic needs. They are, therefore, most abundant in areas of oxidation and storage, or in those in which high energy sources are required. The shape observed most frequently is round or oval, although irregular profiles are not uncommon (Fig. 1–6). For example, mitochondria studied in living cells by phase contrast microscopy, at magnifications approaching several thousands of diameters, have been observed to alter their shape and location. The lengths of mitochondria range from 0.4 μm to several micrometers, although in extremely active cells they may attain lengths up to 10 μm. Such lengths permit their resolution with optical microscopy, either using live cells with the phase contrast microscope or using special stains (supravital), such as Janus green, with the transmission optical microscope.

Ultrastructurally, a mitochondrion is com-

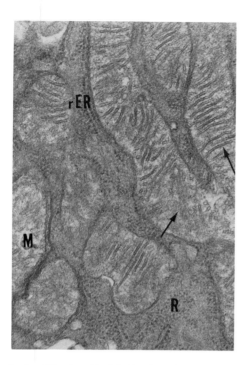

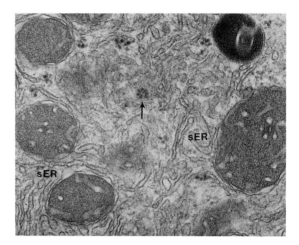

Fig. 1–7. Smooth-surfaced endoplasmic reticulum (sER) with rosette aggregations (arrow) of ribosomes (polysomes). (Uranyl acetate and lead citrate stain; × 53,000)

Fig. 1–6. Mitochondria (M) showing orientation of cristae *(arrows)* and matrix. Also present are free ribosomes (R) and rough-surfaced endoplasmic reticulum (rER). (Uranyl acetate and lead citrate stain; × 16,000)

posed of two membranes separated by a space approximately the width of the membranes. The outer membrane, which is about 7 nm thick, is quite smooth (Figs. 1–4 and 1–6), while the inner membrane, also 7 nm thick, is internally folded to form shelves, or cristae. Cristae are usually oriented transversely across (Fig. 1–6) the mitochondrion. Not uncommonly, however, they are longitudinally, obliquely, or circularly arranged. Furthermore, the number and size of the cristae vary according to cell type. The distance the cristae may extend in any one direction also varies. In very active cells the cristae are densely packed. The interior of a mitochondrion is fluid-filled (the mitochondrial matrix), and round to oval magnesium or calcium matrix granules may be suspended in it. Mitochondria also contain DNA and RNA, that are somewhat different from the intracellular DNA and RNA components.

Endoplasmic Reticulum

This organelle, consisting of squamoid canals, cisternae, sacs, and vesicles of trilaminar unit membrane composition, occurs in networks or parallel arrays. Its membrane is sieve-like in appearance and may be encrusted with *ribosome granules*—hence the name *rough-surfaced (granular) endoplasmic reticulum* (rER) (Figs. 1–4 and 1–5). If ribosomes are absent it is called *smooth-surfaced (agranular) endoplasmic reticulum* (sER) (Figs. 1–4 and 1–7). The walls of the reticular components are between 6 and 7 nm thick, and their luminal diameter is between 10 and 60 nm. The lumina of the rER contain protein, and those of the sER contain steroid. Elaborated proteins of moderate electron density are often resolved within the lumina of the cisternae of the rough endoplasmic reticulum. The elements of the smooth-surfaced system are shorter and more elaborately branched, and do not include cisternae. Some cytologists believe that the ribosomes may become attached or detached from the walls of the reticulum, depending on specific cell needs.

Ribosomes

These are often observed free in the cytoplasm, dissociated from the endoplasmic reticulum (Figs. 1–6 and 1–7). Except for red blood corpuscles, they are located in all cells. At the optical level, especially in actively secreting cells that have been stained with hematoxylin and eosin, they impart a blue hue (basophilia) to the cell segment in which they are concentrated. Ribosomes at this level are called chromidial material, or ergastoplasm. Ribosomes are dense irregular or oval granules that range in diameter from 15 to 25 nm (Fig. 1–7). They may occur singly or may be arranged as rosettes or helices, called polysomes or polyribosomes. Individual ribosomes of a polysomal group may demonstrate specific arrangement of its particle complement on delicate threads of messenger RNA (mRNA). About 80% of the total RNA in cells is present as ribosomal RNA (rRNA), while 16% is found as transfer RNA (tRNA) and 2% as messenger RNA (mRNA). Polyribosomes are believed to be responsible for the production of cytoplasmic proteins used in the division and growth of cells. This is unlike the ribosomes attached to the endoplasmic reticulum, which elaborate proteins channeled to the Golgi apparatus for secretion.

Annulate Lamellae

When present, annulate lamellae are found in developing germ cells and certain epithelial cells, and in the pyramidal cells of neuroblasts in the cerebral cortex. They are associated with the rER and with the leaflets of the nuclear envelope. Some cytologists believe that they originate from the latter. They are comprised of double membranes 7 to 9 nm thick that surround a flattened luminal space about 45 nm in diameter. Serially spaced, at distances of 100 to 200 nm along the length of the lamellae, are found annuli, or pores, a characteristic feature of these organelles (Fig. 1–8). Three possible functions have been assigned to these structures. They may function as conduits for the transfer of information from the nucleus to the cytoplasm, speed up protein synthesis during neuron development, and participate in metabolite transportation, modification, and storage.

Golgi Apparatus

Cells of tissues preserved and prepared using osmium tetroxide or chromium impregnation techniques show a perinuclear network of organelles known as the *Golgi complex (apparatus)*. Electron microscopic examination reveals these to be comprised of units or packets of stacked, flattened, slightly convex membranous saccules or cisternae (Figs. 1–4 and 1–9). The outer convex surface is called the cis, or "forming face," while the concave or trans aspect is the "maturing face" (Figs. 1–4 and 1–10). Small vesicles are often observed at the forming face, while the larger ones are associated with the

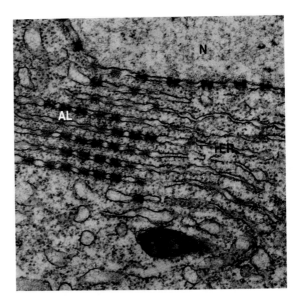

Fig. 1–8. Electron micrograph showing annulated lamellae (AL), rough-surfaced endoplasmic reticulum (rER), and nucleus (N). (×25,000) (Kessel, RG.: Annulate lamellae. J. Ultrastruct. Res., Suppl. *10*:4, 1968.)

edges of the stacked saccules or with the maturing face. Both the saccular and vesicular walls are of unit membrane construction. The diameter of the saccule is about 15 nm. The Golgi complex has been implicated in the modification and packaging of the products synthesized by the ribosomes of the rER (Fig. 1–10). These are subsequently directed to the forming face of the Golgi complexes by the small vesicles, and are released as secretion vesicles (granules) (Figs. 1–9 and 1–10).

Secretory Vesicles (Granules)

These are spheroid cytoplasmic structures that tend to coalesce en route to the secreting surface of the cell, at which they may accumulate for discharge (Fig. 1–10). Tissue fixation procedures effect precipitation of the vesicles' contents so that they may assume granular profiles. The secretory granules may be resolved with optical microscopic techniques in actively secreting cells of the salivary glands, pancreas, and other secretion-producing organs. Electron microscopic study reveals these cytoplasmic structures to be prosecretory vesicles budded from the Golgi apparatus, including the GERL saccules at the external most lamella of the maturing face (Figs. 1–4 and 1–10).

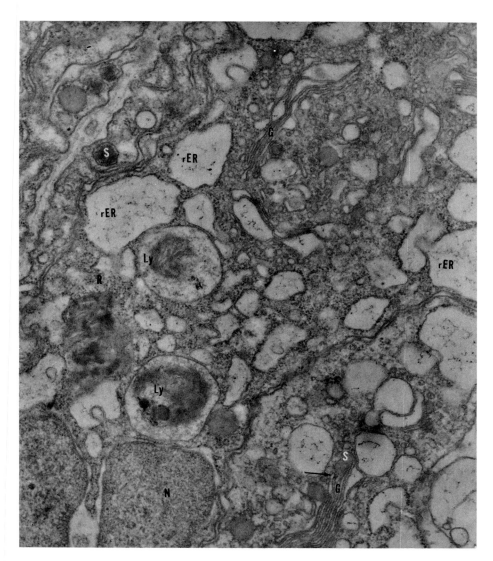

Fig. 1–9. Electron micrograph showing Golgi cisternae (G), with the formation of a secretion granule (S) at its terminal *(arrow)*. Also shown are dilated cisternae of rough-surfaced endoplasmic reticulum (rER), free ribosomes (R), lysosomes (Ly), and nuclear (N) segments intimately associated with elements of the rER. (Uranyl acetate and lead citrate stain; ×42,000)

GERL Saccule

In some cells the outermost saccule at the maturing (secretory) face of a Golgi system differs from its sister saccules in that it houses acid phosphatase, an enzyme normally associated with lysosomes. Furthermore, these saccules are intimately associated with neighboring cisternae of the rER. Because of these features, saccules of this type are designated specifically as *GERL* (*G*, Golgi apparatus; *ER*, rough-surfaced endoplasmic reticulum; *L*, lysosome). Expectedly, GERL saccules produce vacuoles that presage and are followed by lysosomes. The precise relationship of the GERL saccules, if any, to the Golgi complex and/or the rER is unclear.

Lysosomes

These organelles are bound by a single unit membrane and, except for erythrocytes (red blood cells), they may be found in all cells, especially defense and filter cells, such as macrophages and leukocytes (white blood cells). They are usually round or oval. However, because they may attack foreign bodies regardless

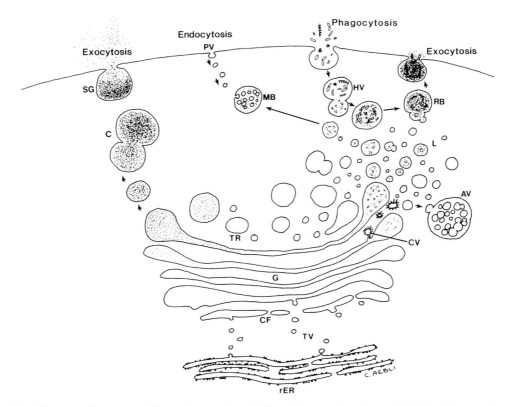

Fig. 1–10. Diagrammatic representation of the secretory *(left)* and lysosomal pathways *(right)* involving rough-surfaced endoplasmic reticulum (rER), transport vesicles (TV), forming (cis) face (CF), Golgi complex (G), maturing (trans) face (TR), coated vesicle (CV), autophagic vacuole (AV), multivesicular body (MB), residual body (RB), heterophagic vesicle (HV), primary lysosome (L), pinocytotic vesicle (PV), secretory granule (SG), and condensing vacuole (C). (Adapted from several sources.)

of dimension or shape, the morphology of the lysosome is often determined by the size and shape of the ingested particle(s). Lysosomes are important in that they contain powerful enzymes capable of selective hydrolysis or digestion of proteins, fats, carbohydrates, and other substances of intra- or extracellular origin. Lysosomes occur as nonactive primary lysosomes and as active secondary lysosomes. The latter are of two varieties, heterolysosomes and autolysosomes (Fig. 1–10). Heterolysosomes, also known as phagolysosomes, are produced by the fusion of a primary lysosome with a phagosome, which is a vacuole formed by the cell during the ingestion of foreign particles. Another type of secondary lysosome is known as a multivesicular body, which is between 0.4 and 3.0 μm in diameter (Fig. 1–10). It is membrane-bound and houses numerous vesicles believed to be enzyme-filled, probably with acid phosphatase. Neither their origin nor ultimate fate is known,

although it is believed that they are produced by pinocytotic vesicles. Autolysosomes are formed by the fusion of primary lysosomes with autosomes which are vacuoles containing various substances, located in the cytoplasm (Figs. 1–10 and 1–11). Following digestion, autosomes are reduced to organelle remnants known as residual bodies. Lysosomes can be identified electron microscopically by cytochemical techniques that localize catabolic enzymes such as alkaline and acid phosphatases. Lysosomes are produced in a manner similar to that of secretory granules. Thus, the catabolizing enzymes synthesized by the rER are directed via vesicles to the cis face of the Golgi complex. As the hydrolytic enzymes contained in and released by the vesicles are channeled through the Golgi apparatus, and arrive at the trans face, they are enveloped by the distended saccules and released at the edge of the uppermost saccules as lysosomes. Other than phagocytic and autolytic (self-destructing)

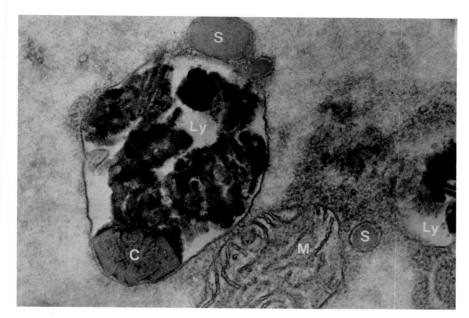

Fig. 1–11. Lysosome (secondary lysosome or autophagic granule) (Ly) containing crystal (C) and other debris. Note the intimate association of the secretion granule with the lysosome (M, mitochondrion; S, secretion (granule) vesicle). (Nylen, M.U., Provenza, D.V., and Carriker, M.R.: Fine structure of the accessory boring organ of the gastropod, *Urosalpinx*. Zoologist, 9:935, 1969.)

activities, lysosomes remove organelle and inclusion excesses in the cell.

Coated Vesicles

Cells in the process of pinocytosis—that is, ingestion of extracellular products—become enveloped by a coating. The cell membrane forms small invaginations known as coated pits. After these cytoplasmic inpocketings or pinocytotic vesicles are pinched off as free cytoplasmic vesicles, the term "coated vesicle" is used (Fig. 1–10). Special electron microscopic study reveals their bristly coats, composed of a network of a protein called clathrin and polygonally arranged over the limiting membrane of the vesicle. Other than the limiting cell membranes, coated vesicles may originate from organelles such as the rER, Golgi complex, and secretory vesicle (Figs. 1–10 and 1–12). Coated vesicles function as carriers, delivering products intracellularly from organelle to organelle.

Microbodies

These organelles, also called peroxisomes, are found in liver parenchyma, epithelial cells of kidney tubules, nonciliated respiratory epithelium of bronchioles, and dentin-forming cells (odontoblasts). They are round to oval in shape (up to 1.5 μm in diameter) and unit membrane-bound, and contain enzymes. They are probably produced by the granular endoplasmic reticulum. The latter may be associated with hydrogen peroxide overproduction, and with fat to carbohydrate conversion.

Microtubules

These are unbranched hollow cytoplasmic tubules 25 nm in diameter and 3 μm or more in length (Figs. 1–4 and 1–13). They are integral components of cilia, flagella, and centrioles, and form the spindles and astral rays of dividing cells. The protein, tubulin, of which they are composed can be identified by immunofluorescent staining. In structures such as cilia, the protein, dynein, is found in the microtubules; this exerts an enzyme action associated with the ATP release of energy. The functions attributed to microtubules include the following: provision of a cytoplasmic skeleton for maintaining cell shape; provision for cell movement, and for chromosome movement during cell division; and acting as cytoplasmic conduits for intracellular flow of water, metabolites, and other molecules. The quantitative aspect of the microtubule population appears to increase with functional demands.

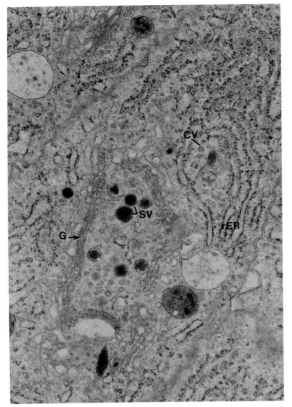

Fig. 1–12. Electron micrograph of organelles associated with the elaboration and liberation of secretory vesicles, showing rough-surfaced endoplasmic reticulum (rER), Golgi components (G), coated vesicles (cv), and secretory vesicle (sv). (Uranyl acetate and lead citrate stain; × 27,500)

Centrioles

In secretory epithelial cells, centrioles are located near the Golgi system. In other cells they may be perinuclear, in a special region of the cytoplasm referred to as the centrosome, or centrosphere. Here they form two cylindric bodies—diplosomes—oriented perpendicularly to one another (Fig. 1–4). Centriole-like structures, known as basal bodies may be found under the limiting membrane of free cell surfaces, associated with cilia and flagella (Fig. 1–14). Optical microscopy shows the centrioles to be dark-staining bodies. Electron microscopic examination reveals them as barrel-shaped, with one end closed. Their length ranges from 30 to 500 nm, and their diameter is about 150 nm. Except for a few dark granules the cytoplasm is clear. In transverse section the centriole is seen to be comprised of a dense matrix in which nine groups of three fused microtubules (triplets) are oriented circularly and at a specific angle to each other. The tubules of each triplet form three concentric circles designated as subfibers a (inner), b (middle), and c (outer). Centrioles have been implicated in cell division, in the movement of cilia and flagella, and possibly in cyclosis.

Filaments

With microtubules, filaments form the cytoskeleton of cells and participate in cell movement. They may be found in almost all cell types of most tissues. Cytoplasmic filaments range in diameter from 5 to 15 nm. They may be aggregated into larger units or fibrils, with diameters of 0.2 to 1.0 μm or more. Thus, they are within the resolution range of the optical microscope. Fibrils may organize into larger bundles called fibers. Filaments are classified as thick filaments (myosin), microfilaments (thin filaments; actin), and intermediate filaments. The latter form five subclasses and include tono- (cytokeratin) filaments, desmin filaments, vimentin filaments, neurofilaments, and glial filaments.

Thick filaments are composed mostly of myosin, a protein of contraction that is associated with muscle cells (fibers) (Fig. 1–15). They are about 15 nm in diameter or about twice the thickness of their counterpart, the thin filament. *Microfilaments*, thin filaments, are between 5 and 7 nm in diameter and contain the protein actin. These filaments are similarly located in muscle cells, and participate in contraction (Fig. 1–15). Actin filaments located beneath the plasma membrane are responsible for the ameboid movement of leukocytes, as well as for their phagocytic activity. Additionally, actin filaments make up the substance of microvilli and extend toward the cell body to the cytoplasmic territory immediately under the cell membrane known as the terminal web (Fig. 1–16). Here, actin filaments form a structural complex with myosin and the tonofilaments. Spectrin, another type of microfilament that is especially prominent in erythrocytes, connects a component of the plasma membrane (ankyrin) to adjacent actin filaments. Spectrin is also associated with microtubules, but the precise interrelationships have yet to be elucidated. Thick and thin filaments are not found exclusively in muscle fibers. Rather, they have been observed in many non-

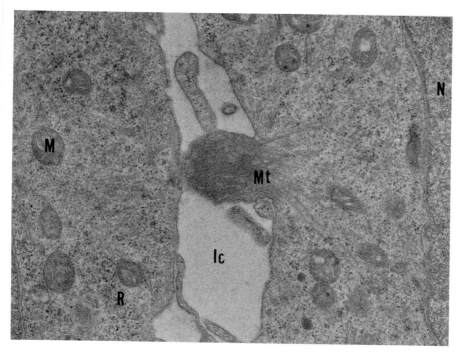

Fig. 1–13. Cell segment showing microtubules (Mt) observed in cell section coursing and converging on the cell process, which projects into an intercellular space (Ic), (N, nucleus; M, mitochondrion; R, free ribosomes). (Uranyl acetate and lead citrate stain; ×6000)

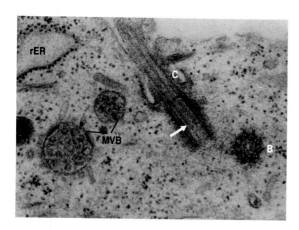

Fig. 1–14. Electron micrograph showing rudimentary cilium (C) and basal body (B). Note the microtubular arrangement in the transverse section of the basal body. In the longitudinal plane, the microtubules of the basal body and cilium appear as dark lines *(arrow)* (MVB, multivesicular body; rER, rough-surfaced endoplasmic reticulum). (×51,500)

muscle cells, in which they are believed to provide for intracellular movement of cytoplasmic structures and for stabilization of the cells' morphology.

Intermediate filaments form a group of structurally diverse filaments, 8 to 10 nm thick; hence, they range in diameter between those of the thick and thin filaments. Included in this group are the following: desmin filaments, associated with muscle cells; glial filaments, associated with the glial (supporting) cells of nerves; neurofilaments, associated with neurons (nerve cells) (Fig. 1–17A); tonofilaments (cytokeratin), associated with epithelia (Figs. 1–3 and 2–6), including cells participating in cornification (Fig. 1–17B); and vimentin filaments, associated with mesenchymal or progenitor cells and their descendants. These cytoplasmic elements mostly provide for cell support, strength, resiliency, and rigidity.

INCLUSIONS

Inclusions are not found in all cells. They are variously associated with specific cells, with a specific stage of cell development, or with spe-

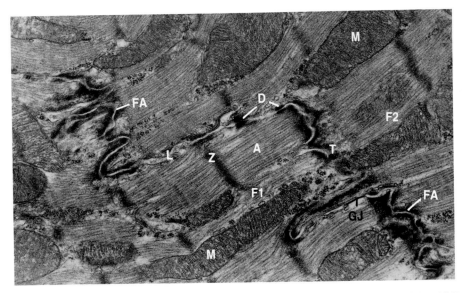

Fig. 1–15. Electron micrograph of cardiac muscle cell segments (F1 and F2). The intercalated disk exhibits a stepwise organization with transverse (T) and longitudinal (L) sections. The transverse portions occur at the Z lines of the myofibrils. The junctional units located at a disk are desmosomes (D), fascia adherens (FA), and gap junctions (GJ). The latter are more prominent at the longitudinal portions. Note the myofilaments in the myofibrils, as well as the typical striations, the A band and Z line, and mitochondria (M). (Uranyl acetate and lead citrate stain; × 20,000)

cific cell activity. Among the classes of materials referred to as inclusions are crystalline bodies, pigments, and nutrients, including lipid deposits.

Crystalline Bodies

These are suspected of being proteinaceous in composition. Iron-containing crystals recovered from worn-out erythrocytes and those found in testicular epithelium occur most frequently. When they are not free cytoplasmic bodies they may also be associated with lysosomes (Fig. 1–11), microbodies, and mitochondria.

Pigments

Pigments imparting color to cytoplasm are of two types—those that are produced by cells, endogenous pigments, and those that are taken up by cells, exogenous pigments. Carotin is a natural pigment of carrots and of certain other vegetables and fruits. This pigment is fat-soluble, and as a component in the lipid complex is called lipochrome. It is responsible for the yellow color in some fat tissue. Carotin-containing food taken in excessive amounts may cause skin to assume a yellow-red color, a condition known as carotenemia. Silver and gold salts employed

in certain heart and blood vessel disorders and in arthritis may accumulate under the skin as gray deposits. Lead poisoning is often detected by a blue streak on the gums, a condition known as the gingival lead line. Coal dust may produce discoloration of the lung, and tattooing procedures introduce pigments subcutaneously.

Lipofuscin, melanin, and hemoglobin are types of endogenous pigments. Lipofuscin imparts a yellow to bronze color in fresh tissue, especially heart, liver, and nerve tissue. The pigment may also be found in aging tissues. In histologic section it is detected with fat stains. Melanin is a nitrogenous pigment that gives tissues a light tan to dark brown color. Its production by cells, melanocytes of neural crest origin, is light-stimulated, producing enzyme conversion of nonpigmented chromogen into melanin deposits (melanosomes) in the cells. Melanosomes may be transferred to other cells (skin, iris of eye), or they may be phagocytized by macrophages (Fig. 1–18).

Hemoglobin is the red pigment of erythrocytes (see Fig. 2–60B). It is distributed more or less evenly throughout the cell, and is responsible for oxygen transport. It remains active for several months. Red blood cells that have become functionally ineffective are removed from

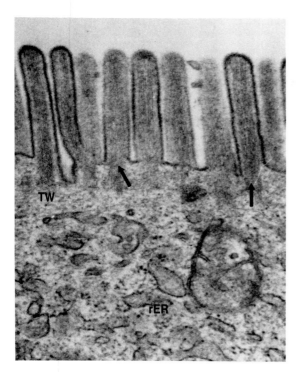

Fig. 1–16. Electron micrograph showing terminal web (TW) and microvilli with their core of actin filaments *(arrows).* Note that the organelles are located below the terminal web (rER, rough-surfaced endoplasmic reticulum). (Uranyl acetate and lead citrate stain; ×28,000)

circulation by macrophages and phagocytes in bone marrow, liver, and spleen. These cells degrade hemoglobin to an iron pigment, hemosiderin, and to an iron-free pigment, hematoidin (bilirubin). The iron component of hemosiderin is recycled in the production of hemoglobin for new erythrocytes. Bilirubin is retrieved by the liver, used in the formation of bile, and stored in the gallbladder. It is yellow-brown in color and, when oxidized, forms biliverdin, which is green in color.

Nutrients

These occur in the cytoplasm as amino acids (proteins), glycogen (carbohydrates), and lipids (fats). Amino acid coupling in protein synthesis occurs on ribosomes. Nutrients are used for the production of secretions or organelles or for metabolism and mitosis.

Glucose, the reduction product of carbohydrates, is stored in cytoplasm as *glycogen.* Except for special histologic techniques—for example, Best's carmine staining—glycogen is rinsed out of the cell so that the cytoplasm appears vacuolated. Electron microscopic study reveals glycogen as small, 15- to 30-nm particles dispersed singly in the cytoplasm as beta particles or in clusters as alpha particles (Fig. 1–17B). Ribosomes and glycogen bodies are similar in size and shape, and require the digestion of one or the other (ribonuclease or amylase, respectively) for their positive identification.

Fat cells, or *adipocytes,* are often found as individual cells or as cell groups in certain connective tissues. If they comprise the dominant cell component, the tissue is referred to as fat (adipose) tissue. In these, the fat (lipid) inclusion may occupy most of the cell volume. This inclusion is not limited exclusively to adipocytes, but may be found in cells such as those of the adrenal cortex, liver, and muscle. As cytoplasmic inclusions lipid droplets are not membrane-bound, and are referred to as liposomes. They range in size from 0.5 to 2.5 μm (Fig. 1–19). They may fuse, occupying increasingly greater volumes of the cell. Cytoplasmic fat may be used by cells as an energy source or for the lipid component in the formation of unit membranes for new or spent organelles.

CELL CYCLE

Somatic (body) cells undergoing regeneration and tissue repair require the production of new (daughter) cells from parent or (stem) cells to replace the spent and injured ones. Daughter cells mature and become specialized so that, generation after generation, cell types are perpetuated. Among the continuously regenerating cells are certain epithelial, blood, and connective tissue cells. Highly specialized cells such as adult neurons and heart muscle do not retain the ability for self-perpetuation through cell division. Growing and maturing cells are in interphase, while those engaged in division are in some stage of mitosis. These two stages, interphase and mitosis, constitute the cell cycle. The periods of the cell cycle include mitosis (M), or cell division, gap 1 (G_1), the period in which the daughter cells engage in synthesis of growth materials, synthesis (S), the period of DNA replication and production preparatory to the subsequent mitosis, and gap 2 (G_2), the period during which the production of all proteins including those of mitosis are completed.

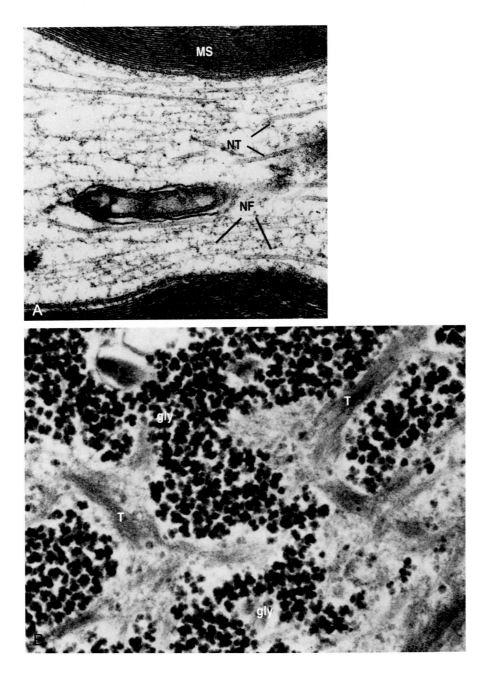

Fig. 1–17. Intermediate filaments. *A,* Electron micrograph of neurotubules (NT) and neurofilaments (NF) in an axon sectioned longitudinally (MS, myelin sheath; ×40,000) *B,* Tonofilaments (T) and glycogen masses (gly). (×20,000) (*A,* Courtesy of R.M. Meszler.)

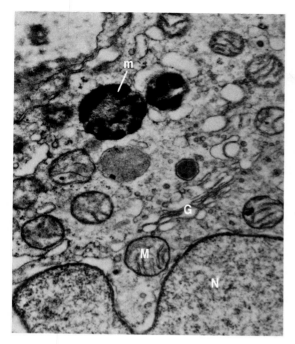

Fig. 1–18. Electron micrograph of cell segment showing melanin pigment (m) among mitochondria (M) and Golgi components (G) (N, nucleus). (× 11,500) (Courtesy of H.S. Sobel and E. Marquet.)

Cell division also occurs in the formation of sex cells, gametes (oocytes and spermatocytes), from stem cells (oogonia and spermatogonia). This process, however, is known as meiosis because, unlike mitosis, there is a reduction (halving) of the chromosome number during cell division of the parent or stem cell. That is, instead of the daughter cell in humans possessing the somatic or diploid number of chromosomes, 46 (23 pairs), each gamete possesses 23, the gametic or haploid chromosome number. Of the 23 pairs, 22 pairs are complementary and are, therefore, homologous. The other pair is dissimilar, consisting of the sex chromosomes (XX, female; XY, male). With fertilization of an ovum by a spermatozoon, the zygote (fertilized egg) will have had the somatic chromosome number restored to 46 chromosomes or 23 pairs, 23 from each parent. From this point on, all subsequent divisions will be mitotic for all cells throughout life, except of course for gametogenesis (development of gametes).

Mitosis

Mitotic division of somatic parent cells succeeds in increasing or maintaining the normal

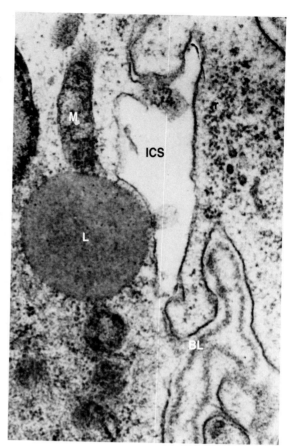

Fig. 1–19. Liposome (L) between mitochondria (M). Note the free ribosomes (r) somewhat below a narrow organelle-poor zone, the ectoplasm (ICS, intercellular space; BL, basal lamina). (× 8,000)

cell complements for tissues. The mechanism involves an equal distribution of cytoplasm, organelles, and nuclear material from parent to the two daughter cells. The division of the nuclear components is known as *karyokinesis*, while the separation of the cytoplasmic components during telophase is *cytokinesis*. These activities usually occur in an ordered sequence, but karyokinesis may occur without cytokinesis. In this event a multinucleate cell will be produced. The number of nuclei present in a single cell will, of course, depend on the number of nuclear divisions. Four stages are involved in mitosis: prophase, metaphase, anaphase, and telophase (Fig. 1–20). Their durations are $1\frac{1}{2}$ hours (prophase), 30 minutes (metaphase), 5 minutes (anaphase) and 45 minutes (telophase). Interphase may take several days, while mitosis occurs within 3 hours.

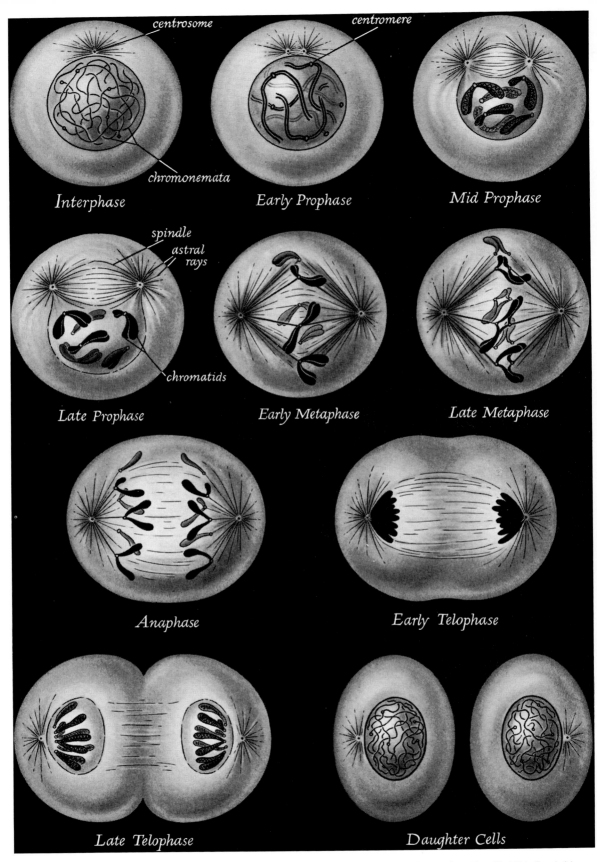

Fig. 1—20.   Diagram showing the various stages of cell division, mitosis. (Winchester, A.M.: Genetics. 2nd Ed. Cambridge, Houghton Mifflin, 1958.)

Prophase. This stage includes disorganization of the nucleolus, disappearance of the nuclear envelope, and organization of the chromatin into chromosomes (Fig. 1–20). Each chromosome is made up of two DNA threads, chromatids, joined only at the centromere sites. The double structure of the chromosomes, begun during the synthesis (S) and completed during the gap 2 (G_2) periods of interphase, become more clearly resolved as organization into chromosomes is achieved. Finally, the centrioles divide and move poleward in the cell as microtubules are formed at their microtubule organization center. They lengthen by end-to-end additions, and succeed in pushing the centrioles further apart until a continuous interpolar system of microtubules, the mitotic spindle system, has been formed. Other shorter microtubules radiate from the centrioles (astral rays) to connect terminally with the polar segments of the plasma membrane.

Metaphase. Total disappearance of the nuclear membrane and nucleolus occurs during metaphase. The chromosomes become aligned at the cell's equatorial (metaphase) plate, at which the centromeres divide and the sister chromatids are disengaged and become attached to the microtubules of the spindle system (Fig. 1–20). The 23 sister chromatids begin to move away from the equatorial plate and toward opposite poles of the cell.

Anaphase. In this stage the chromatids continue to be pulled toward the opposite poles of the cell by their centromere attachment to the spindle system (Fig. 1–20). Although some scientists believe that poleward migration of the chromatids may be due to the interaction of actin and myosin and the microtubules, the precise mechanism involved has yet to be elucidated.

Telophase. Cell elongation continues and, as this stage begins, a cleavage furrow is arranged around the equatorial segment of the cell. Here the cell divides, forming the two daughter cells (cytokinesis) (Fig. 1–20). Other changes involved in telophase include disorganization of the microtubules of the spindle and astral systems, reorganization of the nucleolus and nucleus, and distribution of the cytoplasmic constituents (cytokinesis). Uncoiling of the chromosomes occurs, forming chromatin.

CHAPTER 2

Basic Tissues

WERNER SEIBEL

The body is comprised of four basic tissues: epithelial, connective (supporting), muscular, and nervous. The germ layer derivation (ectoderm, mesoderm, endoderm) of these tissues is called *histogenesis*. Muscle and connective tissues, excluding those derived from the pharyngeal (branchial) arches associated with the head region, are derived from mesoderm. Nervous tissue arises from ectoderm, and epithelium from any of the three (Table 2–1). Sometimes the term "mesenchyme" is used to imply mesodermal heritage; however, mesenchyme as a tissue may originate from any of the three layers and, if derived from a layer other than the mesoderm, may be identified by the appropriate prefix as *ecto*mesenchyme or *endo*mesenchyme. The former is especially important to the student of oral histology, because tissues of the cephalic region, including dental tissue, are derived from the neural crest components (neural ectoderm). These migrate to a mesodermal location, and are appropriately designated as ectomesenchyme (Table 2–1).

EPITHELIAL TISSUE

Epithelial tissues originate from any one of the three germ layers (Table 2–1). In its diverse forms, epithelium may cover the surface of the body and also line tubes, ducts, and various passages leading to the external environment. Among the latter are those of the digestive, respiratory, urinary, and reproductive systems. Skin forms the epithelial covering of the body.

Epithelium also lines the chest cavity as the *pleura*, the heart surface and adjacent serosa as the *pericardium*, the abdominal cavity as the *peritoneum*, and the heart, blood, and lymph vessels as the *endothelium*. Other cavities are also lined, but not by true epithelium. These include the cavities associated with the bursae, joints, brain, and spinal cord. Furthermore, epithelium constitutes the functional or secretory parenchyma of most glands and their ducts. It also forms the sensory epithelium for the organs of taste, smell, hearing, sight, and others. Oral histology students should be particularly knowledgeable regarding epithelial structures such as the skin, linings of the oral and nasal cavities, and the parenchyma of glands, and their ducts, and the specialized epithelia of the taste, smell, and other organs.

General Features

The morphologic features of the surface layer of epithelium determines its subgroup classification as squamous (flat), cuboidal, or columnar. Depending on the number of cell layers present, further distinctions are made, including simple for one layer of cells and stratified for more than one layer. The surfacemost layer of epithelium is free and exposed. The inferiormost, however, lies on a basement membrane, which is composed of a basal lamina and an underlying connective tissue layer, the reticular lamina. The basal lamina may be defined only by electron microscopy but the organized connective tissue layer, or the classic basement membrane, is resolved by optical microscopy. Currently, the term "basement membrane" refers to both layers. The basement membrane may be reflected into the deeper connective tissue as folds, called papillary ridges. Epithelial cells are held together by various junctional complexes, such as

Table 2–1.   Germ Layer Derivations

I. ECTODERM

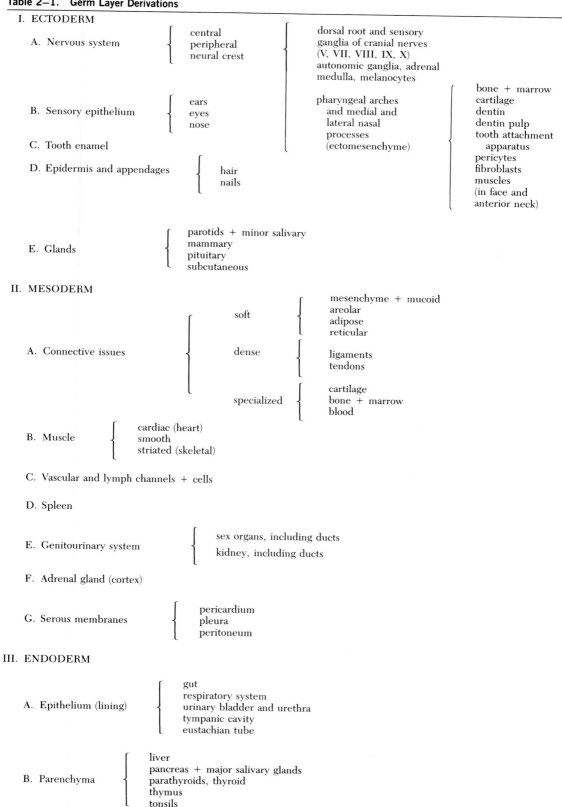

A. Nervous system
- central
- peripheral
- neural crest

dorsal root and sensory ganglia of cranial nerves (V, VII, VIII, IX, X) autonomic ganglia, adrenal medulla, melanocytes

B. Sensory epithelium
- ears
- eyes
- nose

pharyngeal arches and medial and lateral nasal processes (ectomesenchyme)

bone + marrow
cartilage
dentin
dentin pulp
tooth attachment apparatus
pericytes
fibroblasts
muscles (in face and anterior neck)

C. Tooth enamel

D. Epidermis and appendages
- hair
- nails

E. Glands
- parotids + minor salivary
- mammary
- pituitary
- subcutaneous

II. MESODERM

A. Connective issues
- soft
  - mesenchyme + mucoid
  - areolar
  - adipose
  - reticular
- dense
  - ligaments
  - tendons
- specialized
  - cartilage
  - bone + marrow
  - blood

B. Muscle
- cardiac (heart)
- smooth
- striated (skeletal)

C. Vascular and lymph channels + cells

D. Spleen

E. Genitourinary system
- sex organs, including ducts
- kidney, including ducts

F. Adrenal gland (cortex)

G. Serous membranes
- pericardium
- pleura
- peritoneum

III. ENDODERM

A. Epithelium (lining)
- gut
- respiratory system
- urinary bladder and urethra
- tympanic cavity
- eustachian tube

B. Parenchyma
- liver
- pancreas + major salivary glands
- parathyroids, thyroid
- thymus
- tonsils

the zonula occludens, zonula adherens, and macula adherens (desmosomes). Hemidesmosomes (half-desmosomes) attach the basal portion of epithelial cells to the underlying connective tissue. Another cell association is known as the macula communicans, which includes the nexus, or gap junction. The free surfaces of some epithelia are modified for specific functional needs and, therefore, bear cilia, flagella, or microvilli.

Papillary and Epithelial Ridges

Early in development, the ectodermal-mesenchymal interface is straight. Late in the first trimester of development, however, folds develop that are known as primary epidermal ridges and primary dermal ridges of connective tissue. Later, sheets of epithelium descend and divide the primary dermal ridges into two smaller connective tissue segments. These are called secondary dermal ridges (papillary ridges, or papillae), and the epithelial sheets dividing the connective tissue are designated variously as interpapillary pegs, epithelial ridges, or epithelial pegs (Fig. 2–1).

Epithelium does not possess its own vascular or lymph supply; rather, it depends on blood vessels in the subjacent connective tissue. The papillary connective tissue is loose, and is rich in blood capillaries. The deeper reticular connective tissue layer is denser and less vascular. Papillary ridges increase the surface area of the epithelial-connective tissue junction. Here the junctional complexes are more numerous and the blood supply is greater, so that dermoepithelial interactions are enhanced. Papillary ridges and interpapillary pegs are generally longer, thinner, and more numerous at sites of increased friction or possible injury. The basal layer of epithelium that rests on the basement membrane is closest to the blood supply and, therefore, occupies the most advantageous area for mitotic and metabolic activities. The surface cells, which are the furthest from the vascular supply and are more apt to be involved in environmental injuries, have weakened intercellular bonds and tend to desquamate. This characteristic is clinically advantageous in detecting incipient cancers, especially those of oral cavity, lungs, and some female genital organs, as well as the gastrointestinal tract.

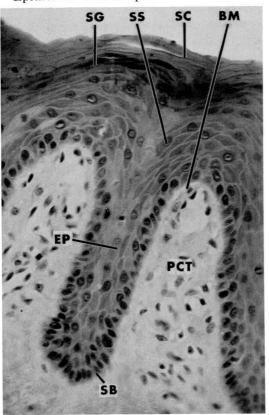

Fig. 2–1. Epithelium and underlying papillary connective tissue from the hard palate, with epithelial pegs (EP) between connective tissue ridges present a papillary effect (PCT). Note the clear thin layer of the basement membrane (BM) on which the low columnar cells of the basal layer rest. The spinous (SS), granular (SG), and cornified (SC) layers are sequentially surfaceward. (Hematoxylin and eosin stain; enlarged from ×415)

Basement Membrane and Glycocalyx

Surrounding epithelial cells either partially or completely is a fuzzy glycoprotein coat, the *glycocalyx* (Fig. 2–2). Fibroblasts, pericytes, and muscle or Schwann cells that are not epithelial may also possess a glycocalyx. In epithelial cells, the glycocalyx is especially prominent over the microvilli of intestinal epithelium, and it is thick enough to be resolved by optical microscopy. Possible functions attributed to the glycocalyx are the following: (1) autocell-type discrimination, essential for tissue development and repair; (2) protection against autolysis; and (3) increase of the cohesive and adhesive bonds between different tissues and among cells.

Before the advent of the electron microscope the basement membrane was interpreted by light microscopy as an amorphous layer intervening between the connective tissue and the base of the epithelial cells. Except for special microtechnical procedures, such as those involved in the periodic acid-Schiff (PAS) reaction and silver staining, this layer could not always be defined. A positive PAS reaction indicates that the basement membrane contains a carbohydrate component. Its positive reaction to silver indicates that it is composed of reticular fibers—hence, the derivation of the term "reticular lamina." Electron microscopic observations have confirmed the presence of reticular (argyrophilic, or argentaffin) fibers, which bear the characteristic 64-nm banding of collagen (Fig. 2–3). Additionally, however, an ultramicroscopic layer the basal lamina (+ 100-nm thick), located between the reticular lamina and the plasma membrane of the basal cell layer, has also been detected. The basal lamina consists of two sublayers—an adepithelial light layer, the lamina lucida, and a denser underlying one, the lamina densa, which abuts the reticular lamina. Although the lamina lucida is similar to the glycocalyx, the lamina densa is densely matted by fine filaments 2 nm in diameter. The reticular lamina and basal lamina collectively form the *basement membrane* (Fig. 2–3). Other than functioning as a supporting cushion for epithelia, basement membranes serve as mechanisms for ultradiffusion and ultrafiltration.

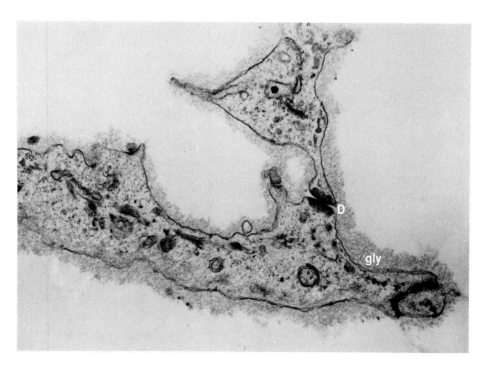

Fig. 2–2. Surfacing epithelial cell segments joined by desmosome (D). (gly, glycocalyx.) (Uranyl acetate and lead citrate stain; × 15,000) (Sisca, R.F., Provenza, D.V., and Fischlschweiger, W.: Ultrastructural characteristics of the human enamel organ in an early stage of development. J. Baltimore Coll. Dent. Surg., 22:8, 1967.)

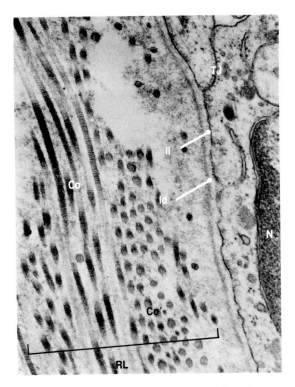

Fig. 2–3. Two epithelial cell segments joined by tight junction (TJ). The basal lamina and its lamina lucida (ll) and lamina densa (ld). Note that the reticular lamina (RL) is cell-free and is composed exclusively of collagenous elements which bear the characteristic banding and are transversely (Co') and longitudinally (Co) sectioned. (N, nucleus.) (Uranyl acetate and lead citrate stain; ×16,000) (Fischlschweiger, W, Provenza, D.V., and Sisca, R.F.: Reorganization of the peripheral layers of the human enamel organ during the bell stage—an electron microscopic study. J. Baltimore Coll. Dent. Surg., 22:28, 1967.)

Intercellular Associations

Formerly it was believed that epithelial cells were cemented together by a dense staining material, which with optical microscopy was seen to be particularly prominent at the cell terminals. These sites were designated as terminal bars (Fig. 2–4). Electron microscopic study of the terminal bar sites has indicated that attachment between cells occurs via structural and functional bonds, or junctional complexes and communicating junctions. The junctional complexes are of three types: zonula occludens (tight or impermeable junction), zonula adherens (intermediate junction), and macula adherens (desmosome). These are arranged sequentially in abutting cells from their apices to their bases (Figs. 2–5 and 2–6).

Cell junctions that interface with the external environment are arranged as tightly sealed circumferential belts or bands, which prevent the passage of material into the intercellular spaces by occluding the entrance. These tight seals or junctions are appropriately called *zonula occludens*, or occluding zones, and are formed by the fusion of the outer leaflets of the plasmalemmas of contiguous cells (Figs. 2–3, 2–5, and 2–6). Special techniques (e.g., freeze fracturing) reveal that the leaflets at these sites are intermittently fused by the interaction of intramembranous protein chains of abutting cell membranes at their contact points. Chelates of calcium and magnesium ions, as well as proteolytic enzymes, cause disruption of the seals.

The *zonula adherens* is also arranged circumferentially about the cells (Fig. 2–5). Because leaflets of adjacent cell membranes are not fused, a microspace, 0.02 μm, intervenes, in which may be found glycoprotein filaments similar to and continuous with those constituting the terminal web that underlies the plasma membrane of columnar epithelial cell apices (Figs. 2–4 and 2–5). Because of their structure, the zonulae adherens probably represent the terminal bars observed with optical microscopy. This junctional complex functions in intercellular adhesions.

Desmosomes, or macula (spot) adherens, are abutting dense plaques that intermittently dot the lengths of continuous cell membranes from the zonula adherens basalward (Figs. 2–2 and 2–5). Because they have no junctional mates in connective tissue cells, those joining the cell bases with the connective tissue on which they rest are comprised of a single plaque, and thus are known as half-desmosomes, *hemidesmosomes* (Figs. 2–5 and 2–7). Each plaque is about 410 nm in length and 25 nm in width, with a 10- to 30-nm interspace. Intermediate filaments, the tonofilaments, about 10 nm thick, course through the substance of the cytoplasm, circle in and out, or terminate at the plaques of the limiting membranes. The space between opposing plaques is divided in half by a dense line, sometimes called the central stratum. Extremely delicate filaments pass from the adjacent cell membranes of the plaques through the microinterspace comingling to form the medial or central stratum. While desmosomes and hemidesmosomes form attachment and anchoring

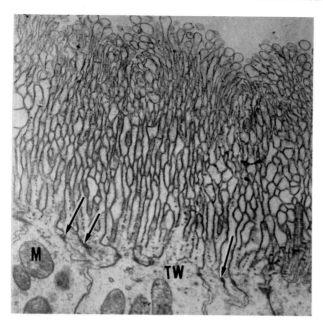

Fig. 2–4. Microvilli of active secretory epithelium. Note the parallel arrangement and length of the microvilli. Also shown are terminal bars *(arrows),* terminal web (TW), and mitochondria (M). (Uranyl acetate and lead citrate stain; ×6500)

sites, the microspace additionally provides for intercellular flow of tissue fluids for metabolic interchanges.

A gap or communicating junction is also known as a nexus. It is spot-like (macular) in profile, and is characterized by conditions in which microconduits extend from abutting limiting cell membranes to span a very narrow intercellular space, or gap. The walls of the conduits are composed of protein particles arranged annularly. Because the conduits are open-ended there is a free exchange of micromolecules and ions flowing directly from cell to cell (Fig. 2–5).

Cell Surface Modifications

The limiting membranes of some epithelial cells may be modified structurally as microvilli and basal infoldings to increase surface area, or as cilia and flagella for motion and locomotion, respectively.

Microvilli. The free surface of some epithelial cells bears outpocketings that are not clearly resolved with optical microscopy; these exhibit a brush-like or striated appearance. Electron microscopic examination reveals these to be minute, finger-shaped processes which are extensions of the apical cytoplasm (Figs. 2–4 and 2–5; also see Fig. 1–4). They are known as *microvilli.* Although in some epithelial cells they are located exclusively on the apical surface, in other cells they may be found on any or all surfaces.

It has been estimated that a single cell may bear several thousands of these projections.

The lengths of the microvilli differ. In some cells, microvilli are similar in length and alignment while in others they may be nonuniform in shape, length, and arrangement. Furthermore, cells in the secretory phase may exhibit elongated microvilli. These differences may be artifacts induced by fixation and processing of the tissue.

The cytoplasm of microvilli is organelle-free, consisting exclusively of bundles of fine filaments (5 to 7 nm diameter), most of which are comprised of actin. These extend longitudinally and become attached to the tip of the microvilli to form distal terminal densities. The proximal terminals of the filamentous bundles connect with the myosin components of the terminal web filaments in the cells' apices. The filaments are believed to affect movement and contraction of the microvilli by actomyosin interaction.

Terminal Web. The terminal web is a thin, organelle-free cell segment approximating the level of the zonula adherens (Figs. 2–4 and 2–5). In addition to actin and myosin filaments, a few intermediate filaments, mostly cytokeratin (tonofilaments), constitute the terminal web. Collectively, these form a dense fibrous network that functions as attachment and support sites for the filamentous core of the microvilli.

Basal Infoldings. The plasmalemma, espe-

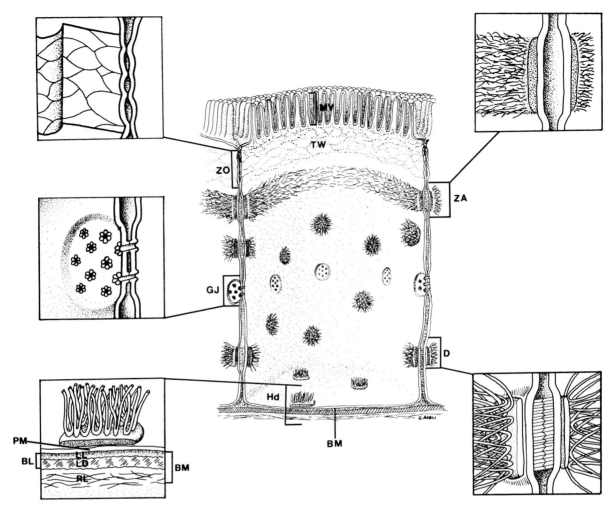

Fig. 2–5. Diagram of cell showing microvilli (MV) and various cell junction types as observed from different planes using the electron microscope. The cell surface is covered with microvilli, subjacent to which is the terminal web (TW). Counterclockwise, the lateral cell membranes of adjacent cells form a tight junction (belt) or zonula occludens (ZO), gap or communicating junction (GJ), hemidesmosome (Hd), macular (spot) junctions or desmosomes (D), and zonula adherens (ZA). The basal cell membrane (PM) rests on the basement membrane (BM), consisting of a reticular layer (RL) and the basal lamina (BL), comprised of a lamina lucida (LL) and lamina densa (LD).

cially of the basal cell membrane of some epithelial cells, exhibits elaborate inpocketing (Fig. 2–8; also see Fig. 9–30). In certain columnar epithelia the inpocketing is limited to the basal cell membrane, and is appropriately referred to as *basal infoldings.* These may be found in the ducts of salivary glands, convoluted tubules of the kidney, and other organs requiring increased surface area for secretion and/or absorption. Cells may also possess lateral projections, which may interdigitate with those of adjacent cells, or more basally, with basal folds of other cells.

Cilia and Flagella. Surfaces of cuboidal or columnar epithelial cells may be modified, bearing structures that either propel substances over their surfaces or that effect locomotion of the entire cell. The latter is exemplified by the male sex cell (spermatozoon), which moves by the whip-like action of its *flagellum.* On the other hand, movement of substances over the cell surface is produced by the beating of *cilia.* For example, mucoid lubricants are spread over cell surfaces and foreign substances, including debris, are moved toward external opening(s) by ciliary action. Cilia may also be found associated with nervous tissue. Here they are nonmotile, and are simple structural modifications. Flagella are up to 15 times longer than cilia; otherwise,

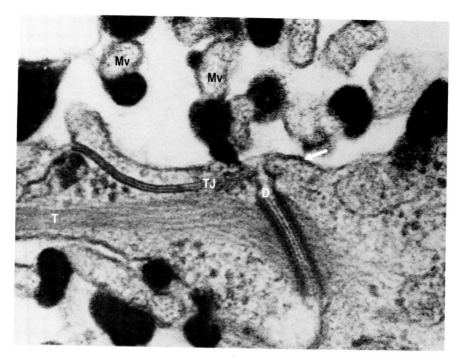

Fig. 2–6. Processes of cells joined by a tight junction (TJ). Note the relationship of tonofibrils (T) and desmosome (D). Round black bodies on the microvilli (Mv) in the intercellular space are reaction products of alkaline phosphatase. Also note the unit membrane *(arrow)*. (Uranyl acetate and lead citrate stain; ×62,000) (Leonard, E.P., and Provenza, D.V.: Alkaline phosphatase activity in sequential mouse molar tooth development. An electron microscopic study. Histochemie, *34*:343, 1973.)

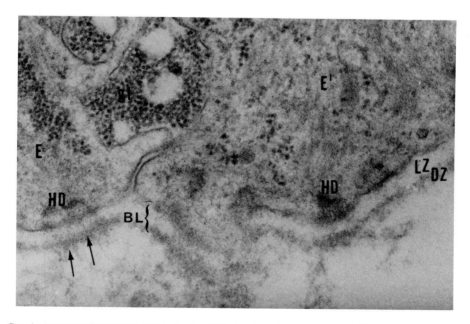

Fig. 2–7. Basal segments of two epithelial cells E and E' showing glycogen accumulations (gl), hemidesmosomes (HD), and anchoring fibrils *(arrows)*. The lucid zone (LZ) and dense zone (DZ) of the basal lamina (BL) are shown. (Uranyl acetate and lead citrate stain; ×10,000)

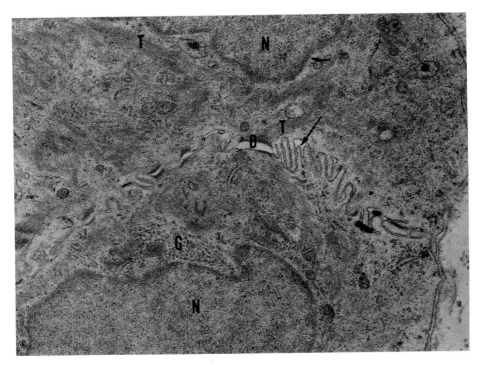

Fig. 2—8. Electron micrograph of several epithelial cells showing interdigitation of their cell membranes *(arrows)* (D, desmosome; N, nuclei; T, tonofilaments (cytokeratin); G, Golgi apparatus). (Uranyl acetate and lead citrate stain; × 3,000.)

the two are structurally the same. Because of this, only the structure of cilia will be discussed here.

Several hundred processes may project from a single cell surface. Electron microscopic examinaiton shows each cilium to be a hair-like process extending from a basal body in the apical cytoplasm (see Fig. 1–14). Each cilium contains a central dark line—the axial filament, or axoneme. At its base is located a basal body; when stained with hematoxylin and eosin it is seen to be dark and refractile. Although both the length and diameter of cilia may vary, they average about 0.03 μm in diameter and between 8 and 10 μm in length. Electron microscopic study reveals that the limiting membrane of the cilia consists of extensions of the plasma membrane of the central cell mass, and contains microtubules longitudinally disposed from the basal body throughout its length. Microtubules are composed of a protein, tubulin, and they are arranged as a circlet of nine pairs with two singlets in the center. This 9 + 2 paired arrangement of microtubules is characteristic of all cilia, including those of sensory perceptors. (Variations do occur, however—for example, olfactory

cilia possess 11 singlets.) A protein, dynein, extends from each microtubule pair. It is believed that the enzyme action of this protein provides the energy for the forward strokes in ciliated movement.

Basal bodies (corpuscles) are barrel-shaped bodies in the apical cytoplasm, with which cilia rootlets are connected and functionally associated (see Fig. 1–14). Structurally, basal bodies are the same as centrioles, and are believed to originate from them. The basal body is about 0.5 μm in length and 0.2 μm in diameter. The parallel arrangement of the basal body microtubules is similar to that of cilia and flagella. It differs in that the central singlets are lacking, and a third half-microtubule is fused to each of the nine doublets to form a circlet complex of nine triplets.

Classification

The profile of the surface and number of cell layers collectively identify the various types of epithelium. Thus, the human body contains squamous, cuboidal, and columnar epithelium. If one layer is present it is classified as simple, and more than one layer is known as stratified.

There is a type of simple columnar epithelium that appears to be stratified because of differences in cell height, cell arrangements, and nuclear position. This epithelium is referred to as *pseudostratified* (falsely layered) *columnar epithelium*. In still another type of epithelium both the number of layers as well as the shape of the surface layer of cells change. For example, the number of layers tends to increase and its surface cells tend to become dome-shaped (cuboidal) when the organ (e.g., the urinary bladder) is not distended. Conversely, when the organ is "stretched," the epithelium becomes thinner as the number of cell layers is decreased and the cells surfaceward become flattened. This tissue is referred to as *transitional epithelium* because the cell shape depends on the functional condition of the organ. Surface transitional epithelium may, therefore, range in shape from cuboidal-like to squamous, depending on the volume of fluid in the bladder or on the duct's contents.

Squamous Epithelium

Simple squamous epithelium consists of a single layer of flattened cells, mosaic in configuration, that forms sheets, tubes, or cups. Squamous epithelium of the latter is observed in Bowmen's capsule, the filtering system of the kidney (Fig. 2–9). In tubules, squamous cells form the endothelial lining of blood and lymph vessels. Squamous epithelium may also form ducts in certain salivary glands. The basal cell membrane is attached to the basement membrane by anchoring a fibril-hemidesmosome complex. Except for the nucleus, the limiting membrane and organelles of these cells are not clearly resolved with optical microscopy without special tissue processing. Squamous cells may be so thin in profile that their thickness may be noted only over the cell center at which the nucleus is located. This is especially true for endothelial and Bowman's capsule cells (Fig. 2–9).

Stratified squamous epithelium forms sheets of varying thicknesses but only the surface cells are squamous (Fig. 2–1). Below the surface, to the basal layer, the cell shapes change from polyhedral to round to cuboidal and/or columnar. The latter rests on a basement membrane. The basal epithelial layer adheres to the underlying connective tissue by hemidesmosomes, and the suprabasal epithelial cells are attached by desmosomes. Shrinkage caused by tissue processing results in the spinous appearance of the intermediate cell layers (Fig. 2–1).

Two varieties of stratified squamous epithelium occur in the body, keratinous (cornified or hornified) and nonkeratinous (mucous or moist) stratified squamous epithelium. Both are found

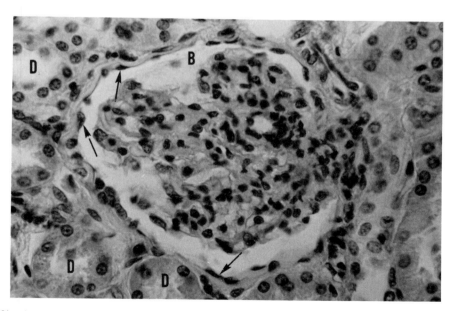

Fig. 2—9. Simple squamous epithelium *(arrows)* lining Bowman's capsule (B) of the kidney. Also shown are ducts (D), comprised of simple cuboidal epithelium. (Hematoxylin and eosin stain; ×320)

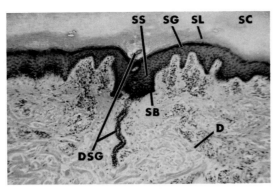

Fig. 2–10. Thick skin showing dermis (D), duct of sweat gland (DSG), stratum basale (SB), stratum spinosum (SS), stratum granulosum (SG), stratum lucidum (SL), and stratum corneum (SC). (Hematoxylin and eosin stain; enlarged from ×215)

in the mouth and oral cavity. Keratinous epithelium, as epidermis, is located in dry areas such as the skin and vermilion (red) border of the lips. Nonkeratinous epithelium is located in moist areas such as the linings of the mouth, oral side of the cheeks, underside of the tongue, floor of the mouth, and soft palate. Oral epithelium may vary in degree of keratinization from nonkeratinized, as in the epithelial lining of the cheeks, to keratinized, as in the gingivae (gums).

The degree of cornification observed in epidermis is not normally observed in the oral epithelium. Accordingly, the terms "orthokeratinous" and "parakeratinous" are used for keratinous epithelium of the oral cavity. The keratinizing process for the former is more developed than for parakeratinous epithelium. Orthokeratinous and parakeratinous epithelia are found in those areas of the oral cavity in which frictional forces created by biting, chewing, and movement of food occur. Clinically, these are known as masticatory epithelia, and are most often found covering the gingivae and hard palate. Because of the presence of taste receptor cells, the epithelium covering the dorsal surface of the tongue is known as specialized epithelium.

Completely keratinous epithelium, such as is found in the skin, contains at least four layers (Fig. 2–10). Beginning with the basalmost, attached to the basement membrane, they are the germinating, spinous, granular, and desquamating cornified layers. Collectively, the germinating (stratum basale, or germinativum) and spinous layers form the malpighian layer. An-

other layer may be interposed between the granular and cornified layers in thick skin or in the skin forming the red border of the lips. This transparent layer, three to four cell layers thick, is known as the stratum lucidum.

Thick skin is found in areas such as the palms of the hands and soles of the feet, or in other areas subject to intense wear. Elsewhere, skin is referred to as thin skin, and the transparent cells are not present as a definite layer. Rather, the transparent cells simply intermingle with those of adjacent layers.

The cells of the stratum basale are similar in shape and internal structure—they are cuboidal or columnar, with their nuclei located in the center of the cell or toward the basement membrane. Mitotic activity occurs more often in this than in the overlying layers. Except for tonofilaments, other cytoplasmic structures are more abundant in the basal layer than in the overlying ones. In this regard, as the cells divide, grow, and mature, and are pushed surfaceward, and the organelle population decreases progressively. Furthermore, in the stratum granulosum, the nuclei undergo degenerative changes: pycnosis, karyorrhexis, and finally karyolysis.

The tonofilaments of the cells in the stratum basale aggregate into fibril bundles in the stratum spinosum. Here the tonofibril bundles increase in number and course through the cell, with many terminating at the desmosomes. Cell shrinkage and its relationship to the tonofibril bundle-desmosome complex results in the "spinous appearance" of this layer (see Fig. 2–35). The keratohyaline granules, which are believed to be produced by ribosomes, and that characterize the stratum granulosum, are irregular in contour and measure from 0.5 to 1.0 μm. Lamellar bodies (membrane-coating granules) produced in the Golgi apparatus appear in the cells of the stratum spinosum. In the overlying layers, the lamellar bodies accumulate at the cell margins and empty their products (mostly glycolipids and sterols) into the intercellular spaces. Here the secretory products are organized into sheets that waterproof or otherwise limit the ingress of materials through the epidermis. In the suprajacent layer the keratohyaline granules probably form a matrix in which the tonofibrils are enmeshed to produce a biostructural keratin complex of filaments 8 to 10 nm thick. As these cells approach the stratum corneum all organelles disappear by lysosomal digestion, including the nucleus, and they undergo dehydration. In the stratum corneum the cells become flattened keratinous plates that can withstand mechanical stresses and the effects of chemical solvents; these are sloughed off. In the process of keratinization, desmosomal bonds are loosened, facilitating desquamation of the dead or moribund cells.

In the oral cavity, epithelia of the gums and hard palate are subjected to harsh abrasive forces. To withstand these insults, the cells undergo keratinization. At these sites, although the stratum germinativum (basal) and spinosum layers are similar, differences are noted in the more superficial layers. In orthokeratinous (keratinous) epithelium, the granular layer although narrow, can be resolved, and the surface desquamating layers, although keratinous, nuclei have disappeared or are in stages of advanced degeneration. In parakeratinous epithelium the process of keratinization has not been completed. Keratohyaline granules are sparser and are not organized into a definite layer. The nuclei of the desquamating cell layer are present in advanced stages of degeneration, exhibiting pycnosis and/or karyorrhexis (Fig. 2–11A).

The layers in nonkeratinous epithelium include the basal or germinating, spinous and overlying intermediate, and superficial or desquamating (Fig. 2–11B). As determined by optical microscopy, nonkeratinous differs from keratinous epithelium in several respects. In nonkeratinous epithelium the keratinocytes of the spinous layer are generally rounder and larger than in keratinous epithelium, the intercellular spaces are smaller and fewer, the desmosome-tonofibril complexes are weakened, with the spinous appearance less prominent, and a granular layer is absent, and in the superficial desquamating layer the nuclei of the keratinocytes are less degenerated. Ultrastructural features distinguishing keratinous from nonkeratinous epithelium will be discussed in Chapter 9.

Cuboidal Epithelium

Simple cuboidal epithelium consists of a single cell layer overlying the basement membrane (Fig. 2–12). In profile, sheets of these cells appear cuboidal. As glandular components (secretory parenchyma or ducts) the cells are trapezoidal, with their broad base juxtaposed to the

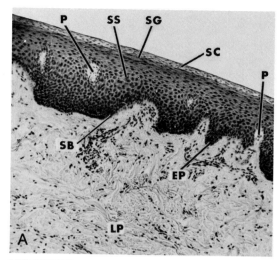

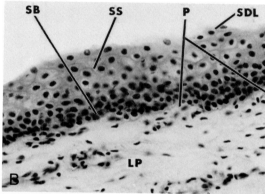

Fig. 2–11. *A,* Parakeratinous gingival epithelium and lamina propria (LP). The former consists of a stratum basale (SB), stratum spinosum (SS), stratum granulosum (SG), and stratum corneum (SC), with a papillary layer showing connective tissue ridges (P), intervened by epithelial pegs (EP). (Hematoxylin and eosin stain; enlarged from ×50) *B,* Nonkeratinous epithelium and lamina propria (LP) with shallow ridges/papillae (P), stratum basale (SB), stratum spinosum (SS), and surface desquamating layer (SDL). In the latter note the presence of nuclei in the cells. (Hematoxylin and eosin stain; ×215)

basement membrane. Surface view reveals them to be hexagonal. The nucleus, the most conspicuous organelle, is round and located centrally. In the oral cavity simple cuboidal epithelium is located exclusively in salivary glands.

Stratified cuboidal epithelium does not occur widely in the body. It is found in ducts of the sweat and sebaceous glands, in which it rarely exceeds a thickness of two cell layers. Although the surface cells are cuboidal, the underlying ones are smaller and polyhedral (Fig. 2–13). Their cytologic features are similar to those seen in simple epithelium.

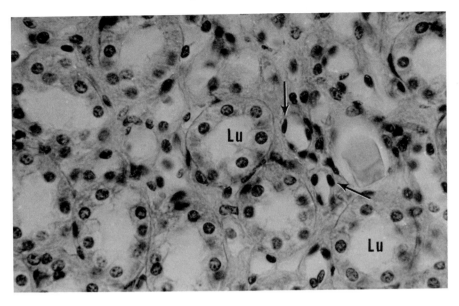

Fig. 2–12. Photomicrograph of kidney tubules. Note that the lumina (Lu) are surrounded by simple cuboidal cells. Some tubule walls *(arrows)* are of simple squamous epithelium. (Hematoxylin and eosin stain; ×320)

Transitional Epithelium

Based on ultrastructural features, transitional epithelium, once thought to be modified stratified cuboidal, is now considered to be a distinct epithelial type. Transitional epithelium lines the urinary bladder, ureters, kidney calyces, and pelvis, which must accommodate volumetric changes. When filled with fluid wastes and distended, the epithelium is squamous and is reduced to less than three layers. In the more evacuated organs, while more contracted, the epithelium increases to six or more cell layers and the surface cells are pear- or dome-shaped. They are often binucleate (Fig. 2–14). Subsurface cell layers are polygonal, cuboidal, or columnar in profile. The basal layer rests on a thin basement membrane.

The unique feature of transitional epithelium is cell maneuverability, which is enhanced by a decrease in desmosomal connections. The cells glide over one another to accommodate volumetric changes. Specializations of the superficial cells are especially noteworthy. For example, the luminal plasmalemma is thicker (up to 12 nm)—that is, the outer leaflet is significantly thickened. Additionally, the apical plasma membrane is plicated when the organ is empty (contracted). The plication (infoldings) are related to the subsurface spindle-shaped vesicles in that they are formed from plications of the plasmalemma. Their membranes are similar to those of the thickened region of the plasma membrane. As the bladder becomes distended, the cell surface becomes smoother and longer by the spindle-shaped vesicles binding to the plications of the plasma membrane. The luminal membrane becomes smoother, as required by bladder distension. Because the apical cell membrane is thicker and less permeable, it probably

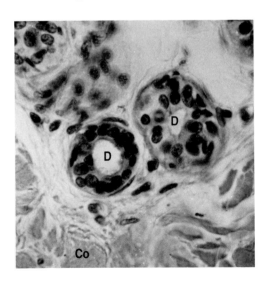

Fig. 2–13. Stratified cuboidal epithelium of sweat gland (D). Note the collagen bundles (Co) surrounding duct components. (Hematoxylin and eosin stain; ×80)

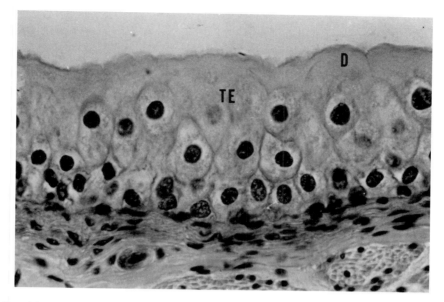

Fig. 2–14. Transitional epithelium (TE) from urinary bladder. The superficial pear-shaped cells (D) are characteristic of the empty (nondistended) organ. (Hematoxylin and eosin stain; ×320)

functions as an osmotic barrier between tissue fluids and urine wastes.

Columnar Epithelium

Simple columnar epithelium, viewed on the free surface, forms a mosaic of hexagons. In profile, the cells are rectangular and vary in height from low to tall columnar (Fig. 2–15). The bases of the cells are joined to the basement membrane. Their nuclei are oval, linearly located at the same level in the basal cell segment. Columnar epithelium lines the digestive tract, from the stomach to the large intestine. It also comprises certain ducts of the respiratory and genitourinary systems, as well the secretory and duct portions of glands. Thus, columnar cells are found in areas of absorption and secretion. A specialized columnar cell found in the respiratory, digestive, and reproductive systems, the goblet cell, produces a lubricant, *mucus* (Fig. 2–16). These cells may be interspersed among other cells, which are ciliated. The sweeping action of cilia spreads the secretion over the surface.

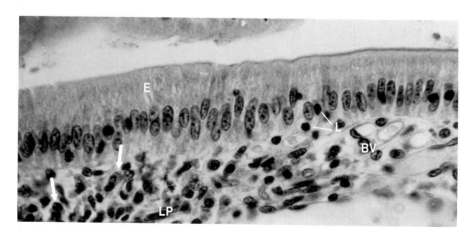

Fig. 2–15. Tall simple columnar epithelium (E) surfacing the intestinal lamina propria (LP). The latter is well supplied with blood vessels (BV). Note plasma cells *(arrows)* and lymphocytes (L) infiltrating the region. (Hematoxylin and eosin stain; ×320)

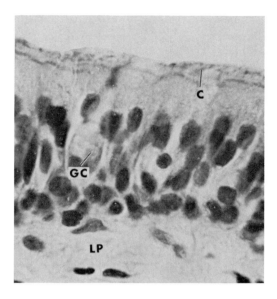

Fig. 2–16.   Goblet cell (GC)/unicellular gland interspersed among the pseudostratified ciliated columnar epithelium (C) of the nasal passage (LP, lamina propria). (Hematoxylin and eosin stain; enlarged from ×430)

Stratified columnar epithelium occurs infrequently. When present, it is found in certain areas of the pharynx and larynx but rarely in the nasal cavities. Only the surface cells are columnar and possess features similar to those of their monolayered counterpart, especially regarding orientation and location of the nuclei. The linear arrangement of the nuclei is important, because it distinguishes stratified columnar epithelium from pseudostratified columnar epithelium (Fig. 2–17). The cells of the intermediate and basal layers are small and polygonal, with centrally located nuclei.

Pseudostratified Columnar Epithelium

These cells are found lining the ducts of the reproductive and respiratory systems. Pseudostratified epithelium consists of two cell components, *columnar* and *basal*, which rest on a basement membrane. Because the nuclei are not located at the same level, cell stratification is simulated—hence, the designation "pseudostratification" (Fig. 2–17). Interspersed among the columnar cells are goblet cells (Fig. 2–16). The columnar cells are tall, and the surface bears cilia. Basal cells are spindle-shaped or polyhedral, intermingling with the columnar type. They are believed to be the replacement source for the spent or injured columnar cells.

Specialized Epithelia

The special functions associated with the organs of hearing, sight, smell, and taste are accommodated by epithelial sensory receptors (see below, Nervous Tissue). Another specialized type possesses contractile properties similar to those of muscle, and is known as *myoepithelium.* This type of cell embraces the secretory end-pieces and their draining ducts in some salivary glands. Contractile activity of these cells expels the secretions into ducts. Myoepithelial cells will be described in Chapter 9.

Functions

Epithelium may function in the processes of absorption, excretion, lubrication, protection, reproduction, secretion, sensory reception, and transport. Epithelia associated with absorption and resorption, secretion, and lubrication are generally simple epithelia. They are often squamous, facilitating transport—for example, the endothelium of vessels and of Bowman's capsule of kidney (Fig. 2–9). The epithelium of ducts or tubules of the kidney and lungs may be simple low columnar or cuboidal, providing for efficient exchanges and/or extraction of materials for excretion. The lubricating secretions of the mesothelium and epithelium of the lower bowel and of certain organs of the genitourinary system reduce frictional resistance. Male and female gonads produce and secrete cells, the *gametes.* Similarly, enzymes and hormones synthesized by glandular epithelium are released and distributed by ducts (exocrine glands), or absorbed and distributed by blood vessels (endocrine glands). Some types of epithelia are specialized to function as sensory receptors. Others may be modified to sweep secretion-trapped debris toward the external body openings for expulsion. The external body covering (skin) and cavity linings (e.g., of the mouth and nose) function as barriers against physical and chemical injuries.

Ependyma of the central nervous system and certain types of connective tissue cells—osteoblasts (bone), cementoblasts (cementum), and odontoblasts (dentin)—are epithelium-like in morphology, arrangement, and function. Because they resemble epithelium, they are designated as "epithelioid."

Study of the functions attributed to epithelium suggests that all materials entering, trav-

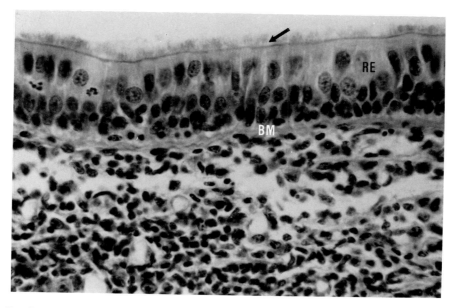

Fig. 2–17. Respiratory tract epithelium (RE) showing cilia at the surface *(arrow)* of ciliated pseudostratified columnar epithelium. The lobulated nuclei seen in the epithelium are leukocytes about to enter the respiratory tube to be expelled with mucus. Note the thickness of the basement membrane (BM). (Hematoxylin and eosin stain; ×320)

eling through, and leaving the body are subject to epithelial action and/or direction. These actions occur primarily under the control of the cells' plasmalemmas, and to a lesser extent the basal lamina and glycocalyx, with which they are in intimate contact.

Glands

The function of glandular parenchyma, which is comprised mostly of cuboidal or columnar epithelium, is the synthesis of secretions. Glands are variously classified as endocrine or exocrine, unicellular or multicellular, simple or compound, tubular or alveolar/acinar or (mixed), serous or mucous or (mixed) and holocrine, apocrine, or merocrine.

Exocrine and Endocrine Glands. Secretory portions of glands that empty their contents into a system of ducts leading to external openings are known as glands of external secretion—exocrine or ducted glands. Conversely, glands with elaborations (hormones) that are taken up by capillary plexuses and distributed throughout the body by vessels are known as glands of internal secretion—endocrine or ductless glands. The secretory segment of endocrine glands forms cords or spherical masses, known as follicles. These are not found as paraoral or paranasal organs. On the other hand, exocrine

glands are found associated with both the oral and nasal cavities; these are classified on the basis of cell number, shape, complexity of arrangement, and the nature of secretion.

Unicellular and Multicellular Glands. Unicellular glands are the simplest, and consist of a single cell, the *goblet cell* (Fig. 2–16). Multicellular glands are composed of many cells, which may be in the form of covering sheets, or they may dip down into underlying connective tissue as simple glands and assuming the shape of tubules (straight, coiled, or branched) or of sacs (alveoli or acini—single or branched) (Fig. 2–18B–D). They may also be complex in organization, compound glands. The ducts of compound glands branch and terminate in secretory segments (units, or end-pieces) of varying shapes: tubes, sacs, alveoli, or acini (Fig. 2–18). Some glandular units are composed of both tubes and acini; these are known as tubuloacinar (tubuloalveolar) (Fig. 2–18E).

The secretory end-pieces communicate directly with the terminals of ducts. The variety and number of both the ducts and secretory end-pieces depend on gland size and organizational complexity. The largest, the excretory duct, communicates with interlobar ducts located in the tissue that separate the lobes of the gland. Interlobar ducts collect secretions from the in-

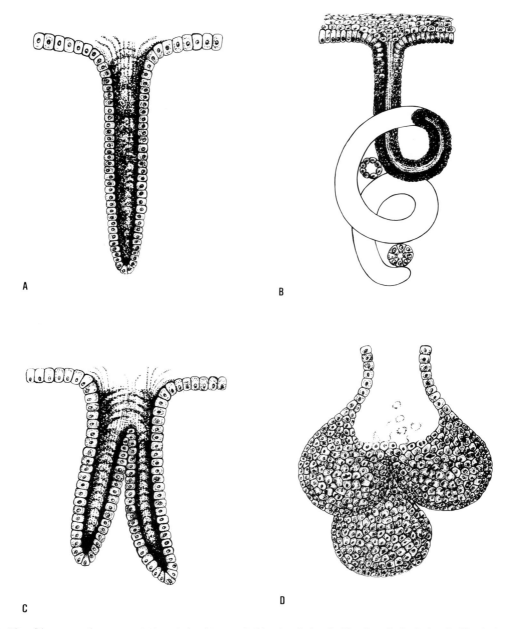

Fig. 2–18. Diagrammatic representation of gland types. *A,* Simple tubular. *B,* Simple coiled tubular. *C,* Simple branched tubular. *D,* Simple branched acinar.

tralobular (striated) ducts located within the lobules, and the latter communicate with the ducts that drain the secretory end-pieces of the gland. These small ducts are known as intercalated ducts. Duct epithelium can be either simple or stratified columnar. The cells shorten as they decrease in diameter, so that a gradual transition occurs from columnar to squamous epithelium.

Regarding the type of secretion, some glandular parenchyma, serous cells, may produce a thin, transparent, albuminous product. Others, mucous cells, may synthesize a viscous, sticky substance. Organs comprised of cells that produce only the former are called serous glands, and those elaborating only the latter are known as mucous glands. Some glands contain both cell

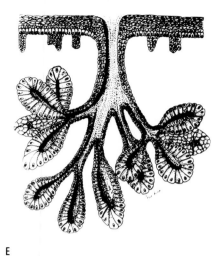

E

Fig. 2–18. Continued. *E,* Compound tubuloacinar. (Adapted from several sources.)

types, synthesizing both mucous and serous secretions; these are appropriately known as mixed glands. A more detailed discussion is provided in Chapter 9.

Merocrine, Apocrine, and Holocrine Glands

Secretory products are liberated by glandular parenchyma in various ways. In some glands, secretions are released into the lumen of the secretory end-piece by exocytosis without any injury or discontinuity of the limiting membrane of the secreting surface. This method of secretion, in which the surface membrane remains intact, is characteristic of merocrine glands; these include such organs as the salivary glands and the pancreas. Electron microscopy shows that a unit membrane encapsulates the secretory vesicles. The membrane of the vesicle fuses with the surface membrane in releasing the secretion, so that the integrity of the surface plasmalemma is maintained (Figs. 1–4 and 1–10).

The secretory mechanism of *apocrine glands* (e.g., sweat glands of the skin) was once believed to involve detachment of the surface membrane or apical cytoplasm, thereby providing for expulsion of the secretory product. This observation, which has since proven to be incorrect, was based on light microscopic studies. Electron microscopic investigations indicate that minimal, if any, damage occurs in the surface membrane of the cells during the release of the se-

Table 2–2. Classification of Connective Tissue Types

I. Connective tissue proper
   A. Embryonal connective tissue
      1. Mesenchymal
      2. Mucous
   B. Regular connective tissue
      1. Loose connective tissue
      2. Dense connective tissue
         a. Irregular
         b. Regular
   C. Special forms of connective tissue proper
      1. Adipose connective tissue
      2. Reticular connective tissue
      3. Elastic connective tissue
II. Specialized connective tissue
   A. Fluid connective tissue
      1. Blood
   B. Rigid connective tissue
      1. Nonmineralized
         a. Cartilage
            (1) Hyaline
            (2) Elastic
            (3) Fibrous
      2. Mineralized
         a. Bone
            (1) Compact
            (2) Spongy (cancellous)
         b. Dentin
         c. Cementum
            (1) Acellular
            (2) Cellular

cretions. Hence, these glands are merocrine-like.

In *holocrine glands* (e.g., sebaceous glands of the skin), the secretory process involves the total cell. In these, the cells accumulate sebum as they are pushed further away from the basal cell layer. The synthesis and storage of sebum, which lubricates the skin, occurs at the expense of the organelle population. As a consequence the cells die and are expelled as the secretion. The male gonads (testes) are also holocrine. Spermatocytes are secreted into the spermatic ducts as male gametes (spermatozoa).

CONNECTIVE TISSUES

These tissues give form to the body while supporting and connecting the diverse organs and systems. Connective tissues are classified in Tables 2–1 and 2–2. The three basic components of connective tissue include an intercellular amorphous ground substance and tissue fluid (matrix complex), formed intercellular protein fibers (reticular, collagenous, and elastic) and diverse cell types (mesenchymal, fibroblast, fat, histiocyte, mast, plasma, reticular, macrophage, and leukocyte).

Functions

Except for those of the head region that are related to the neural crest and designated as ectomesenchyme, connective tissues are of mesodermal origin. Other than specific functions performed by specialized connective tissue (e.g., bone, cartilage, cementum, dentin, ligaments, tendons, aponeurosis), some connective tissues serve more general functions—for example, insulation (fat), partitioning (septae), and encapsulation (capsules and fascia). The cells of some connective tissues serve a protective function through phagocytic and antibody-producing actions. Additionally, connective tissue matrices serve as a medium through which nutrients and the metabolic products are dispersed and exchanged selectively by way of tissue fluids, blood, and lymph.

Components

Extracellular Matrix

As indicated, extracellular matrices of connective tissue are composed of an amorphous ground substance, including tissue fluid, and formed protein fibers.

Ground Substance. The cellular and fibrous intercellular components of connective tissue are embedded in a homogeneous and virtually transparent matrix, or ground substance. Depending on such factors as age and injury, the ground substance varies in consistency from fluid to gel. Early in embryonic development the ground substance is almost fluid; in mature tissue, however, it becomes semifluid or gelatinous, which provides the cementing action necessary for the formation and organization of tissues. Fluidity, which facilitates the transport of materials through tissues, tends to be increased by the enzyme hyaluronidase.

In addition to water and salts, ground substance contains glycosaminoglycans (nonsulfated and sulfated) and glycoproteins. The most common of these is hyaluronic acid. Mesenchymal cells and fibroblasts are principally responsible for the elaboration of ground substance. Other than serving as a matrix, ground substance provides for transportation of metabolites to and from vascular channels. Furthermore, it assists in maintaining tissue electrolyte balance. It also prevents the rapid movement of foreign materials and the spreading of pathogens.

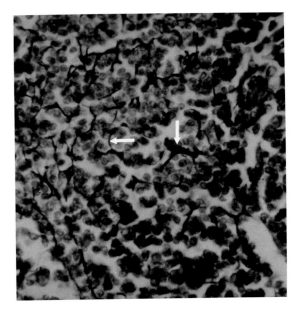

Fig. 2–19. Reticular fibers *(arrows)* of splenic stroma. (Silver stain; ×320)

Tissue fluid, the liquid arriving in tissues from plasma in the blood vessels, constitutes up to 30% of the body fluid. It holds crystalloids, gases, metabolites, and proteins in solution or suspension. Under special conditions, tissue fluid occurs interstitially as a free liquid. It quickly accumulates in injured and inflamed tissues.

Intercellular Fibers. Three varieties of intercellular fibers are recognized by histologists— *reticular, collagenous,* and *elastic.* Some oral histologists suggest a fourth variety, the *oxytalan fiber.*

Reticular Fibers. These are fine fibers with diameters smaller than 1 μm, that are aggregates of smaller elements, the *fibrils.* Reticular fibers have been demonstrated to be composed mainly of collagen type III (see below). Stained with hematoxylin and eosin, reticular fibers are almost imperceptibly pale pink; when silver techniques are used, they stain brown-black. Because of their affinity for silver they are known as *argyrophilic* (Gr. *argyros,* silver) or *argentaffin* (L. *argentum,* silver) fibers (Fig. 2–19). They are PAS-positive and mildly birefringent.

Reticular fibers differ from the classic collagen fiber (type I) in that they are smaller and richer in carbohydrates. Ultrastructurally, both reticular and collagen fibrils show cross markings at intervals of 64 nm, suggesting that their molecular arrangements are similar. Developmentally, reticular fibers appear first and are probably produced by mesenchymal cells. Reticular fibers are also the first produced in wound repair and later are replaced by collagen fibers (type I). Reticular elements comprise the framework of hemopoietic organs (lymph nodes, red bone marrow, and spleen) and stroma of glands. They also make up a latticework as the perithelium of capillaries, endomysium of smooth muscle fibers, and endoneurium of nerve cells. In teeth, reticular fibers form the developing matrix of the dental pulp.

Collagen Fibers. Collagen fibers are present in all connective tissues in which support and strength are required (Fig. 2–20). They are composed of collagen (type I) (see below). Collagen accounts for at least one-third of the protein in the body. Arrangement and orientation of the fibers vary with the functional demands of the tissues. For example, in tendons, the fibers are parallel to provide maximum strength with limited stretching ability (Fig. 2–21). Because of the support and strength potential of collagen, the fibers are numerous in tendons, ligaments, fascia, capsules, and sheaths of muscles and nerves (Fig. 2–21). Collagen fibers are also the chief component in fibrocartilage, dermis, and matrices of bone, cementum, and dentin. Collagen is believed to engage selectively in permeability, especially in the walls of vascular channels.

In the unpreserved state, collagen is colorless and is known as *white fiber*. When stained with hematoxylin and eosin the fibers are pink; with Masson's trichrome they are green, and with Mallory's trichrome they are blue. Collagen fibers are the most birefringent because of the ordered arrangement of the tropocollagen molecules. When boiled or denatured by strong chemicals, the fibers are reduced to a gel of polypeptides. Less harsh treatment involving mild acids or neutral salts reduces the fibers to *tropocollagen* molecules, the basic unit of a collagen molecule.

The fibril unit of collagen as determined by electron microscopy is not of uniform thickness; rather it varies between 20 and 100 nm. These units aggregate into fibril bundles or *fibers* that are 1 to 12 μm thick (Fig. 2–22). Fibers are sheathed and collected into fiber bundles that vary in thickness. Although fibrils of a given fiber do not branch, elements of one bundle can become separated to join a neighboring bundle. Later, they may recombine with the parent bundle.

The organization into fibrils and fibers depends on the type of collagen composition. Collagen types I and III organize into fibrils and fibers, type II into fibrils only; types IV and V

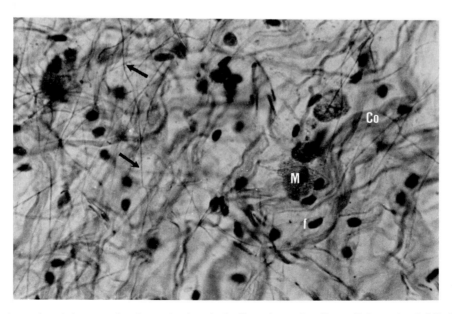

Fig. 2–20. Loose (areolar) connective tissue showing elastic fibers *(arrows),* collagen (Co), mast cell (M), fibroblast (f), and an assortment of other cells. (Hematoxylin and eosin stain; ×80)

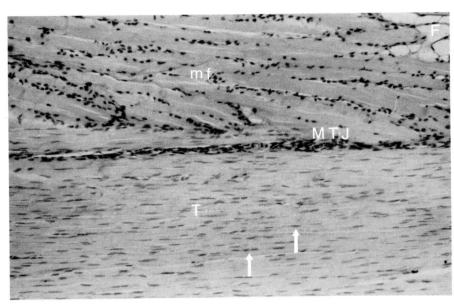

Fig. 2–21. Muscle and tendon (T) junction (MTJ). Note the flattened fibrocyte nuclei *(arrows)* compressed among the longitudinally oriented collagen bundles of a tendon (F, fat cells; mf, muscle fibers). (Hematoxylin and eosin stain; ×80)

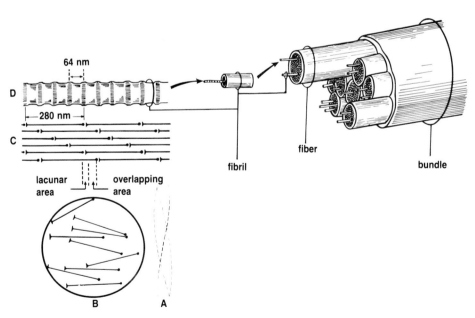

Fig. 2–22. Diagrammatic representation of collagen bundle and its subunits: polypeptide (α) chains, tropocollagen molecules, fibrils, and fibers. The protein subunits, tropocollagen molecules *(B)*, are composed of three polypeptide chains helically arranged *(A)*. Rod-shaped tropocollagen molecules (280 nm in length) combine in an ordered stepwise overlapping pattern to form collagen fibrils *(C)*. At the electron microscopic level fibrils appear striated, with repetitive dark and light bands at 64-nm periods *(D)*. The more intensely staining dark bands in negative stained preparations are gaps or lacunar areas between tropocollagen molecules. The light bands are overlapping sites of the ends of tails and the heads of the tropocollagen molecules. (Modified from Junqueira, L.C., and Carneiro, J.: Basic Histology. 3rd. Ed. Los Altos, CA, Lange Medical Publications, 1980).

form neither fibrils nor fibers. The organization of tropocollagen molecules into fibrils and fibers may involve proteoglycans and structural glycoproteins.

As mentioned above, ultrastructural observations reveal that fibrils possess repetitive cross banding at 64-nm intervals (Fig. 2–22D). This is a result of the stepwise orientation of the 280-nm-long rod-shaped tropocollagen molecules, which produce a lacunar space between molecules and overlapping areas. The lacunar segments stain more intensely and are darker (Fig. 2–22D).

Collagen Types. Many types of *collagen molecules* have been isolated which differ in the composition of tropocollagen molecules. Types I, II, III, IV, and V have been more completely characterized as discussed below.

The various collagen types occur in differing combinations and quantities. Tropocollagen, which is 280 nm long and 1.5 nm wide, is comprised of three polypeptide chains arranged in a triple helix (Fig. 2–22A). For example, the structure of collagen type I, that of the classic collagen fiber, includes one alpha-2 (α2) and two alpha-1 (α1) peptide chains dextrally coiled so that the interval between complete turns of the chains is 816 nm. The various types of collagen are due to differences in amino acid composition and sequences in the peptide chains (alpha-1, alpha-2). The alpha chains contain an abundance of *glycine* and two amino acids, *hydroxyproline* and *hydroxylysine* which are generally not present in large amounts in other proteins. The amounts of these amino acids in tissues are indicative of their collagen content. The conversion of proline to hydroxyproline requires vitamin C. Deficiency of vitamin C produces scurvy.

COLLAGEN TYPE I. Type I collagen is the typical collagen fiber in all connective tissues, in which support, strength and resistance to tension are required (Fig. 2–20). It not only possesses the most general distribution of the collagens, but it is the most abundant. With collagen types II and III, type I collagen fibers are known as *interstitial fibers*. The cells responsible for synthesis of type I fibers are cementoblasts, chondroblasts, fibroblasts, odontoblasts, osteoblasts, and their progenitor cells. The molecular formula of collagen type I is $[\alpha 1(I)]_2 \alpha 2(I)$ because it consists of two similar alpha-1 chains and one alpha-2 chain.

COLLAGEN TYPE II. Type II collagen occurs only as fibrils and is found in the matrices of hyaline and elastic cartilage and in the vitreous body of the eye, where it is organized as a loose network of delicate fibrils surrounded by an abundance of ground substance. The function of these fibrils is principally to resist intermittent pressure. They are observed with polarizing microscopy or with light microscopy, when the tissue is picro-Sirius stained. These fibrils are produced by chondroblasts, and they possess type II molecules with a molecular formula of $[\alpha 1(II)]_3$. Hence they consist of three identical alpha-1 chains.

COLLAGEN TYPE III. Type III collagen molecules are the main component of reticular fibers. Cells involved in the synthesis of type III fibers include pericytes, hepatocytes, fibroblasts, mesenchymal cells, reticular cells, smooth muscle cells, and Schwann cells. They possess the molecular formula $[\alpha 1(III)]_3$, because they consist of three identical alpha-1 chains.

COLLAGEN TYPE IV. Type IV collagen molecules are components of basement membranes and eye lens capsules. At the optical level, collagen appears as a thin diffuse coat or membrane that is weakly birefringent. It has been suggested that this type of collagen is mostly unpolymerized or partially polymerized procollagen molecules that are not organized into fibrils or fibers. The molecular structure reported is $[pro\alpha 1(IV)]_2 \, pro\alpha 2(IV)$ or $\alpha 1(IV)$. This type of collagen is produced by endothelial and epithelial cells, muscle cells, and Schwann cell. The functions suggested are those associated with basement membranes; hence they include support and selective filtration between epithelial and connective tissues.

COLLAGEN TYPE V. The alpha chains of some collagens are different from those of others, at least in basement membranes, and these collagens are collectively grouped as type V. They are found in placental basement membranes, fetal membranes, blood vessels, and other tissues. Neither fibers nor fibrils are produced, and the molecular formula suggested is $[\alpha 1(V)]_2 \alpha 2(V)$. Definitive data are meager and await further study.

Elastic Fibers. These fibers are homogeneous and elastic ranging from 0.2 to 4.0 μm in diameter, and can stretch twice their length (Figs. 2–20 and 2–23). They are not organized into bundles and branch freely, often forming a meshwork. Their color in living tissue is responsible for the term "yellow fibers." Stained with hematoxylin and eosin the fibers are bright red. With polarized microscopy a birefringence is noted, which increases with stretching.

Electron microscopy reveals that the core of the fiber is comprised of an electron-lucid amorphous material, the protein *elastin*. Aperiodic microfibrils, 13 nm in diameter, surround the core. Elastin is composed of the amino acids glycine and proline, with generous amounts of valine. The hydroxyproline content is sparse, and hydroxylysine is absent in elastic fibrils. Two distinct amino acids form elastin—desmosine and isodesmosine—and, with elastin polymerization, these cause cross-linkage of the polypeptides. The peripheral microfibrils are carbohydrate-rich but contain no hydroxyproline or hydroxylysine. Accordingly, they are considered to be glycoproteins.

Proelastin, which is organized extracellularly into elastic fibrils, is synthesized by mesenchymal cells, fibroblasts, and smooth muscle cells of vascular channels. The mechanism of fibril production is similar to that of collagen. That is, fibrillogenic cells synthesize proelastin, which is converted to *tropoelastin* extracellularly. In the

intercellular spaces enzymes promote linkage of the tropoelastin molecules to form elastin. The difference in the processes of elastic fiber formation is primarily in the microfibrils that are produced initially. These extracellular structures provide the framework on which elastin organization into fibers is based. In the process some of the microfibrils become incorporated or embedded into the elastin core. Cross-linkage of elastin tends to progress with age and, when excessive in arterial walls, reduces their elasticity, with potential harmful effects.

Oxytalan Fibers. Electron microscopic examination indicates that the ultrastructural features of the oxytalan fiber are similar to those of the elastic fiber. Accordingly, some oral histologists believe them to be developing elastic fibers. In oral tissues the fibers occur in the developing dental pulp, gingivae, and periodontal ligaments.

Connective Tissue Cells

There are two types of connective tissue cells, fixed and transient.

Fixed Cells. These include fat cells, fibroblasts, and fixed macrophages (histiocytes) and mesenchymal cells.

Mesenchymal Cells. These are the dominant cell type in the early embryo (Fig. 2–24). In definitive tissues, however, they are scarce and

Fig. 2–23. Elastic fibers *(arrows)* constituting the middle layer (media) of a large artery. (Verhoeff's stain; ×80)

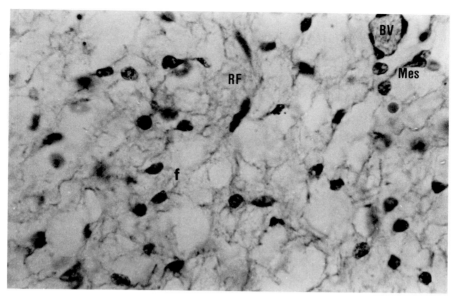

Fig. 2—24. Mesenchymal cell nuclei (Mes) and fibroblasts (f) in embryonal (mesenchymal) connective tissue (BV, blood vessel; RF, reticular fibers). (Hematoxylin and eosin stain; ×320)

are located mostly in capillary beds. They are small star-shaped cells and, because the stainable cytoplasmic components are so few and the nuclei so large, with coarse chromatin, only the latter are resolved at the optical level. The organelles include strands of endoplasmic reticulum and a few mitochondria. Numerous intercommunicating processes present a network

appearance. Though smaller, they can be mistaken for fibroblasts.

Fibroblasts. Of the fixed connective cells, fibroblasts are the most numerous. They are large and fusiform, with branching processes of differing dimensions. The nucleus, the most prominent organelle, is light-staining, oval, and centrally located. Mature *fibrocytes* are regular

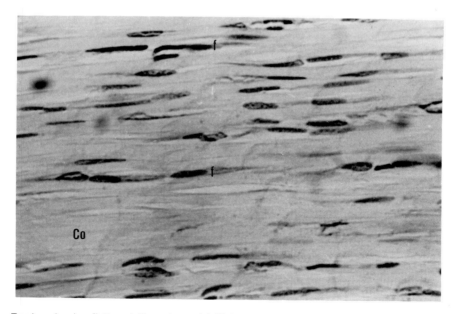

Fig. 2—25. Tendon showing flattened fibrocyte nuclei (f) between collagen fibers (Co). (Hematoxylin and eosin stain; ×320)

components of tendons and ligaments. Only their flattened nuclei may be seen (Figs. 2–21 and 2–25). Fibrocytes do not produce new fibrils; rather, they maintain the collagen.

Electron microscopic study shows that fibroblasts are rich in organelles, particularly rER and Golgi complexes, which are active in fiber synthesis (Fig. 2–26). Lysosomes remove spent and surplus organelles. Fibroblasts are functionally competent and capable of mitotic activity; these features are valuable in wound healing.

The main functions of fibroblasts are synthesis of ground substance and fibril production and maintenance. It should be noted that the fibroblast is not the exclusive fibrillogenic cell. Odontoblasts (dentin-forming), cementoblasts (cementum-forming), chondroblasts (cartilage-forming), osteoblasts (bone-forming), mesenchymal cells, and even epithelial cells (in early embryogenesis) produce fibrils for their matrices. Other cells engaged in the synthesis of collagens have been discussed above. It should be noted further that the fibrillogenic mechanism is the same, irrespective of the cell type involved. The production of the procollagen alpha chains is initiated in the rER and, in minutes, the material

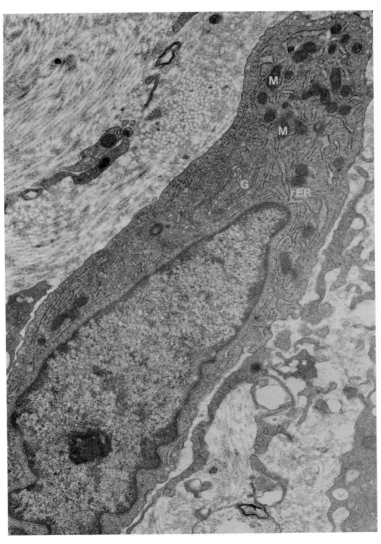

Fig. 2–26. Electron micrograph of fibroblast showing organelles involved in collagen synthesis: rough endoplasmic reticulum (rER), Golgi apparatus (G), and mitochondria (M). (Urnayl acetate and lead citrate stain; ×53,000) (Courtesy of Aaron Bernstein.)

is directed by vesicles to the saccules of the Golgi system (Fig. 2–27). In the latter, the alpha chains are organized first into parallel threads and then into rods. Within minutes the procollagen rods are extruded by Golgi saccules in secretory vesicles. These are released from the cell surface into intercellular spaces, in which they are converted by enzyme action to tropocollagen and then to collagen fibrils.

Fat Cells (Adipocytes). Fat cells occur in most connective tissues (Figs. 2–21 and 2–28). In areas in which they are the dominant cell population, they form *adipose tissue.* Fat cells do not reproduce—rather, they are replaced or replenished by mesenchymal cells and by adventitial cells in yellow marrow. Young cells, lipoblasts, contain fat droplets (liposomes); as cells mature to adipocytes, the liposomes accumulate and coalesce to form a single droplet. In the process the nucleus is pushed against the plasmalemma, where it is surrounded by a halo of cytoplasm. The organelle populations diminish proportionately with liposome formation. The

maturation of the adipocyte described represents *unilocular* fat storage; in humans this type of lipid is called *white fat.* This is in contrast to *brown fat,* which occurs mostly as fetal (newborn) fat, in which the liposomes do not coalesce but remain independent and form *multilocular* deposits. Brown fat cells are smaller, with larger organelle populations and larger mitochondria.

Lipoblasts are transformed into adipocytes during fat synthesis and storage and, if the stored lipid is consumed, they revert to lipoblasts.

Histiocytes (Fixed Macrophages). These are soft tissue scavengers that destroy debris, dead cells, and foreign bodies (phagocytosis). Normally they are stationary, but with severe inflammations they can migrate to affected sites.

Histiocytes are star- or spindle-shaped cells with intense-staining oval, centrally located nuclei (Fig. 2–29). With electron microscopy microvilli are observed on the cell surface. The centrosome is located near the nucleus; it is surrounded by a dense population of mitochondria and the Golgi complex. Lysosomes, residual

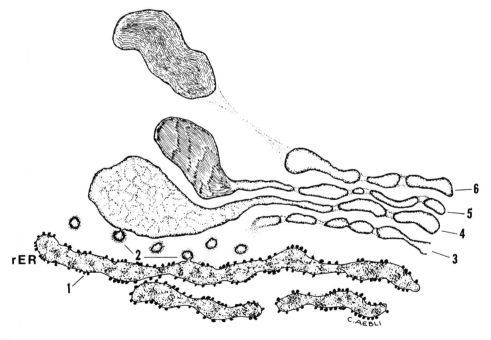

Fig. 2–27. Diagrammatic representation of the mechanism of collagen synthesis. Stage 1 shows rough-surfaced endoplasmic reticulum (rER) with budding transfer vesicles. In stage 2 the transfer vesicles departing from the rER fuse with a dilated saccule of the Golgi apparatus. Stage 3 shows the first Golgi saccule of a stack. Stage 4 shows a dilated terminal in a Golgi saccule, the second of the stack with its associated transfer vesicles. Stage 5 shows a dilated saccule housing threads in parallel array. Stage 6 shows saccules and a freed secretory vesicle, derived from the saccule. (Redrawn from Weinstock, M., and Leblond, C.P.: Synthesis, migration and release of precursor collagen by odontoblasts as visualized by radioautography after ³H proline administration. J. Cell Biol., 60:121, 1974, by copyright permission of Rockefeller University Press.)

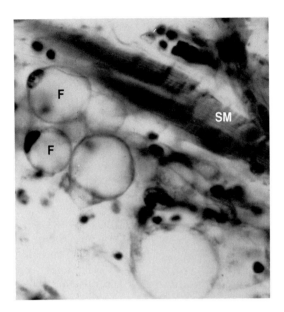

Fig. 2–28. Fat cells (F) showing "signet ring" appearance. Note that the empty space intervening between the nuclei and the cell-limiting membrane is caused by the dissolution of the fat droplet, which occurs during histologic tissue processing (SM, striated muscle fibers). (Hematoxylin and eosin stain; ×320)

bodies, and vacuoles are abundant. In addition to phagocytosis, histiocytes are involved in immunologic defense.

Free Cells. Nonstationary cells include leukocytes, mast cells, plasma cells, and some chromatophores, as well as macrophages, including reticulocytes. (Cells such as the lymphocytes, monocytes, neutrophils, basophils, and eosinophils will be described below.)

Mast Cells. In the oral cavity, mast cells are abundant in gingival connective tissue. They are also found in vascular beds, in which they produce histamine (a stimulant for increasing capillary permeability) and heparin (a blood anticoagulant).

The cells are large and round or oval in profile, and contain large, red-staining cytoplasmic granules (Figs. 2–20 and 2–30). The nuclei are centrally located. Ultrastructural features include short blunt processes and cytoplasm that contains numerous mitochondria and a well-developed Golgi complex. Neither smooth nor rough endoplasmic reticulum is extensive. The granules are membrane-bound, and vary in diameter from 0.2 to 0.8 μm. Vacuoles, products of discharged granules, and disintegrating granules are also present (Fig. 2–31). The Golgi apparatus

is responsible for the production of heparin and its packaging as granules.

Plasma Cells. These antibody-producing cells are the body's first line of defense; their progenitors are B-type lymphocytes. In lymphoid tissue and intestinal mucosa they are abundant, but are scarce elsewhere. Plasma cells are unlike lymphocytes in that they contain more basophilic cytoplasm. Also, they show a pale-staining juxtanuclear area. The eccentric nucleus contains dense chromatin masses that are radially arranged, similar to spokes in a wheel (Fig. 2–32). Young cells contain acidophilic *Russell bodies.*

Electron microscopic examination reveals an area adjacent to the nucleus that contains the diplosome and Golgi apparatus. The rER consists of flattened cisternae and saccules among helical arrangements of polyribosomes. Proteins produced by the rER may be incorporated with immunoglobulins immediately, or subsequently in the Golgi complex, in which they are packaged in secretory vesicles. Russell bodies observed by electron microscopy have been determined to be extra-large accumulations of secretion vesicles (granules), which probably indicate imminent cell death.

Reticular Cells. These are cells that synthesize the reticular fiber matrix for glands. The numerous cell processes are joined by desmosomes, forming a network. The cytoplasm is abundant, and weakly basophilic, and the nucleus is large and pale-staining (Fig. 2–33). Electron microscopy confirms these features, suggesting that these are developmental cells. They may be related to the mesenchymal cell and, therefore, may be the immediate precursors of other connective tissue cells, including blood cells and yellow marrow. In hemopoietic (blood-forming) tissues the reticular cell is known as an adventitial, or primitive reticular cell. It has been proposed that, as needed, reticular cells engage in immunologic activities and become phagocytic.

Macrophages. These are defense and scavenger cells that function by lysosomal (enzyme) digestion. Monocytes of the red bone marrow migrate to connective tissues, where they become macrophages. They are strategically located near blood vessels in lymphoid tissue and liver sinusoids (Fig. 2–34A). The shape of macrophages varies with functional activity, except

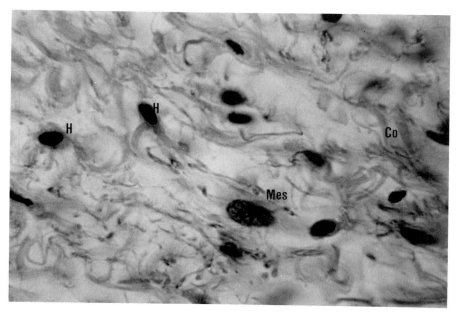

Fig. 2–29. Loose connective tissue showing histiocytes (H), a mesenchymal cell (Mes), and collagen fibers (Co). (Hematoxylin and eosin stain; ×970)

during quiescent periods. Normally they resemble fibroblasts, except for their decreased size and condensed nuclei. Macrophages are ameboid. Their short blunt processes probe surfaces for attachment and engulf material for lysosomal digestion. If the substance to be removed exceeds the capacity of one cell several macrophages may fuse to execute the task, and a *foreign body giant cell* is formed.

At the ultrastructural level, folds and microvilli-like structures occupy the cell surface. Except for phagosomes, the organelle complement is reduced (Fig. 2–34B).

Leukocytes. These white blood cells leave the vascular channels, wandering among connective tissues. They return to blood vessels by entering and traveling through successively larger lymph vessels, eventually arriving at the thoracic or

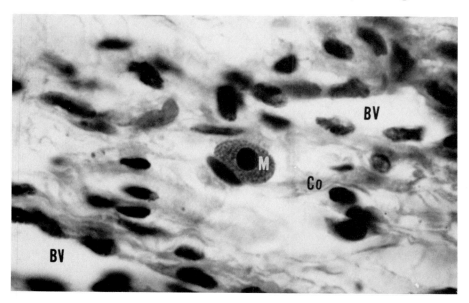

Fig. 2–30. Gingival connective tissue showing mast cell (M) between two blood vessels (BV) (Co, collagen). (Hematoxylin and eosin stain; ×800)

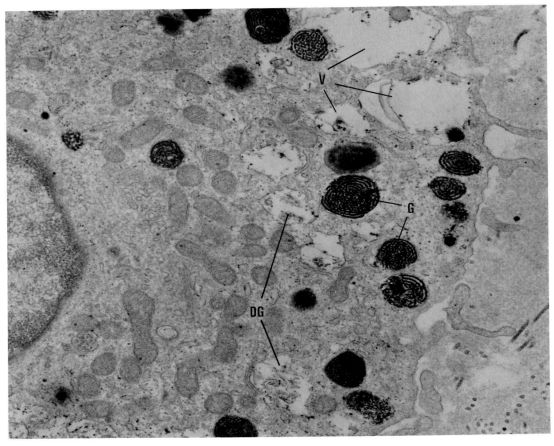

Fig. 2–31. Electron micrograph of mast cell in stomach connective tissue showing nucleus and mitochondria, as well as granulated (G), degranulated/vacuolated (V), and disintegrating granules (DG). Note the short cell processes. (Steer, H.W.: Mast cells of human stomach. J. Anat., *121*:395, 1976, Cambridge University Press.)

right lymphatic ducts, which drain lymphatic circulation and empty its lymph into venous circulation. The agranular leukocytes (monocytes and lymphocytes) are more abundant in connective tissues than the granular ones (eosinophils, basophils, neutrophils).

Connective tissue *lymphocytes* average 7 μm in diameter and consist of a thin rim of basophilic cytoplasm surrounding the round nucleus. The nucleus is slightly cleft, and stains intensely because of the condensed chromatin. There are few other organelles, except for free ribosomes. Lymphocytes are found most frequently at strategic sites in the digestive and respiratory systems, where they produce antibodies and combat inflammation. Some lymphocytes of the lamina propria and certainly those in epithelium represent moribund cells that have migrated to the lumen to be excreted with wastes (Fig. 2–15).

Connective tissue *eosinophils* are found mostly in the gastrointestinal and respiratory tracts and in lactating breasts. They are especially abundant in persons undergoing parasitic and allergic attacks and are morphologically similar to eosinophils in circulating blood. They are round, and the nucleus is bilobular. With acid dyes, the cytoplasm reveals distinct brick-red granules of uniform diameter and distribution. With electron microscopy the enzyme-containing granules are found to be membrane-bound, with a granular protein matrix and a few flat crystals.

Chromatophores. These pigment-containing cells are located mostly in the skin, lining of the mouth, hair, pia of the brain, and choroid coat of the eye. In regard to melanin pigment, chromatophores may exist as pigment producers (melanocytes) or as pigment phagocytes (melanophores).

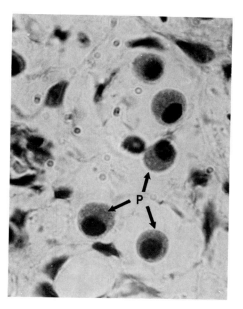

Fig. 2–32. Plasma cell showing eccentric nuclei (P). (Hematoxylin and eosin stain.)

Melanocytes are pigment synthesizers that originate from the neural crest. After the third month of embryonic development they travel as melanoblasts to the dermoepithelial junction of the skin. Here they differentiate into melanocytes, with processes extending into the deeper

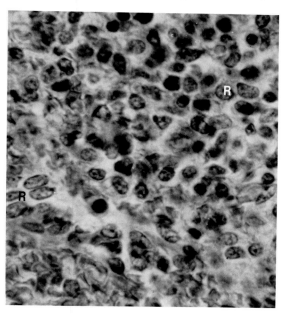

Fig. 2–33. Reticular cells (R) and assorted hemopoietic cells in splenic parenchyma. Note that reticular fibers are not observed in this preparation. (Hematoxylin and eosin stain; ×970)

layers of the skin (Fig. 2–35). Except for infants, with Down syndrome, the dermoepithelial junction is the exclusive location for these cells. Melanophores envelop and remove melanin from melanocytes. In the skin they are components of the stratum spinosum (see Fig. 1–18), and in the dermis they are probably macrophages.

The round cell bodies of melanocytes in the skin occupy positions in the deeper layers as clear cells, with processes extending surfaceward between epithelial cells. It has been suggested that about one-fourth of the cells of the basal epithelial layer are melanocytes. Because they are nonpigmented they are difficult to distinguish from epithelial cells, except when a histochemical procedure that blackens the cells, the dopa reaction, is used.

Electron microscopic examination shows these cells to differ from their epithelial neighbors by an absence of tonofilaments and desmosomes. Synthesis of melanin is normally stimulated by ultraviolet light and x-rays. Trauma and infection may also initiate pigment production. Melanin synthesis begins in the endoplasmic reticulum as a protein product, which is converted into protyrosinase-containing membrane-bound packets in the Golgi apparatus. The packets are changed first to *premelanosomes* and later to tyrosinase-containing oval *melanosomes*. Melanosomes are about 0.5 μm long, and reveal an internal concentric lamellar structure. Both the structure and enzyme content of the melanosome are lost as pigment production progresses. Melanin is transferred to epithelial cells, a mechanism that is not totally understood. It is believed that the epithelial cells of the skin phagocytize the pigment contained in the processes of the melanocyte by endocytosis.

Pericytes. These are found generally around capillaries, often girdling the endothelium by their processes. Because of their location, they have been designated variously as adventitial, perivascular, and undifferentiated mesenchymal cells. Because pericytes probably participate in wound-healing activities such as fibril and ground substance production, vascular bud invasion, and blood vessel formation, it has been proposed that pericytes are precursors of fibroblasts and smooth muscle cells. Pericytes do not possess contractile properties. Their cytomorphologic features are similar to those of mesenchymal cells.

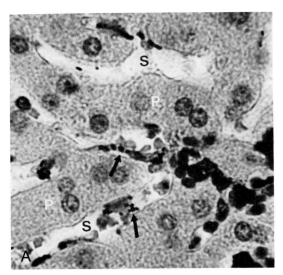

Fig. 2–34. *A,* Liver parenchyma (P) and sinusoids (S) of an animal injected with India ink. Particles (arrows) are phagocytized by sinusoidal macrophages. (Hematoxylin and eosin stain; ×970) *B,* Electron micrograph of rat macrophage showing eccentric nucleus (N), dense bodies (D), and scant rER. (Courtesy of C.P. Leblond.)

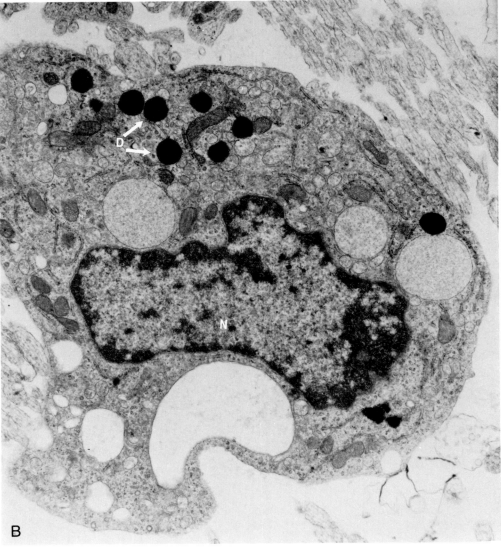

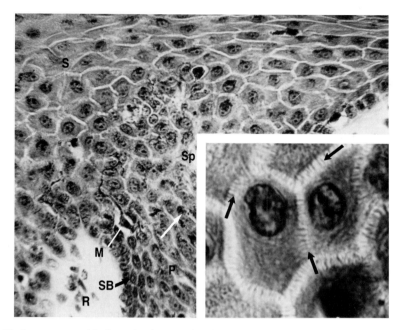

Fig. 2–35. Stratified squamous epithelium showing an epithelial peg/ridge (P). External to the connective tissue (papillary) ridges (R) is the basal epithelial layer (SB). Note that the spinous layer (Sp) contains melanocytes or their processes (M). Also note that the superficial cells are more squamous in appearance (melanin pigment, *white arrow*). (Hematoxylin and eosin stain; ×80) *Inset,* Note the intercellular bridges *(arrows),* artifacts formed by adjacent cell segments attached at desmosomes.

Classification

Connective tissues are identified by the presence of few cells, with an abundance of intercellular material (amorphous ground substance and fibers). In some organs connective tissues are characteristically distinct, while in others they are indistinguishable. Connective tissues include two categories, connective tissue proper and specialized connective tissue (Table 2–2).

Classification of regular connective tissue involves either the number or type of the fibers or cells present. For example, if many fibers are present, the term "dense" is used. If the number of fibers is relatively few and diffuse, the tissue is characterized as "loose." Cell or fiber preponderance in the tissue results in such designations as elastic, reticular, adipose, or mesenchymal connective tissue.

Connective Tissue Proper

Embryonal Tissue. These are more prominent in early embryonic development, and include mesenchymal and mucous connective tissues. After birth these tissues are normally found in the dental pulp; otherwise they are present in healing wounds or tumorous lesions.

Mesenchymal Tissue. This is the progenitor connective tissue, and is found mostly in the young embryo. It has the potential to differentiate into other connective tissues and into muscle (Table 2–2). This primitive connective tissue consists of networks of stellate or spindle-shaped cells in a gel-like environment that is abundant in reticular fibers and ground substance. The nuclei of the mesenchymal cells are large, round, and pale. Optical microscopic examination shows the tissue to stain weakly with hematoxylin and eosin (Fig. 2–24).

Mucous Tissue. Some histologists suggest that this tissue is intermediate between mesenchyme and the adult connective tissues. As a definitive tissue it comprises *Wharton's jelly* in the umbilical cord (Fig. 2–36), the vitreous humor of the eye, and the dental pulps. Other than a few fibroblasts, mesenchymal cells, macrophages, and fibrils (mostly collagenous), the bulk of the tissue is comprised of a gelatinous matrix. Except in dental pulps, blood vessels, lymphatics, and nerves are usually not constituents of this connective tissue. When present, they are en route to neighboring tissues.

Fibers of the umbilical cord and dental pulp

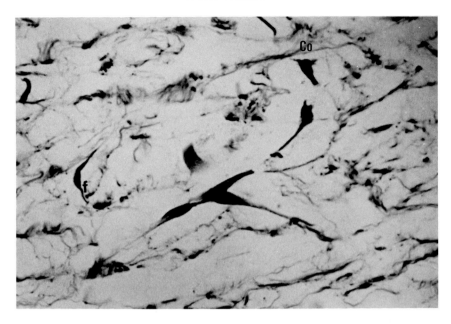

Fig. 2–36. Umbilical cord. This mucoid connective tissue is mostly composed of viscous matrix, collagen (Co), and a few fibroblasts (f). (Hematoxylin and eosin stain; ×320)

tend to increase with age. In the umbilical cord, mucous connective tissue serves to protect the umbilical vessels, the life-sustaining link between mother and embryo. In the dental pulp it provides for tooth vitality, and in the vitreous humor it accommodates fluid exchanges with the aqueous humor.

Regular Connective Tissue. Loose and dense connective tissues are the two main subtypes of regular connective tissue. They are more differentiated connective tissues, as evidenced by the different cell types present.

Loose Type. In general, loose connective tissue is distinguished from the dense type in that the fibers are fewer and diffusely arranged, while cells and ground substance are more plentiful. Loose (areolar) connective tissue is found in organs requiring mobility, packing, padding, fastening, stretching, and sheathing. Loose connective tissue is a necessary component of nerves, blood, and lymph vessels, as well as of the walls of tubes in the respiratory and digestive systems. The subepithelial connective tissue of the digestive and respiratory tubes is loosely arranged, and is called the *lamina propria*. Collectively, the two layers (epithelium and connective tissue) comprise the *mucosa*. The connective tissue beneath the mucosa is known as the *submucosa*, and it is generally separated

from the lamina propria by a muscle layer, the *muscularis mucosa*. In some areas, as in the nasopharynx, the separating tissue is an elastic fiber layer, the *elastic lamina*.

Other than the functions indicated above, loose connective tissue may serve to support and protect the organs in which they are contained. The repair process is performed mostly by mesenchymal cell/pericyte reserves in the capillary beds. These are not only fibrillogenic but may differentiate into connective tissue cells.

Of the cells, fibroblasts and macrophages are the most numerous in loose connective tissue (Figs. 2–20 and 2–29). Mesenchymal and mast cells, when present, are restricted mostly to capillary beds (Fig. 2–30), while fat and other types of cells occur throughout the tissue, singly or in groups.

Dense Type. The increase in fiber content in dense connective tissue is accompanied by a decrease in both cells and ground substance. This requires fewer nerves, blood, and lymphatic vessels. Fibers may be arranged regularly or irregularly, and thus form dense regular or dense irregular connective tissues, respectively.

Body parts requiring strong, tough sheets or cords are constructed of dense, regularly arranged fiber bundles, as in ligaments connecting bones and supporting organs in the abdominal cavity, tendons attaching muscles to bones (Fig. 2–21), and aponeuroses linking muscles.

Membranes, protectively surrounding or internally partitioning organs, are generally irregularly arranged, dense, fibrous connective tissue. For example, the kidneys and certain glands are protectively covered by *capsules*. Membranes known as *fascia* separate organs of the abdominal cavity (Fig. 2–37). *Sheaths* of connective tissue cover muscles as epimysium, nerves as epineurium, cartilage as perichondrium (see Fig. 2–40), and bone as periosteum (see Fig. 2–46A). Capsular connective tissue extensions called *septa* partition the substance of glands; these divide the glandular parenchyma into large territories known as lobes or into smaller segments called lobules (see Figs. 9–17, 9–33, and 9–34). A lobule is a tissue unit accommodated developmentally by a single arteriole.

Adipose Tissue. This classification is reserved exclusively for tissues in which adipocytes make up most of the cell population. Among the regions of the body in which adipose tissue may be found are under the skin (panniculus adiposus), surrounding the kidneys and suprarenal glands, filling the surface grooves of the heart, and in mesenteries. Originally it was thought that the fat component of yellow bone marrow was due to adipocytes. The current belief is that adventitial (reticular) cells engage in fat accumulation in hemopoietic (myelocytic) tissue.

Fat tissue is generally surrounded by a dense fibrous sheath, and its mass is partitioned into lobules by septa. Loose connective tissue originates directly from mesenchyme, fat tissue arises from loose connective tissue. Additionally, although loose connective tissue cannot revert to mesenchyme, adipose tissue can revert to loose connective tissue.

Reticular Connective Tissue. The dominant fiber type is the reticular fiber. This type of tissue is found in organs requiring a fine supporting latticework as in bone marrow, liver, and lymphoid organs, including the spleen (Fig. 2–19). In these, the fibers are arranged as a diffuse framework, the stroma, which supports the parenchyma (functional cell components) of the organ while simultaneously providing for percolation of fluids and cells through the gland. As components of the basement membrane, the fibrils serve to support, anchor, and connect the epithelium to the underlying connective tissue, the lamina propia.

Reticular cells, mesenchymal cells (pericytes), macrophages of the reticuloendothelial system, and fibroblasts are normally found in this tissue. Although reticular fibers are also found in numerous structures, such as fat aggregations, endomysium, and capillary perithelium, they are not abundant enough to warrant classification as a tissue.

Elastic Tissue. This classification is based on the principal fiber components, the elastic fiber, which is composed of the protein elastin. Under conditions of sustained tension, its properties of elasticity and resiliency make this tissue most suitable for structures such as the following: (1) the ligamentum flavum, which connects the laminae of adjoining vertebrae; (2) the stylohyoid ligament, which connects the styloid process of the temporal bone and the lesser horn of the hyoid bone; (3) the suspensory ligament of the penis; and (4) the true vocal cords. Although not present in amounts large enough to be considered elastic tissue, the fibers are abundantly present in the walls of large arteries (Fig. 2–25) and in other tubes requiring elasticity and resiliency.

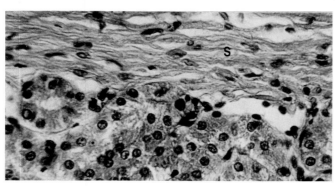

Fig. 2–37. Dense irregular connective tissue forms the capsular sheath of the kidney (S), surrounding tubules consisting of cuboidal and low columnar epithelium. (Hematoxylin and eosin stain; ×450)

Specialized Connective Tissue

These may be either fluid or rigid.

Fluid Connective Tissue. There are two main categories, blood and lymph.

Blood. This fluid constitutes about 7% of the body weight, and is about 5 quarts in volume. It is a connective tissue that consists of plasma, ground substance, fibrin fibrils (present in clotting), and cells, which are either white (leukocytes) or red (erythrocytes) blood cells (Fig. 2–38). There are between 5000 and 9000 leukocytes per cubic millimeter and, depending on the presence or absence or cytoplasmic granules, are designated as granular or agranular. Plasma also contains hemoconia, chylomicrons, and platelets, which are fragments of cytoplasm liberated by large cells called megakaryocytes that are found in red bone marrow, a blood-producing (hemopoietic) tissue (see Fig. 2–39). Hemoconia are cell debris, and chylomicrons are fat droplets 1 μm in diameter.

AGRANULAR LEUKOCYTES. These include monocytes and lymphocytes.

Lymphocytes make up between 20% to 35% of the total white blood count. They may be small, medium, or large; those in circulating blood are almost exclusively small and are about the size of an erythrocyte. Most of the cell volume is occupied by a round, slightly indented nucleus that stains intensely because the chromatin material is condensed. The cytoplasm is scant, and may contain azurophilic granules. Medium and large lymphocytes are found mostly in lymphatic tissue, and probably represent the less mature lymphocytes. Three types of lymphocytes have been identified by special cytochemical methods, and have been named according to their embryologic site of activation. These are the T lymphocytes (thymus-dependent lymphocytes), B lymphocytes (bursa-dependent lymphocytes), and null cells. Lymphocytes are active in the immune response and are phagocytic. The T lymphocytes are involved with the cell-mediated immune response, while B lymphocytes function in the humoral immune response.

Monocytes comprise about 5% of the total white blood count and are about 16 μm in diameter. The cytoplasm contains nonspecific azurophilic granules, probably primary lysosomes. The nucleus, which occupies about half of the cell volume, is composed of less condensed chromatin than lymphocytes, so it stains lighter. It is somewhat eccentric, oval, kidney-shaped, or deeply notched. Monocytes migrate into the tissues of the body and are transformed into phagocytic macrophages, which are involved in the immune system.

GRANULAR LEUKOCYTES. Most leukocytes of circulating blood which includes neutrophils, eosinophils, and basophils, possess distinctly lobed nuclei that are connected by chromatin strands. The cytoplasm contains distinctive granules that have an affinity for specific stains.

Neutrophils make up about 65% of the white blood cells. In dry smears they are 10 to 12 μm in diameter. There are two to five nuclear lobes; nuclear lobulation and phagocytic potential increase with age. Of the specific and nonspecific granules in the cytoplasm, the latter are fewer and azurophilic, while the former stain poorly with neutral dyes. The azurophilic granules contain enzymes, suggesting that they are lysosomes. Because neutrophils are involved in defense against pathogenic organisms, they occur most often in areas of infection.

Eosinophils in humans possess bilobed nuclei, and the cytoplasm contains coarse specific granules. In dry smears eosinophils are about 12 μm in diameter. The eosinophilic granules are rounded and contain lysosomal enzymes similar to those of the nonspecific granules of the neutrophils. Because these cells participate in combating allergic reactions and parasitic infections, they occur mostly in sites subjected to allergic irritation, such as the nasal and respiratory passages. They are thought to phagocytose antigen-antibody complexes.

Basophils are not observed frequently in circulating blood (0.5% of leukocytes). In dry smears their diameter is about 10 μm, and they contain coarse basophilic round granules that stain blue to purple. When present in great numbers they may obscure the pale-staining, S-shaped, bilobed or trilobed nucleus. Basophils become phagocytic when necessary. They normally contain histamine and heparin, and are functionally similar to mast cells.

Erythrocytes are biconcave disks, 7 to 8 μm in diameter and devoid of nuclei (Fig. 2–38). As mature cells they circulate in blood vessels for their life span of about 120 days. There are between 4.5 and 5.0 million erythrocytes in

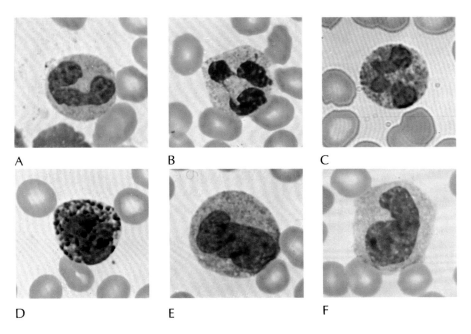

A

B

C

D

E

F

Fig. 2–38. Blood smear of human blood. *A* and *B,* Neutrophil. *C,* Eosinophil. *D,* Basophil. *E* and *F,* Monocytes. (Wintrobe, M.M., et al.: Clinical Hematology. 8th Ed. Philadelphia, Lea & Febiger, 1981.)

women, and between 5.0 and 5.5 million in men per cubic millimeter of blood. As prospective erythrocytes differentiate in red marrow, the nucleus and other organelles are removed to provide space for the accumulation of hemoglobin. This pigment provides for the exchange of gases (oxygen and carbon dioxide) in the capillary beds. The biconcave shape of the erythrocyte allows it to fold so as to pass through the smallest capillaries. Its shape further provides for increased hemoglobin content.

Thrombocytes, or blood *platelets*, are colorless cytoplasmic extrusions of megakaryocytes that are about 3 μm in diameter. Surface view reveals them to be round, while in profile they are spindle- or rod-shaped (Fig. 2–38). In circulating blood, in which they live for about 9 days, they number between 200,000 and 350,000/mm³. Hematoxylin and eosin staining reveals two areas: a lighter peripheral area, the hyalomere, and a darker center, the chromomere. Blood platelets function in blood clotting by adhering to the traumatized site of the blood vessel to produce a white thrombus.

PLASMA. This fluid transports material of the blood contains various substances, including amino acids, antibodies, carbohydrates, chylomicrons, enzymes, fats, gases, hormones, hemoconia, and proteins.

HEMOPOIESIS (BLOOD FORMATION). The life span of blood cells is short, and they must be continously replenished by hemopoietic (blood-forming) tissues. Hemopoiesis (also known as hematopoiesis) occurs in lymphatic and myeloid tissues, such as red bone marrow. Lymphatic tissues are responsible for the formation of lymphocytes; thus, they are developed in lymph nodes and nodules, tonsils, and spleen. The developmental lymphocytes, large and medium, are normally found only in lymphatic tissue. The adult functional form (small lymphocyte) is found in circulating blood.

Red bone marrow is the site for the production of erythrocytes, granular leukocytes, monocytes, and platelets. It is also believed to be the site of origin for lymphocytes. Active sites for myeloid hemopoiesis in adult humans include the red marrow of flat bones such as those of the cranium, pelvis, ribs, scapula, and sternum. Additional sites include the proximal ends of long bones such as the humerus and femur.

It has been demonstrated that all blood cells arise from a pleuripotential stem cell, which can proliferate as well as differentiate into more limited, unipotential stem cells. The *colony forming unit spleen* (CFU-S) cell is considered to be the multipotential stem cell that gives rise to several colony-forming units (CFU), which differentiate into the specific blood cell types. The CFU-S cell thus produces colony-forming units giving rise to erythrocytes (CFU-B), to eosinophils (CFU-Eo), to neutrophils (CFU-G), to monocytes (CFU-M), and to platelets (CFU-Meg). Neutrophils and monocytes may also originate from another stem cell, the colony-forming unit—granulomonocyte (CFU-GM). Basophils may arise from CFU-G cells, but their exact origin is unknown. Lymphocytes arise from unipotential lymphoblasts, which in turn arise from the CFU-S cell without a CFU as an intermediary. The CFU cells are not morphologically identifiable using the usual blood staining techniques.

The first morphologically identifiable cell type in the erythrocyte series, the *proerythroblast*, arises from the colony-forming unit—erythroid (CFU-E or E-CFU) cell. The latter takes its origin from the colony-forming unit—burst (CFU-B or E-BFU) cell, the first unipotential cell in the series. The progressive stages in the sequential development of the erythrocyte from the proerythroblast are basophilic eythroblast, polychromatophilic erythroblast, orthochromatophilic erythroblast (normoblast), reticulocyte, and mature erythrocyte. During the various developmental stages, the size of the developing erythrocyte decreases, the chromatin condenses, organelle populations (including the nucleus), decrease and eventually disappear, hemoglobin increases, and cell morphology changes.

In the granulocytic series, the *myeloblast* is the first morphologically recognizable cell to arise from the unipotential CFU cells. The sequential stages from myeloblast to granulocyte include the promyelocyte, myelocyte, metamyelocyte, band (stab) cell, and mature granulocyte. Beginning with the myelocyte, specific granules are present so that the cells can be distinguished histologically as basophilic, neutrophilic, or eosinophilic types. During differentiation cell size diminishes, chromatin condenses, the nucleus becomes lobulated, and the cell acquires its full complement of specific cytoplasmic granules.

In lymphocyte formation, lymphoblasts arise from CFUs (stem cells). The latter abandon the marrow (site of origin) and migrate first to the thymus, where they proliferate and differentiate into programmed T lymphocytes; then they proceed to the spleen for further modification and finally enter circulation. B-lymphocyte production in mammals occurs primarily in bone marrow, although the spleen and gut lymphoid tissue are also implicated. Their sites of maturation and activation have not yet been clearly defined. In avian hemopoiesis, however, it has been determined that B-lymphocyte stem cells from the marrow migrate to the bursa of Fabricius, where they develop into mature (functional) cells. B lymphocytes survive only a few days, while T lymphocytes may live for months or years.

Both B and T lymphocytes are active participants in the immune response system of the body. B lymphocytes react to antigens, and they are believed to differentiate into plasma cells, which synthesize antibodies. T lymphocytes do not produce antibodies; rather, they are programmed to identify and respond to specific foreign cells (antigens). They are, therefore, part of a cell-mediated immune response in which foreign cells are destroyed by direct contact with the T-lymphocyte.

Monocytes have been demonstrated to develop in red bone marrow from the same stem cell (CFU-GM) as that from which granulocytes may also develop. Monocytes originate from monoblasts, which are formed from the CFUs.

Platelets originate from megakaryocytes, which in turn arise from the megakaryocytoblast derived from CFU-Meg cell. The megakaryocyte is a giant cell 100 μm in diameter and located in red marrow. It contains multilobed nucleus surrounded by an abundance of cytoplasm (Fig. 2–39). The latter is extruded as free cytoplasmic fragments, the platelets.

Lymph. This material is collected into specialized vessels from interstitial or extracellular tissue fluid. Lymph contains electrolytes, lipid globules, proteins, water, and other substances, including leukocytes (mostly lymphocytes). En route to the general circulation lymph is filtered by lymph nodes, which are arranged in protective chains at strategic sites.

Rigid Connective Tissue. Of the rigid types

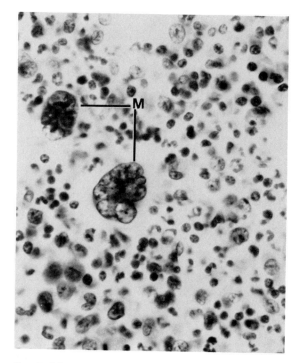

Fig. 2–39. The megakaryocytes (M) are multinucleate, and are the largest of the hemopoietic cells in the marrow. (Hematoxylin and eosin stain; ×330)

of connective tissue, cartilage is normally nonmineralized, and bone, cementum, and dentin have their matrix calcified by crystal apatite (Table 2–2).

Cartilage. Cartilage is comprised of cells and fibers embedded in ground substance. In cartilage, as in bone and dental hard tissues such as dentin and cementum, the fibers and ground substance constitute the intercellular matrix. Unlike these tissues, however, the matrix of cartilage is not mineralized. Cartilage cells, chondrocytes, are found in matrix niches called lacunae. Hyaline and elastic cartilage and fibrocartilage make up the three types of cartilage, a classification based on the quantitative and qualitative aspect of the matrix fibers. Cartilage does not normally have its own vascular, lymphatic, and nerve supplies. Nutrients, oxygen and metabolic by-products pass through the matrix by diffusion.

HYALINE CARTILAGE. In the unpreserved state, hyaline cartilage is porcelain in appearance. Its elastic and flexible properties are necessary requirements for its presence in anterior rib terminals, articulating surfaces of long bones, nose,

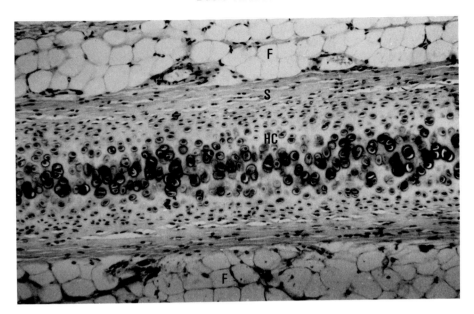

Fig. 2–40. Hyaline cartilage (HC), perichondrial sheath (S), and external deposit of fat cells (F). (Hematoxylin and eosin stain; ×80)

larynx, trachea, and other respiratory tubes. In the respiratory system cartilage is found as rings, plates, or irregular masses. Hyaline cartilage produced during embryonic development, especially that associated with the fetal skeleton, becomes mineralized, degenerates, and is replaced by bone. The characteristics of hyaline cartilage, as observed by optical microscopy, include a perichondrial sheath surrounding a matrix containing chondrocytes housed in lacunae (Figs. 2–40 and 2–41).

Perichondrium in the definitive stage is composed of a fibrous layer of irregularly arranged dense fibrous tissue, in which the chief com-

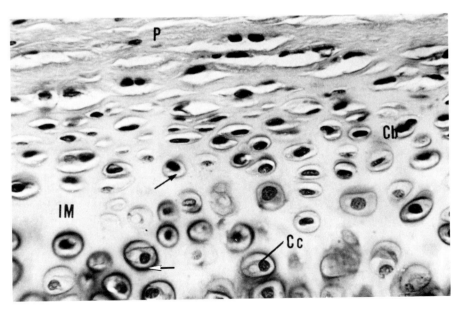

Fig. 2–41. Hyaline cartilage segment showing the external perichondrial sheath (P), chondroblasts (Cb), chondrocytes (Cc) in lacunae *(black arrow)*, interterritorial matrix (IM), and capsular matrix *(white and black arrow)*. (Hematoxylin and eosin stain; ×320)

ponent is collagen maintained by fibroblasts. That portion of this layer abutting the cartilage contains fewer fibers and more cells. In growing cartilage, the internal surface of the sheath consists of a developmental, or chondrogenic, layer of chondroblasts (Fig. 2–41). Perichondrium of mature hyaline cartilage is thinner, and is composed exclusively of the fibrous layer. Its internal cartilage-abutting surface is gradually transformed from regular connective tissue to cartilage. Although nerves and blood and lymphatic vessels are absent in the substance of cartilage (except en route elsewhere), they are present in the perichondrial sheath, particularly in the deeper layers. It should be noted that, as articulating cartilage surfaces are developed, the perichondrium is lost.

Chondrocytes, formed initially from the chondrogenic layer, are squamoid and biconcave. As the cells mature and become more deeply situated in the matrix, they become rounded and polygonal if compressed by adjacent chondrocytes sharing the lacuna. They attain diameters up to 40 μm and, although they may be present in lacunae singly, they mostly form isogenous groups (cell nests). Because the cells shrink during microtechnical processing, chondrocytes may be washed out, to leave empty lacunae that are often observed on histologic slides. Immature chondrocytes are mitotically active, accounting for the multiple cell occupation or formation of isogenous groups. Optical microscopy reveals chondrocytes to possess small nuclei, with diffuse chromatin and conspicuous nucleoli. The cytoplasm is abundant and basophilic, a result of the well-developed system of rER. Lipid and glycogen inclusions removed in tissue processing are reflected in the vacuolated appearance of the cytoplasm (Fig. 2–41).

Electron microscopic examination reveals that the surface membranes of chondrocytes possess short truncated processes extending from shallow ridges. These conform to the furrows on the internal surface, or *capsule* of the lacunae. Mitochondria are relatively few; otherwise, the organelle populations are consistent with their secretory role.

Cartilage growth occurs appositionally and/or interstitially. Appositional growth is a product of chondroblasts arising from the chondrogenic layer. Chondroblasts elaborate matrix, become invested in their secretion, and develop into chondrocytes. With interstitial growth, chondrocytes in the lacunae engage in mitotic activity to form isogenous groups. These synthesize matrix, which is secreted into the intercellular spaces so that adjacent cells are pushed further apart, thereby increasing the intercellular matrix. Interstitial growth increases the chondrocyte population and matrix mass internally. Appositional growth, which is on the external surface, adds to the circumference. Little or no growth occurs in definitive hyaline cartilage.

Cartilage matrix components include fibers, which make up 40% of its dry weight, and ground substance. The fibers are masked, because fibers and ground substance possess similar refractive indices. With hematoxylin and eosin the matrix does not stain homogeneously throughout; rather, the lining of the lacuna, known as the capsular or territorial matrix, is strongly basophilic, while the intervening areas, or interterritorial matrix, stain less intensely (Fig. 2–41). The ground substance, synthesized by the chondrocyte, is composed of proteoglycans, which are sulfated glycosaminoglycans attached to noncollagenic proteins. The basophilia imparted to the matrix is due to the glycosaminoglycans, which are made up predominantly of acidic chondroitin 4- and 6-sulfates. The territorial matrix possesses more proteoglycans, and thus has more chondroitin sulfate than the interterritorial matrix. This accounts for its increased receptivity to basic stains. Thus, the greater the number of cells constituting the lacunar cell nest, the more intense the basophilia of the capsular matrix.

ELASTIC CARTILAGE. Because of its elastic properties, elastic cartilage is found in the pinna (external ear) and epiglottis, as well as in certain cartilages of the larynx and auditory tube requiring resiliency. Because of the predominance of yellow fibers, this cartilage has a yellow hue in the unpreserved state. Its perichondrium, cells, and matrix, except for the presence of elastic fibers, are similar to those of hyaline cartilage (Fig. 2–42).

FIBROCARTILAGE. This tissue provides firm support while simultaneously permitting stretching of body parts. Because of this property, fibrocartilage is an essential component of intervertebral disks (Fig. 2–43), joints, and other sites at which ligaments and tendons are attached to bones. Fibrocartilage is found either

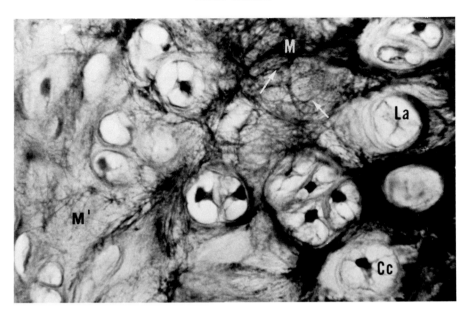

Fig. 2–42. Elastic cartilage. The density of the elastic fibers *(white arrows)* in the matrix (M) is greater than that in M'. Note the vacuolated appearance of the chondrocytes (Cc) (La, Lacuna). (Hematoxylin and eosin stain; ×320)

with hyaline cartilage or with dense fibrous connective tissue. It is unique in that it has no perichondrium, and the fibers are not masked. With light microscopy the chondrocytes are seen to occur singly or in groups of cell rows among the parallel fiber bundles (Fig. 2–44).

    Bone. The *structural components* of bone (osseous tissue) include cells (osteocytes) that are housed in matrix niches, lacunae. Bone, too, is covered by a connective tissue sheath, the periosteum (Figs. 2–45 and 2–46). Unlike cartilage, however, the matrix is calcified, and growth only occurs appositionally. It is the mineral component of bone that provides the structural strength to protect and support the various organs of the body (e.g., brain, heart, lungs).

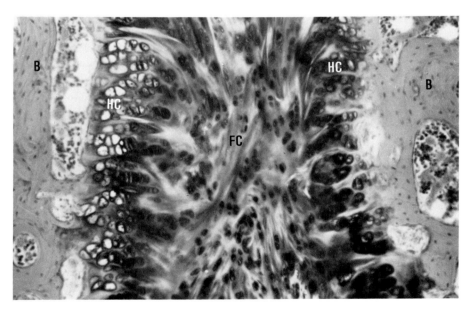

Fig. 2–43. Segments of bone (B) and intervertebral disk composed of fibrocartilage (FC). Hyaline cartilage (HC) interfaces fibrocartilage and bone. (Hematoxylin and eosin stain; ×80)

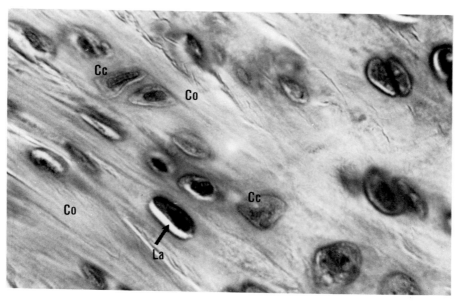

Fig. 2—44. Fibrocartilage at higher magnification than that in Fig. 2—43 showing the collagen (Co) in parallel array and the intercalated linear arrangement of the chondrocytes (Cc) in lacunae (La). (Hematoxylin and eosin stain; ×450)

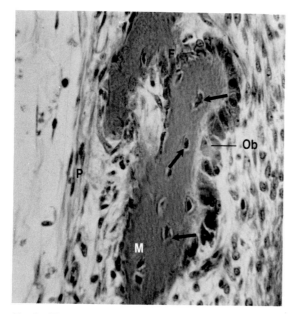

Fig. 2—45. Developing bone showing two osseous spicules surfaced by osteoblasts (Ob) in the process of fusion (F). The bone matrix (M) does not exhibit lamellae. Osteocytes in lacunae *(black arrows)*. Note that a fibrous periosteal sheath (P) is forming on the external aspects of the developing bone. (Hematoxylin and eosin stain; ×320)

Bones are also repositories of mineral salts, blood-forming organs, and levers for body movements. As an organ, bone contains osseous buttresses called trabeculae. The spaces between the trabeculae, as well as the central cavities of bones, contain a formative connective tissue, the marrow, which is of two types, yellow or red, depending on age and location (Fig. 2–46B). The importance of red marrow as the developmental source of blood cells of the myeloid series has been discussed. Yellow marrow is mostly fat, and is not active in hemopoiesis. The marrow cavity and the intertrabecular areas are lined by a thin nonfibrous layer of connective tissue, the endosteum (Fig. 2–46B).

CLASSIFICATION. Regarding shape and size, bones may be grouped into *long* (limb), *short* (wrist and ankle), *flat* (ribs and skull), and *irregular* (vertebrae and certain bones of the skull) types. Short, flat, and irregular bones consist of three layers, two outer compact layers sandwiching a spongy or cancellous layer (Fig. 2–47). In the flat bones of the skull the spongy layer is called the diploe; it is surrounded by two tables of compact bone (inner and outer). Limb bones possess an ivory-textured shaft (diaphysis), a tube of compact bone surrounding a marrow cavity. The terminals of limb bones, the epiphyses, are composed of a thin outer layer of compact bone, which covers a cancellous mass in which

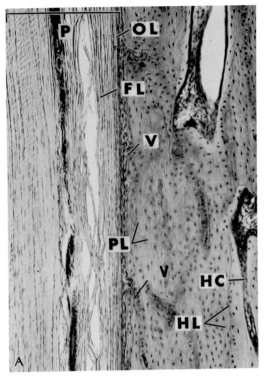

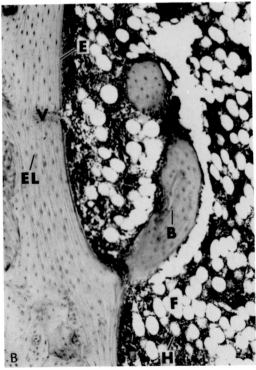

Fig. 2–46. External *(A)* and internal *(B)* bone segments decalcified and longitudinally sectioned. *A,* Periosteum (P), with fibrous (FL) and osteogenic (OL) layers (V, Volkmann's canal; PL, periosteal lamellae; HL, haversian lamellae, surrounding haversian canal, HC). *B,* Inner (endosteal) lamellae (EL), Volkmann's canal (V), bone spicule (B), fat (F), and hemopoietic (H) tissues in marrow cavity.

Note that the inner fibrous layer (FL) is less dense than the outer layer, and that the periosteal lamellae are longitudinally oriented (in transverse sections these form concentric circles around the external bone). Volkmann's canals pass through the periosteal lamellae (PL) and then extend from haversian canal (HC) to haversian canal to arrive at the endosteal lamellae and enter the marrow cavity by another Volkmann's canal (V). Note that the endosteum (E) is not fibrous. (Hematoxylin and eosin stain; ×50)

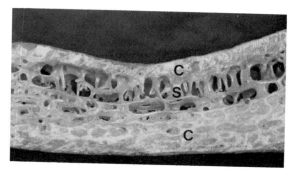

Fig. 2–47. Section of cranial (flat) bone showing the spongy diploe (S) between the tables of compact bone (C).

the trabeculae are oriented so as to accommodate functional stresses (Fig. 2–48). The articulating surface of the epiphysis is covered by hyaline cartilage.

Thus, there are two structural bone types—compact and spongy. Spongy bone is more prevalent in developing bone, and is comprised of trabeculae intervened by marrow spaces (Fig. 2–49). Compact bone, on the other hand, is solid and is composed of layers, lamellae, with no intervening soft tissue (Fig. 2–50).

PERIOSTEUM. The periosteum is a connective

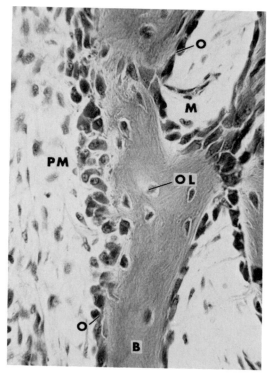

Fig. 2–49. Membranous osteogenesis of mandible showing environment of mesenchyme (PM) in which alamellar bone spicules develop and fuse to form a lattice work (O, osteoblasts; OL, osteocyte in lacunae; B, bone matrix; M, primitive marrow). (Decalcified section, hematoxylin and eosin stain; enlarged from ×450)

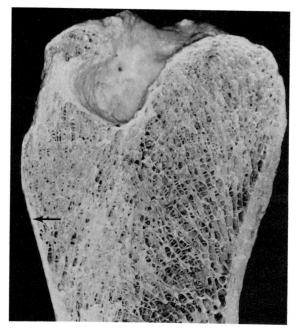

Fig. 2–48. Long bone showing epiphysis in longitudinal section. Note the extremely thin layer of compact bone *(arrow)* on the external surface. The rest of the epiphysis is comprised of functionally oriented trabeculae with intervening spaces that housed marrow.

tissue sheath that covers the external surface of all bones, except at articulations and sites of muscle attachment. The periosteum consists of an outer layer of dense fibrous connective tissue, and an inner layer of looser connective tissue (Fig. 2–45). During periods of growth and development the inner layer contains an osteogenic layer consisting of osteoblasts (bone producers) and, in some areas, osteoclasts (bone destroyers) (Figs. 2–46A and 2–51). The periosteum is attached to bone by collagen fiber bundles that extend from the outer layer through the inner layer to anchor in the periosteal lamellae as Sharpey's fibers (Fig. 2–52A). Terminals of Sharpey's fibers are splayed and fan out to provide larger areas for insertion and anchorage (Fig. 2–53).

CELLULAR COMPONENTS. The cells associated with developing and definitive bone include osteoblasts (bone formers), osteoclasts (bone destroyers or resorbers) and osteocytes (bone-maintaining cells). Marrow, although intimately

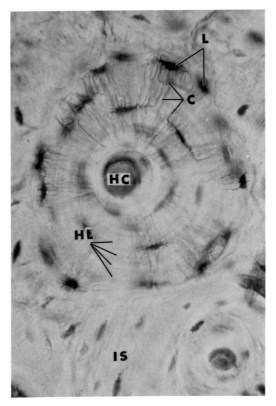

Fig. 2–50. Compact bone showing osteon (haversian system) consisting of haversian lamellae (HL), haversian canal (HC), and lacunae (L), from which canaliculi (C) radiate. These appear to communicate. Interstitial lamellae (IS) are probably an osteon remnant. (Transverse ground section; enlarged from ×430)

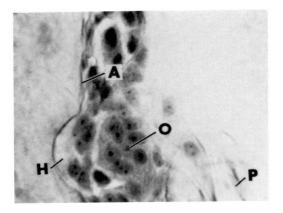

Fig. 2–51. Osteoclasia (bone resorption) showing resorption line (A) and Howship's lacuna (H) containing osteoclast (O) (P, periosteum). (Decalcified section, hematoxylin and eosin stain; enlarged ×970)

associated with bone, is actually a hemopoietic tissue.

Osteoblasts are especially numerous during bone formation and remodeling. They originate from mesenchymal cells, and occupy the inner (osteogenic) layer of the periosteum (Fig. 2–46A). They also surface trabeculae of developing spongy bone (Fig. 2–49). During active periods of development, osteoblasts are round to cuboidal, forming a monocellular layer. When inactive, as the immediate precursors of osteoblasts (preosteoblasts), the cells are squamoid in shape (Fig. 2–49). Preosteoblasts are found in all osteogenic areas associated with osteoblasts. In marrow preosteoblasts occur mostly in the vascular beds.

Osteoblasts exhibit ultrastructural features similar to those of secretory cells. The rER and mitochondria are numerous and their nuclei are eccentric, ovoid, and large. The Golgi complexes are highly organized, and are associated with diverse vesicles. Organelles do not exhibit preferential orientation; rather, they are generally located throughout the cell. Cell surfaces bear microvilli; with differentiation from osteoblasts to osteocytes these become longer. Osteoblasts produce collagen fibrils and ground substance for the bone matrix; during this period they contain large quantities of alkaline phosphatase. After secretion and calcification of the matrix, the osteoblasts are transformed into osteocytes.

Osteoclasts are associated mostly with bone formation, remodeling, and repair processes. They are large multinucleate cells (with up to 30 nuclei) that are generally found in small concavities called Howship's lacunae, located on bone surfaces (Fig. 2–51). They are active in the removal and resorption of both the organic and inorganic components of bone. Osteoclasts were traditionally thought to have been derived either from mesenchymal cells or from the fusion of osteoblasts; now, however, it is thought that they result from the fusion of macrophages, possibly monocytes.

At the ultrastructural level osteoclasts show several specialized regions. The cell membrane is greatly folded adjacent to the resorbing bone; this surface is known as the ruffled border. Circumscribing this area is a clear zone that is organelle-poor, which separates the ruffled border from the adjacent deeper layer that contains

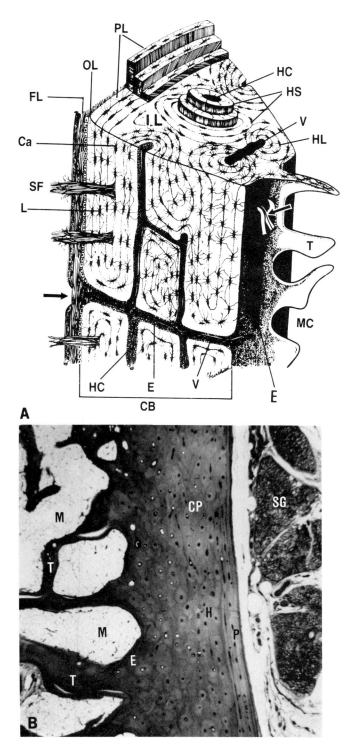

Fig. 2–52. Lamellar systems. *A,* Diagrammatic representation of longitudinal and transverse aspects of compact bone (CB). Trabeculae of spongy bone (T) project into the marrow cavity (MC). The periosteum consists of fibrous (FL) and osteogenic (OL) layers. Note that the Sharpey's fibers (SF) are inserted into the periosteal lamellae (PL). A haversian system (HS) is composed of concentrically arranged lamellae (HL) about the haversian canals (HC). The latter intercommunicate by Volkmann's canals (V) and eventually enter the marrow cavity (MC) by passing through the endosteal lamellae. Blood vessels and nerves *(arrows)* enter Volkmann's canals and the marrow cavity. The osteocytes are metabolically accommodated by lacunae (L) and by an intercommunicating system of canaliculi (Ca) (IL, interstitial lamellae; E, endosteum). (Modified from Elias, H., and Pauly, J.E.: Human Microanatomy. 1st Ed. Chicago, John Wiley and Sons, 1960.) *B,* Cortical plate (CP) of alveolar process consisting of lamellae forming compact bone. The haversian systems (H) are found between the periosteal (P) and endosteal (E) lamellae. Trabeculae (T) of bone project into the marrow (M) (SG, lobules of minor salivary glands). (Hematoxylin and eosin stain; ×8)

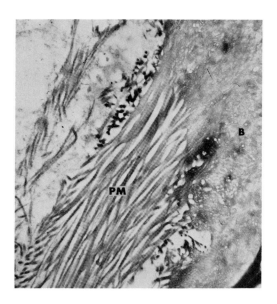

Fig. 2–53. Insertion of collagen fiber bundle (Sharpey's fibers) from the periodontal ligament (PM) into alveolar bone (B). (× 19,000) (Courtesy of K.A. Selvig.)

many membrane-bound vesicles (coated and smooth vesicles, phagosomes, and residual bodies). Even deeper and more removed from bone is an organelle-rich zone in which are found nuclei, numerous mitochondria, Golgi complexes, free ribosomes and polysomes, and a modest population of rough endoplasmic reticulum. The ordered arrangement of the cytoplasmic constituents is related to cell function. For example, the osteoclast is believed to adhere closely to the bone by the action of the filaments in the clear zone. Additionally, the clear zone is thought to confine the resorbing segment of the osteoclast. The ruffled border has been implicated in bone destruction, because dissolved bone and mineral apatite have been observed in its folds and intracellularly in lysosome-related organelles. Although much is known regarding the mechanism of bone resorption, more information is needed.

Osteocytes are the definitive bone cells. They are contained singly in lacunae of both lamellae and trabeculae. The lining of the lacunae, which measures under 2 μm, consists of noncalcified collagen matrix. Canaliculi are tunnels that penetrate the lacunar wall and progress into the matrix to communicate with canaliculi of adjacent lacunae (Figs. 2–50 and 2–52). A canalicular network is produced throughout the bone.

Canaliculi accommodate osteocyte processes. Separating the cell membrane of the cell processes from the mineralized matrix is a microspace of about 0.1 μm, which provides for movement of fluids and nutrients from lacuna to lacuna, hence from cell to cell. With optical microscopy the osteocyte appears biconvex, with numerous processes extending from its surface. The nucleus is small, and the cytoplasm is scant. Electron microscopic studies show that the cell processes of adjacent osteocytes contact one another by gap junctions. Whether or not the processes persist throughout maturity is questionable. Some investigators believe that the processes retract, and that the osteocyte resides totally in the lacuna. Orientation of the processes is mostly toward the nutritional source. Although young osteocytes bear cytoplasmic features similar to those of the osteoblasts, the mitochondria, rER and Golgi complexes become less organized and reduced in number as they mature and become more deeply embedded in the calcified matrix. It is believed that osteocytes no longer synthesize collagen and ground substance; however, they do maintain the matrix. Furthermore, it has been suggested that osteocytes can affect bone dissolution by the liberation of calcium from the matrix.

Bone matrix is composed of organic and inorganic components. The former consists of collagen fibers and ground substance comprised of glycoprotein or glycosaminoglycans. The inorganic component constitutes about 65% of bone (dry weight); it is made up of 85% calcium phosphate in the form of crystalline hydroxyapatite. The crystals are aligned along the collagen fiber length. The matrix formed during embryonic development is composed of spicules (spines), in which the collagen fibers are interwoven in an irregular meshwork (Fig. 2–49). Matrix produced after birth forms more slowly and is organized into definite layers called lamellae. Each lamella ranges from 3 to 7 μm in width, and is composed of osteocytes surrounded by the calcified matrix. The collagen fibers are in parallel arrangement, but are oriented differently than those of adjacent lamellae. This variation in fiber orientation from lamella to lamella renders them more visible with optical microscopy. Lamellae occur mostly in parallel arrangements that form straight, wavy, or circular patterns.

In compact bone, lamellae underlying the

periosteum; thus, encircle the exterior of long bones and are known as periosteal or outer circumferential lamellae (Fig. 2–46A). Those of the interiormost portion of the shaft, which surround the marrow cavity, are known as endosteal or inner circumferential lamellae (Figs. 2–46B and 2–52).

Between the periosteal and endosteal lamellae are longitudinally arranged *haversian systems,* or osteons (Figs. 2–46A, 2–50, and 2–52). These consist of a central channel, the haversian canal, which is surrounded by four to twenty concentrically arranged haversian lamellae. Haversian canals are lined by endosteum and are filled with loose connective tissue containing blood vessels, lymphatics, and nerves. Haversian canals communicate with adjacent ones, with the periosteum, and with the marrow cavity by transverse or oblique channels known as Volkmann's canals (Figs. 2–46A and 2–52). These also contain endosteum and loose connective tissue with nerves and blood and lymphatic vessels. Volkmann's canals, however, are not surrounded by concentric lamellae.

Irregularly shaped lamellar groups located between haversian systems are called interstitial lamellae (Figs. 2–50 and 2–52). These are remnants of other lamellar types, which in the reshaping or remodeling of bone from birth to maturity are only partially destroyed. Pronounced *cement lines* separate developing from resorbing bone sites. Matrix lines stain differently, possess much less collagen, and do not contain canaliculi.

Spongy bone consists of thin trabeculae, spicules, or plates of bone arranged in a latticework (Figs. 2–47 and 2–48). Although they do not contain blood vessels, trabeculae are in direct contact with marrow spaces that contain vascular and lymph vessels. Trabeculae of postnatal spongy bone are lamellar, but in developing bone they lack lamellae and are covered by osteoblasts (Fig. 2–49) and osteoclasts engaging in bone formation and remodeling to arrive at the definitive structure. In the latter, the intertrabecular spaces contain endosteum and red bone marrow.

MUSCULAR TISSUE

Body movements are produced by muscles. For example, gross body movements necessitate the action of skeletal muscles attached to bones of the skeleton. Chewing is accomplished by the action of skeletal masticatory muscles of the cheeks attached to the upper jaw (maxilla) and mandible (lower jaw). Food is driven through the gut by smooth muscles in the walls of the digestive tube. Blood is pumped into arteries (vessels leading blood away from the heart) by the cardiac muscle (heart). Blood is circulated through arteries and veins by the action of visceral (smooth) muscles in the vessel walls. Skeletal muscles can be controlled at will, so they are also known as voluntary muscles, while those of the viscera (organs of the abdominal and chest cavities) are involuntary muscles because they are not willfully controlled. Cardiac muscles are also involuntary.

Classification

Three criteria for muscle classification have been employed: location, functional control, and the presence or absence of muscle fiber stripes or striations.

Location	Structure	Functional Control
Visceral	Smooth	Involuntary
Skeletal	Striated	Voluntary
Cardiac	Striated	Involuntary

Smooth Muscle

Smooth (visceral) muscles are located in the walls of blood and lymph vessels, walls of gut (esophagus, stomach, small and large intestines), walls of respiratory tube (trachea, bronchi, bronchioles), walls of some tubes of reproductive and urinary systems, ducts of glands, skin, and eyes.

Features. Smooth muscle cells are mostly long and fusiform, with tapered or bifurcated terminals (Fig. 2–54A). In some organs they are stellate. Their size depends on functional requirements. For example, in small blood and lymph vessels, and in excretory ducts of glands, in which work requirements are small, the fibers are short (less than 20 μm). On the other hand, in the walls of large vessels and in pregnant uteruses, which require greater work effort, the fibers may be 0.2 mm or longer. Their diameters range from 3 to 8 μm, with the widest part toward the center of the cell in which the nucleus is located. In relaxed fibers, nuclei are elongated; in contracted fibers, they are twisted or wrinkled (Fig. 2–54A). The limiting membrane

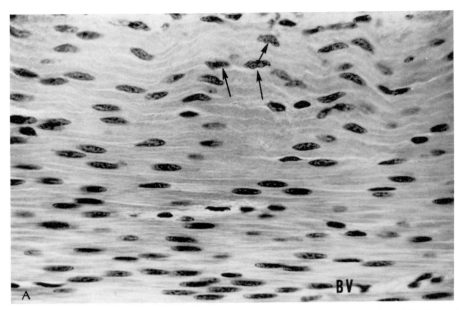

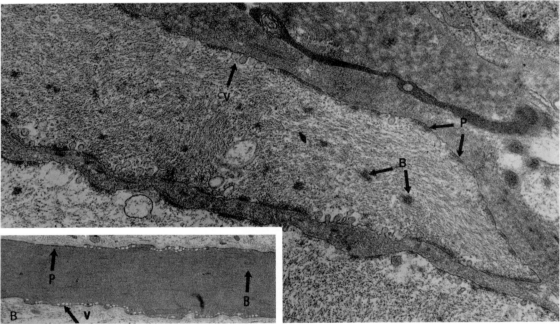

Fig. 2–54. Smooth muscle. *A,* Photomicrograph of smooth muscle coat surrounding duodenum. The nuclei occupy the cell center, the widest portion of the cell. The cells overlap so that the thickest or nuclear segment rests on the attenuated tips of adjacent cells. Several nuclei *(arrows)* appear twisted, indicating that the cells are contracted. Blood vessels (BV) are found between the muscle fibers as part of the endomysium. (Hematoxylin and eosin stain; ×320) *B,* Electron micrograph showing myofilaments sectioned in different planes. Note the dense bodies (B) to which the myofilaments may be attached. Dense regions (plaques, P) seen on the sarcolemma are related to anchoring or attachment of myofilaments (V, pinocytotic vesicles). (Uranyl acetate and lead citrate stain; ×27,000) *Inset.* A low-power electron micrograph of a portion of a longitudinally sectioned smooth muscle cell. Note micropinocytotic vesicles (v) and densities along cell surface (P) and among myofilaments (B). ×6,300.

(sarcolemma) is not visible, except with electron microscopy. Light microscopic examination of the sarcoplasm (cytoplasm) stained with hematoxylin and eosin appears homogeneously pink, but myofibrils may be seen using special stains.

Electron micrographs reveal the myofibrils to be composed of two types of thread-like myofilaments (thick and thin), oriented lengthwise in the fiber (Fig. 2–54B). The filaments are composed of proteins, actin and myosin, which are similar to those found in striated muscle. Although other organelles are perinuclear, most are found near the fiber terminals.

Fiber Organization. Each muscle fiber is surrounded by a basal lamina and a meshwork of reticular fibers, in which elastic and collagen components occur in limited quantity. Smooth muscle fibers may be organized into bundles or fascicles—for example, the erector muscles of the hair. More often, however, they are found as interlacing networks or sheets. When in sheets, the fibers are arranged in overlapping layers in which the narrow tapered ends of one layer are adjacent to the thicker central cell segment (Fig. 2–54A).

Growth and Repair. After maturity, increases in the size of smooth muscle fibers is a result of functional demands. In conditions in which greater muscle action is required, and this cannot be accomplished by fiber hypertrophy, mitosis of smooth muscle does not normally occur. Rather, smooth muscle fibers are added by differentiation of mesenchymal cells in the area. In cases of serious injury, muscle cells are replaced by connective (scar) tissue.

Skeletal Muscle

The muscles form the body flesh. The bone-attaching terminals of skeletal muscles are intercalated with the collagen of tendons (Fig. 2–21). These, in turn, are inserted into bone.

Features. The length of striated muscle fibers or cells ranges from 1 to 40 mm, and in width from 10 to 100 μm. Individually the fibers are cylindric but, when packed into fascicles, they are prismatic. The fibers are multinucleate, with the oval nuclei occupying areas subjacent to the indistinct sarcolemma.

In transverse section the sarcoplasm appears granular because of the cut ends of the myofibrils. These may be seen as distinct areas, Cohnheim's areas, which are artifacts of histologic tissue preparation. Cross striations (light and dark bands) are observed most distinctly along the length of relaxed fibers and their corresponding myofibril subunits (Figs. 2–55, 2–56, and 2–57). The light zone is known as the I or isotropic band, and the dark zone as the A or anisotropic band (Figs. 2–56 and 2–57). A thin dark zone, the Z line, is seen in the center of the I band. A lighter zone, the H band, is found in the center of the A band.

With the electron microscope, a very thin dark line, the M band, is seen inside the H band. The smallest unit showing these striations is the myofibril; this is composed of sarcomeres, which are the fundamental contractile units of the striated muscle fiber. Sarcomeres are arranged successively along the length of the myofibril. A sarcomere is that segment of the fiber between two adjacent Z lines (Fig. 2–56). Accordingly, a sarcomere is made up of an A band and the halves of two I bands that are adjacent to the A band.

Electron microscopic studies have demonstrated that the myofibril, and thus the muscle fiber, is composed of myofilaments within the cytoplasm (Fig. 2–57A). The specific organization of the myofilaments produces the pattern of cross striations observed at the optical and electron microscopic levels. Two types of myofilaments are recognized: thick (myosin) and thin (actin) filaments. Myosin filaments extend the full length of the A band, and are about 10 nm in diameter and 1.1 μm in length. The thinner actin filaments are 5 nm in diameter, and extend from the Z line for about 1 μm through the I band and into part of the A band. Hence, the I band has only actin filaments, while the A band has both actin and myosin filaments. In the central lighter region of the A band—the H band—the actin filaments do not overlap the thick filaments. Both the I and H bands become smaller during contraction, while the A band maintains a constant length. It is thought that muscle contraction is caused by the sliding of the actin filaments over the myosin filaments toward the M line. This hypothesis is known as the *sliding filament theory.*

The Golgi system is perinuclear and poorly developed. The mitochondria, called sarcosomes, are large and exhibit well-developed cristae; they are located between the myofibrils and under the sarcolemma. In the vicinity of the A-I

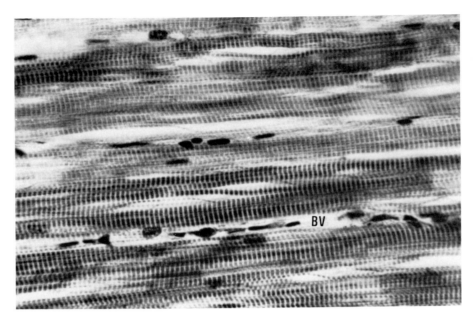

Fig. 2–55. Longitudinal section of striated (skeletal) muscle showing light I bands and dark A bands. Blood vessels (BV) are found between the muscle fibers within the endomysium. (Iron hematoxylin stain; ×320)

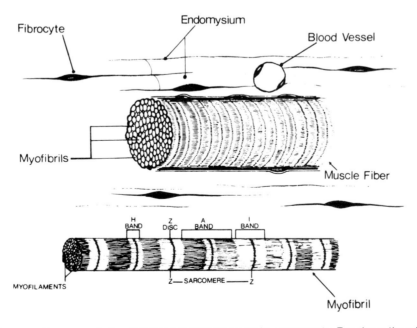

Fig. 2–56. Diagrammatic representation of striated muscle fiber and its components. Top shows the relationship of the fiber to its endomysial covering. The fiber shows the sarcoplasm, marginally located nuclei, myofibrils and striations. Bottom diagram shows an individual myofibril and its markings (bands). In the center of the H band is a thin dark line, the M line.

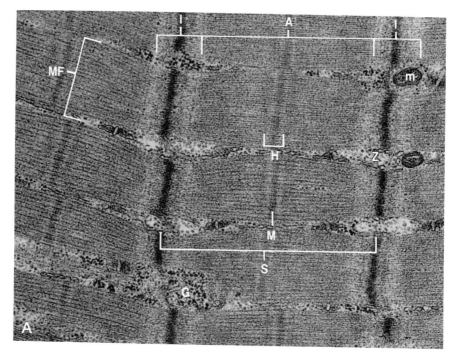

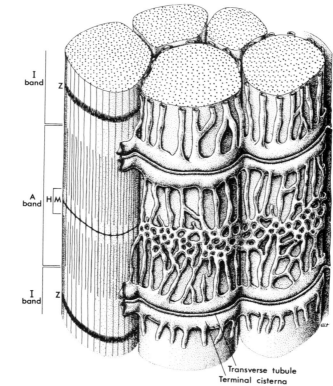

Fig. 2–57. Skeletal muscle. *A,* Electron micrograph of skeletal muscle sectioned longitudinally to show the I band with the Z line and the A band with the central H band and M line. The sarcoplasm between myofilbrils (MF) contain mitochondria (m) and glycogen accumulations (G). The longitudinally oriented dark striations in the I bands and A bands are myofilaments. Note the composition of the sarcomere (S). (Uranyl acetate and lead citrate stain; ×26,700) *B,* Diagram of five myofibrils of mammalian skeletal muscle fiber viewed longitudinally and transversely. The myofibrils are surrounded by the sarcoplasmic reticulum, which consists of a network of sarcotubules near the H band and of terminal cisternae connected by longitudinally oriented sarcotubules. Note the transverse T tubules are invaginations of the sarcolemma at the junction of the I band and A bands and extending among the myofibrils. The T tubule abuts two terminal cisternae, forming a triad. Each sarcomere contains two triads. The various bands in skeletal muscle are shown in a myofibril. Note the granular appearance of the myofibrils cut in transverse section. (Courtesy of C.P. Leblond.)

band junctions the sarcolemma invaginates, extending among the myofibrils as a tubular channel, the T (transverse) tubule (Fig. 2–57B). The T system of tubules increases the surface area of the cell that is in contact with the external environment. Around the myofibrils is found an interconnecting tubular network of smooth endoplasmic reticulum, the sarcotubules of the sarcoplasmic reticulum. Near the A-I junction, adjacent to the sarcolemma of the T tubule, the sarcotubules are continuous with dilated sacs of the sarcoplasmic reticulum known as the *terminal cisterna*. These are oriented transversely to the fiber. At the junction of the A and I bands two terminal cisternae occur, interposed by a T tubule. This complex is called a triad, and two are found in each sarcomere (Fig. 2–57B). The triad and the sarcotubules function in muscle contraction.

Inclusions (e.g., myoglobin pigment, glycogen, lipid) are found in abundance in striated muscle. The myoglobin is responsible for the pink to red color of striated muscle.

Fiber Organization. The sarcolemma of individual fibers is enveloped by an endomysium, which consists of a covering glycocalyx and a fine reticular network containing capillaries (accommodating the cells' metabolic needs) and a few collagen and elastic fibers (offering support) (Fig. 2–56). The fibers are organized into bundles or fascicles, which are held together by a fibrous sheath known as the perimysium (Fig. 2–58). The sheath contains nerves and vascular and lymphatic channels. Muscle, as an organ, is organized from fascicles and is bound by a denser fibrous sheath, the epimysium. The three muscle sheaths, of decreasing densities, interconnect.

Growth and Repair. Skeletal muscle fibers may be ten times longer at maturity than at birth. Exercise, with its accompanying metabolic activity, may increase the volume and hence the diameter of the fibers by 25%. Male hormones increase the thickness of the fibers. Minor injuries to striated muscle fibers may be repaired by new muscle growth; however, extensive injury causing muscle death results in the production of scar tissue.

Cardiac Muscle

The markings in cardiac muscle, although striated, are less distinct than in skeletal fibers.

Muscle action is involuntary. Some action is said to be automatic because the muscle continues to contract even when nerve connections have been severed and the heart removed from the chest cavity. Bathed in an appropriate medium, the heart may remain active for an extended period. Although the structural features of cardiac muscle are similar to those of skeletal muscle, it is more like smooth muscle in nerve supply, function, and embryonic development.

Although located principally in the heart, cardiac muscle fibers are found extending from the heart into the walls of the great vessels.

Features. Cardiac muscle fibers branch to form an intercommunicating network. Individual muscle cells of 10 to 20 μm in diameter are joined terminally by specialized junctional units (intercalated disks) to form a fiber. With the optical microscope, intercalated disks are seen as dark-stained bands at the level of the Z line (Fig. 2–59A). The disks appear in the fibers immediately before birth, and increase in number and patency with age. At the ultrastructural level, intercalated disks appear scalloped where the limiting membranes of adjacent cells are joined by desmosomes, fascia adherens, and gap junctions (Fig. 2–59B; also see Fig. 1–15). Disks, possibly via gap junctions, are also implicated in the transmission of electrical impulses.

In heart muscle, the composition (actin and myosin) and orientation of the myofilaments, the tubular sarcoplasmic reticulum, and the striations (A, I, Z, H, and M bands) are similar to those in skeletal muscle. Also, the sarcoplasm is abundant in both, with the Golgi systems occupying perinuclear positions. Inclusions are distributed similarly. Skeletal muscle cells, however, do not branch to form a network. Cardiac muscle cells have a centrally located nucleus (sometimes two), and mitochondria are more numerous than in skeletal muscle cells.

In mammals, the sarcolemma of cardiac fibers invaginate at the Z lines to form transverse tubules, the T tubules. Adjacent to these are small sacs of cisternae, which are confluent with the tubular network of the sarcoplasmic reticulum. The function of these structures in cardiac muscle is the same as in skeletal muscle. Transverse terminal cisternae are not present.

Purkinje Fibers. These are specialized muscle cell groups that conduct impulses through the heart. Purkinje fibers are important in that they

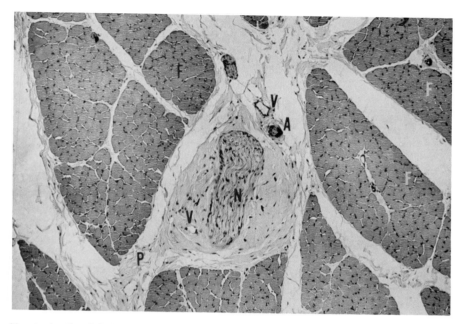

Fig. 2–58. Muscle sheaths of skeletal muscle in transverse section. The muscle fascicles (F) are bound by the perimysium (P) tissue, containing arteries (A), veins (V), nerves (N), and connective tissue fibers and cells. (Hematoxylin and eosin stain; ×50)

transmit impulses regulating and coordinating the sequence of contractions associated with the heart chambers. They form an extensive network from the inner heart lining (endocardium) well into the substance of the heart (myocardium). Fiber diameters may exceed 50 μm. The staining properties are less intense than in the cardiac fiber (Fig. 2–60). The sarcoplasm is abundant and rich in glycogen. The myofibrils are less regularly arranged than in cardiac fibers, and intercalated disks are rare.

Organization. Separating the cardiac fibers is diffuse connective tissue similar to endomysium, which consists of reticular fibers interspersed with a few collagenous and elastic elements. These invest nerves, lymphatics, and blood vessels for the coronary circulation. The heart is covered by a serous membrane, the visceral pericardium or epicardium, and is lined by epithelium, the endocardium. Cardiac muscle does not regenerate by mitotic division. When seriously injured or destroyed it is replaced by scar (connective) tissue. Scar tissue decreases heart efficiency and compensation is then attempted by fiber enlargement, resulting in the total enlargement of the heart.

NERVOUS TISSUE

This basic tissue comprises the central nervous system (CNS)—that is, the brain and spinal cord—and it is also located in other portions and organs of the body, in which it constitutes the peripheral nervous system (PNS) (Table 2–1). Conductivity, the ability of cells to transmit stimuli, and irritability, the ability to respond to stimuli, are most highly developed in nerve cells.

Neurons

The neuron is the basic unit of structure and function of the nervous system. Neurons are ectodermally derived from neuroblasts, and each neuron has two parts: the perikaryon (cell body), in which is located the nucleus, and the cell processes, which are extensions of the cell body. Cell processes are of two types, dendrites and axons (axis cylinders). At junctional points, called synapses, the ends of axons assume intimate relationships with membranes of dendrites, perikarya, or axons. Within the central nervous system clusters of similar perikarya constitute a nucleus; in the peripheral nervous system clusters of perikarya form a ganglion (plural, ganglia) (Fig. 2–62).

Perikaryon and Cell Processes

The central cell mass housing the nucleus is the *perikaryon* (Figs. 2–61 and 2–62). Its morphology varies, depending on the number and

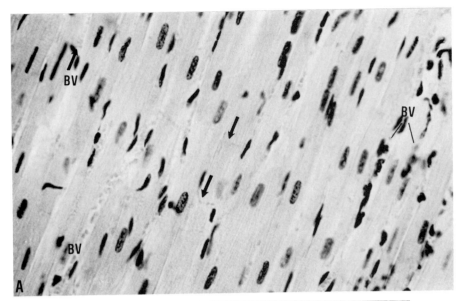

Fig. 2–59. Cardiac muscle. *A,* Photomicrograph of longitudinal section. Shown are centrally located nuclei and intercalated disks *(arrows).* Note that the I and A bands and other markings are not as clearly demonstrated as in skeletal muscle. Heart muscle is more highly vascularized, as indicated by the many blood vessels (BV). (Hematoxylin and eosin stain; ×320) *B,* Electron micrograph showing two intercalated disks (ID) and the stepwise orientation of the junctional complexes. Each disk complex joins two cardiac muscle fibers end to end (F1 and F2, F3 and F4). Similar to skeletal muscle, each cardiac muscle fiber consists of numerous myofibrils embedded in sarcoplasm. Note the numerous mitochondria (m) between the myofibrils (MF). The characteristic bands of striated muscle are present. Surrounding the fibers is connective tissue containing capillaries (C) with erythrocytes (E). (Uranyl acetate and lead citrate stain; ×12,800)

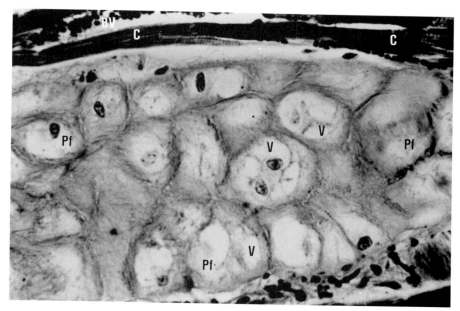

Fig. 2–60. Purkinje fibers (Pf) surrounded by other cardiac fibers (C) and blood vessels (BV). The vacuolated (V) cytoplasm is an artifact—that is, glycogen is rinsed away during histologic processing. (Hematoxylin and eosin stain; ×320)

arrangement of *cell processes* associated with it. For example, those having one process (unipolar neurons) possess round cell bodies. True unipolar neurons are rare. Most, so-called unipolar perikarya, are pseudounipolar (Fig. 2–62). These are exemplified by the cells of the sensory ganglia of the peripheral nervous system. Pseu-

dounipolar perikarya have a single process extending from the spherical cell body. A short distance from the perikaryon the process branches into two similar processes, each traveling in different directions. Neurons, such as those of special sense organs, possess two distinct processes extending directly from a spin-

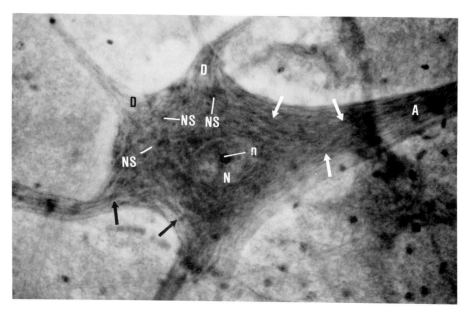

Fig. 2–61. Perikaryon of multipolar neuron showing nucleus (N), nucleolus (n), dendrites (D), Nissl substance (NS), and neurofibrils *(arrows)*. The diameter of the axon segment (A) associated with the perikaryon is greater than any of the other processes. (Hematoxylin and eosin stain; ×800)

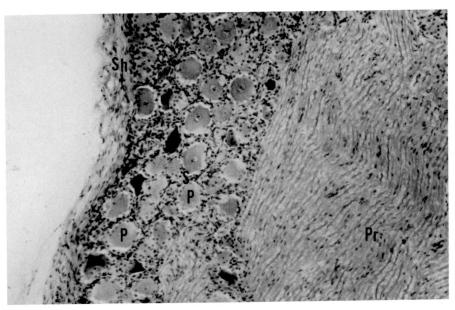

Fig. 2–62. Ganglion section showing a sheath (Sh) surrounding the dorsal root ganglion, a group of pseudounipolar perikaryons (P) and their surrounding satellite cells (S), and myelinated nerve processes (Pr). (Hematoxylin and eosin stain; ×80)

dle-shaped perikaryon, and are bipolar. Most neurons are multipolar, having more than two processes. Their perikarya may have a pear, pyramidal, or stellate shape (Fig. 2–61). The diameters of perikarya range from 4 to 135 μm.

The cytoplasm of neuron cell bodies is basophilic and contains neurofibrils, coarse, granules called Nissl bodies (rER aggregates), and a large round nucleus with a conspicuous nucleolus (Fig. 2–61).

Ultrastructurally the perikaryon is surrounded by a plasmalemma. Contained in the abundant cytoplasm are ribosomes, rER, mitochondria, Golgi components, lysosomes, microfilaments and microtubules, and various inclusions, including glycogen and pigments such as melanin or lipofuscin. With age lipofuscin tends to increase. Neurofibrils, which are aggregates of neurofilaments and neurotubules, are located in the perikaryon and its processes. They extend in the cell from terminal to terminal(s). The nuclear envelope is well defined, and chromatin forms a loosely arranged network.

Dendrites. A perikaryon may possess more than one dendrite, which transmits impulses toward the perikaryon. These are attached to the central cell mass by a trunk-like base. At varying distances from the perikaryon, dendrites arbor-

ize and become brush-like, with numerous spinous projections, the gemmules, whose ends are areas of synaptic contacts. Dendrites are unsheathed. Their cytoplasm is perikaryon-like and contains neurofibrils, endoplasmic reticulum, mitochondria, and other organelles and inclusions.

Ganglia. Groups of perikarya in the PNS are called ganglia (Fig. 2–62). There are two types, sensory and autonomic. Sensory ganglia house pseudounipolar or bipolar neurons and form the sensory components of the spinal nerves (dorsal root ganglia) and of some cranial nerves. The autonomic ganglia house multipolar neurons of the visceral motor system.

Each perikaryon is surrounded by a squamous satellite cell, which is similar developmentally to the Schwann cell (Fig. 2–62) in that it originates from the neural crest (neuroectoderm). Perikarya of ganglia and nerve fibers are invested by diffuse connective tissue that is successively continuous with the endoneurium, perineurium, and epineurium.

Satellite cells are supporting elements for the neuronal components of autonomic and craniospinal ganglia. When functioning specifically as coverings for axons, they are called Schwann cells (see above). Elsewhere, as support or investing cells for perikarya, they are known as capsule or satellite cells.

Axons. Neurons have only a single axon, which is generally the longest process of the cell, and that may attain a length up to several feet. Its diameter is quite constant along its length, and varies widely from less than 1 μm to greater than 20 μm. Impulse transmission occurs more slowly in thinner axons. In most cases the axon carries nerve impulses (action potentials) away from the cell body. In peripheral sensory neurons—that is, pseudounipolar and bipolar neurons—the dendritic processes that convey the nerve impulses toward the cell body resemble those of axons structurally and functionally. Axonal branches, called collaterals, are not uncommon; these are thread-like processes that project at right angles from the axon. The tip or end of the axon bears a brush-like terminal named the telodendrion because of its similarity to the dendrite. An axon is joined to its perikaryon by an elevated implantation cone, or axon hillock. Neither the implantation cone nor the axon contain Nissl substance (Fig. 2–61).

The cell membrane of the axon, the axolemma, is continuous with that of the perikaryon. Other than neurotubules and neurofilaments, the axoplasm or neuroplasm contains mitochondria and elements of smooth endoplasmic reticulum. Multivesicular bodies occur occasionally, but other organelles are not generally present in the axonal process. Axoplasm originates in dendrites and perikarya, and is synthesized by rER. From these sites the cytoplasm streams through the axon toward its terminal, carrying substances needed for growth and maintenance of the process. It also transports organelles and neurosecretory substances to and through the axon. Axoplasmic flow occurs at the rate of about 1 mm/day.

Axon Coverings. Neuron processes generally occur in bundles. In the central nervous system they are called tracts, but in the peripheral nervous system they are known simply as nerves. In a peripheral nerve the term "nerve fiber" refers to both the axon and its sheath, both of which are of ectodermal origin. A nerve fiber in the PNS is encased by a thin layer of diffuse connective tissue, the endoneurium. Nerve fiber bundles or fascicles are bound together by a denser fibrous layer, the perineurium (Fig. 2–63). Fascicles are bound into a nerve by a dense fibrous connective tissue sheath known as the epineurium. Thus, the density of the sheathing tissue progressively decreases as the quantitative aspect of the neuronal process groups are diminished from epineurium to endoneurium.

Individual axons are sheathed by Schwann cells (in the PNS) or by oligodendrocytes (in the CNS), which are types of neuroglia. These cells are strung along the length of the axon, similar to beads on a string. The sheath formed by the Schwann cell is also referred to as the neurolemma. Axons may be myelinated or unmyelinated. In myelinated axons, the plasmalemma of the Schwann cells or oligodendrocytes are wrapped like a jellyroll in concentric layers around a segment of the axons to form a multilayered coating rich in lipoprotein called myelin (Figs. 2–64, 2–65, and 2–66).

As a result of this wrapping in peripheral nerve fibers, the nucleus and cytoplasm of the Schwann cell are squeezed to the end of the cell at the outer rim of the myelin (Figs. 2–63 and 2–64). Consequently, when viewed with the light microscope there appears to be a separate cellular sheath external to the myelin, which has been called the neurolemma. It is important, therefore, to remember that myelin is formed by the cell membrane of the Schwann cell.

Oligodendrocytes, on the other hand, possess processes that may form myelin wrappings on segments of one or more axons. The myelin sheath is formed as a result of a continuous series of axon segments being covered by the myelin-forming cells. Sites at which myelin segments abut exhibit a small gap in the coating, thereby exposing the axon. These sites are known as nodes of Ranvier (Figs. 2–64 and 2–65). The myelinated interval between adjacent nodes is designated as an internodal segment. In the PNS one Schwann cell forms myelin for an internodal segment for one axon, while an oligodendrocyte produces internodal myelin for several axons. In unpreserved tissue the protein-lipid complex imparts a glistening cream color to myelinated fibers, so that massive accumulations of these fibers in the CNS have suggested the term "white matter."

At the light microscopic level, only the squamoid nuclei of the myelin-forming cells may be observed; the cytoplasm external to the myelin cannot be easily resolved. Furthermore, with

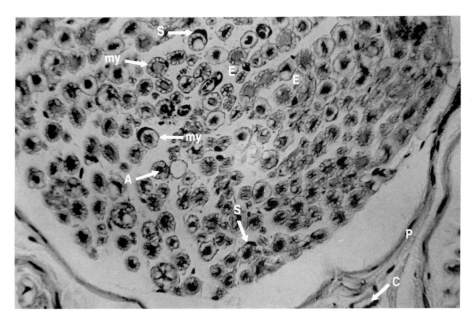

Fig. 2–63. Myelinated nerve fibers in transverse section showing perineurium (P), capillary (C), Schwann cell (S), endoneurium (E), and myelin (my) encasing a central axon (A). (Hematoxylin and eosin stain; ×450)

routine histologic techniques the fatty material of the myelin complex dissolves, leaving behind an artifactual space containing a fine protein network (Fig. 2–63). Special techniques, however, preserve the lipid material so that the sheath can be resolved. Specially stained nerves exhibit diagonal markings, the Schmidt-Lantermann clefts, with optical microscopy (Figs. 2–64, 2–65).

Electron microscopic studies reveal the lipid-protein complex of myelin to be a unit membrane, a continuation of the Schwann cell plasma membrane surrounding the axon that results in alternating dark and light areas (Fig. 2–66). The

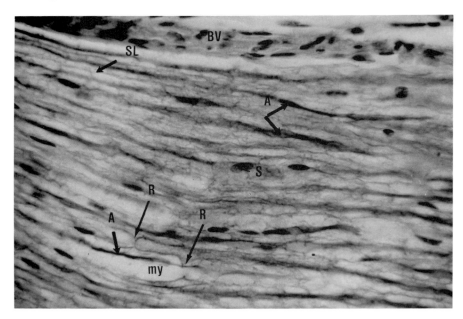

Fig. 2–64. Fascicle of myelinated nerve fibers in longitudinal section showing blood vessel (BV), Schmidt-Lantermann cleft (SL), axis cylinder (A), Schwann cell (S), nodes of Ranvier (R), and myelin (my). (Bodian stain; ×350)

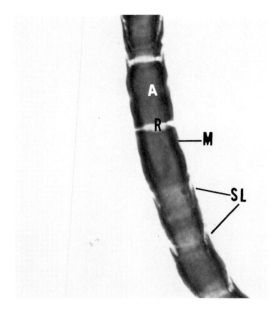

Fig. 2–65. Myelinated fiber in longitudinal section showing axis cylinder (A), node of Ranvier (R), myelin sheath (M), and Schmidt-Lantermann clefts (SL). (Picrocarmine and OsO_4 processed and stained; ×800)

light or electron-lucid zones represent the lipid component of the plasma membrane. Fused unit membranes produce a regular pattern of dense lines among the alternating light layers. The darker of the two electron-dense lines (major dense line) represents the fusion of the inner (cytoplasmic) leaflets of two abutting membranes. The less dense, or intraperiod line, is formed by the fusion of the outer leaflets of two adjacent unit membranes. At the nodes of Ran-

vier processes of adjacent Schwann cells interdigitate but do not contact. Thus, this is a naked zone of a myelinated fiber. Schmidt-Lantermann clefts examined with the electron microscope reveal them to be sites of separation, or at least of diminished fusion, of the spirally arranged Schwann cell membrane.

Unmyelinated fibers are also ensheathed by Schwann cells or oligodendrocytes, but they do not have the concentric wrapping of the plasmalemma. A single Schwann cell or oligodendrocyte may ensheath more than one axon. Because these fibers are so thin, with no myelin coating, they are often difficult to resolve with the light microscope.

Neuroglia

Just as the components of neurons of the PNS are surrounded by nonneural supporting elements such as satellite and Schwann cells, so are the components of the CNS. Here the supporting cells or neuroglial cells are of three types—astrocytes, oligodendroglia, and microglia.

Astrocytes

These are star-shaped, with oval nuclei and numerous cytoplasmic processes. They occur in two forms, according to the type and number of processes. Protoplasmic astrocytes, located in the gray matter, are recognized by processes that are thicker than those of fibrous astrocytes (Fig. 2–67). These are usually located in the white matter, and bear fewer, longer, and thinner processes. The diameter of astrocytes ranges

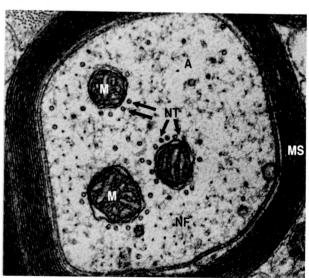

Fig. 2–66. Electron micrograph of transverse section of axon (A) surrounded by myelin sheath (MS) and composed of fused unit membranes of a Schwann cell. The axoplasm contains neurofilaments (NF) of about 10 nm, mitochondria (M), and neurotubules (NT) with a diameter of about 26 nm. (×53,000) (Courtesy of R.M. Meszler.)

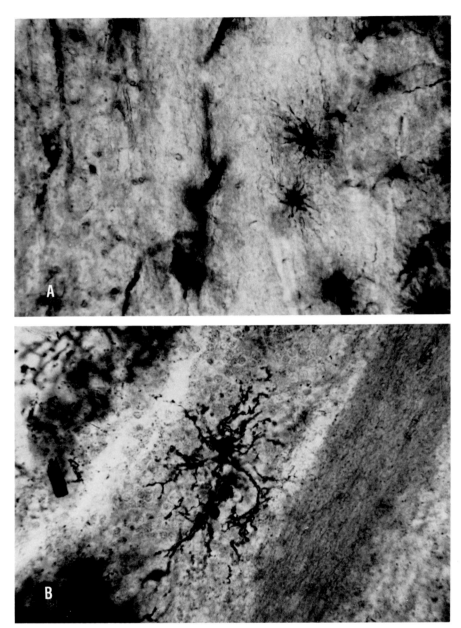

Fig. 2–67. Astrocytes. *A,* Fibrous astrocytes. *B,* Protoplasmic astrocyte. (Golgi-Cajal stain; ×70)

from 7 to 10 µm. Their nuclei are oval and pale-staining. At the ultrastructural level, it is noted that there is a paucity of free and attached ribosomes, as well as of mitochondria. The Golgi complex is more extensive than that of the oligodendroglia, and lysosomes are a regular component of the organelle population. Astrocytes encapsulate axons and function as satellite cells for perikarya. They also participate in the sup-

port of nerve tracts. At some sites the processes of astrocytes abut blood vessel walls, where they are called perivascular feet and are believed to be associated with the blood-brain barrier.

Oligodendroglia

The most numerous neuroglial components are these small angular cells, with diameters of 6 to 8 µm. Oligodendroglia are of ectodermal origin and are found most often in the white matter of myelinated axons. A given oligodendrocyte may produce myelin simultaneously for more than one axon. In the gray matter they act as satellite cells surrounding perikarya. Oligodendroglia are smaller and possess fewer processes than astrocytes. Their nuclei are small, with compact chromatin. The perinuclear cytoplasm is sparse, and contains free and attached ribosomes. The mitochondria and Golgi components are present in limited amounts, as are the microtubules.

Microglia

It is believed that microglia originate from the mesoderm, and on this basis they may be called mesoglia. Impregnation techniques have indicated the presence of microglia in both white and gray matter, although they are most numerous in the latter. They are associated most frequently with vascular channels, and bear a relationship to them similar to that of pericytes in other tissues. These ameboid cells are fusiform, and exhibit a few short processes bearing spines. The nucleus is elongate and stains intensely, and a thin rim of perinuclear cytoplasm is present. The ultrastructural features are similar to those of fibroblasts or macrophages and their function is one of phagocytosis.

Ependyma

The central canal of the spinal cord and the lumina of the ventricles of the brain are lined with cuboidal epithelial cells, which are joined together by tight junctions. During embryonic development the ependymal cells are ciliated. With maturation, however, the cilia disappear, although a profusion of microvilli on the luminal surface persists. It appears that these cells are involved in the exchange of metabolites between the cerebral spinal fluid and the extracellular spaces of the CNS. Indeed, in the region of the hypothalamus, specialized ependymal cells called tanycytes are implicated in the discharge of neurosecretions from that region of the brain. Ependymal cells also cover the luminal surface of the tela choroidea (choroid plexus) in the ventricular system of the brain. The choroid plexus secretes cerebrospinal fluid (CSF) into the ventricles. The choroidal epithelium is important in the formation of the CSF, because they act as part of the barrier between the blood and CSF

and help to control the chemical composition of CSF.

Nerve Endings and Impulse Transmission

Effectors

Neurons influence the activities of other neurons, muscle fibers, and visceral tissues via specialized membrane appositions known as *synapses*. Synaptic contacts between neurons may occur between an axon and dendrite (axodendritic), axon and soma (axosomatic), axon and axon (axoaxonic), dendrite and dendrite (dendrodendritic), and soma and soma (somasomatic). Axonal endings appear as numerous terminal twigs that vary in shape called terminal arborization, or telodendria. Some are basket-shaped, while others are balloon-like with end-bulbs called boutons (end feet). These are expansions of the terminal twigs, and are apposed to the plasma membrane of the adjoining neuron. For example, at an axodendritic synapse, the plasmalemma of the axon terminal (presynaptic membrane) is separated from the plasmalemma of the dendrite (postsynaptic membrane) by a space of 10 to 20 nm known as a synaptic cleft. Most synapses use a chemical trigger called a neurotransmitter that is released from the presynaptic terminal to activate receptor elements in the postsynaptic membrane. Thus, chemical synapses transmit in only one direction. The presynaptic terminals contain mitochondria and synaptic vesicles of various densities and sizes (20 to 100 nm) (Fig. 2–68). These synaptic vesicles contain neurotransmitters, such as acetylcholine and noradrenalin.

When a neuron terminates (synapses) on striated muscle, forming a myoneural junction, the axon branches to innervate many muscle fibers. A motor fiber (neuron) and all the muscle fibers it innervates is referred to as a motor unit. The site of contact of an axon terminal results in an elevation on a muscle, forming a complex known as a motor end-plate. The axon terminal forming the motor end-plate is sheathed by a Schwann cell but not by myelin.

Electron microscopic observations reveal that the terminals arborize and form many smaller endings that abut the muscle fiber. They appear as small swellings submerged in depressions of the sarcolemma. The terminals possess mitochondria and synaptic vesicles containing ace-

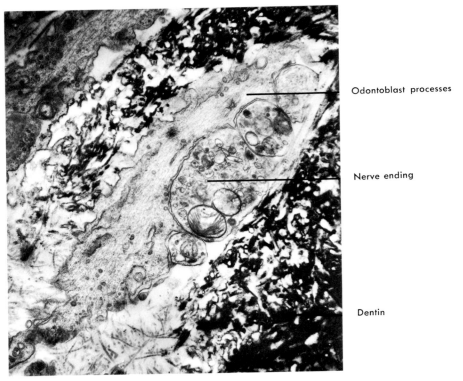

Odontoblast processes

Nerve ending

Dentin

Fig. 2–68. Dentopulpal junction showing nerve ending in dentin tubule. Vesicles in nerve ending house neurotransmitter material. (×18,000) (By permission from Avery, J.K.: Dentin. *In* Orban's Oral Histology and Embryology, 9th Ed. Edited by S.N. Bhaskar. St. Louis, C.V. Mosby, 1980.)

tylcholine. The axolemma, at the trough, is separated from the sarcolemma by a synaptic cleft or gap, 20 to 30 nm deep, which contains a basal lamina. The sarcolemma of the trough is thrown into many folds known as subneural clefts or folds (junctional folds). The sarcoplasm underlying the subneural clefts contains numerous mitochondria and nuclei.

Nerve terminals in visceral tissues form plexuses in and around the smooth muscle fiber, cardiac muscle fiber, and/or secretory components of glands. These terminals may occupy small troughs in the plasmalemma of these tissues, but no specialized endings have been identified by either light or electron microscopic observations.

Receptors

There are two types, free and encapsulated receptor endings. All types are associated with structures of the oral and nasal cavities. Special nerve endings, those related to taste and smell,

found on the tongue and in the nose, respectively, will be discussed later.

Free Endings. The axon endings of sensory nerves may terminate as a branching network of free endings between cells. They may also be a constituent of a specialized receptor that reacts to specific stimuli, such as those associated with touch, pressure, heat-cold, proprioception, vision, hearing, equilibrium, taste, and olfaction. The most numerous receptor endings are free endings which may be found in such tissues as the epithelium of the skin and cornea of the eye, as well as mucous membranes (Fig. 2–69A). They are thought to be receptors of cold, heat, and pain.

Encapsulated Endings. Some specialized receptors are enveloped by a connective tissue, and are called encapsulated endings (Fig. 2–69). These are distributed variously among the connective tissues of the body; they include receptors for touch (Meissner's corpuscle) and deep pressure (pacinian corpuscle) as well as other mechanoreceptors, such as the Krause end-

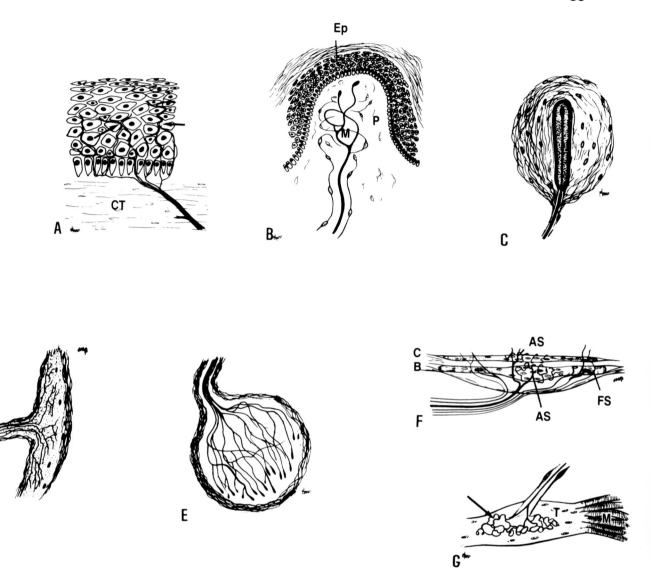

Fig. 2–69. Diagram of sensory receptors. *A,* Naked free nerve endings *(arrows)* in epithelium for pain and temperature reception (CT, connective tissue). *B,* Meissner's corpuscle of touch (M), in connective tissue papilla (P) (Ep, epidermis). *C,* Pacinian corpuscle of pressure, an elliptic body consisting of circumferentially oriented lamellae of squamoid cells about a central core with terminals of a single sensory neuron. *D,* Ruffini mechanoreceptor consisting of collagen fibers surrounded by a network of axonal terminals. *E,* Krause end-bulb mechanoreceptor consisting of a nonmyelinated bulbous nerve terminal. *F,* Neuromuscular spindle with nuclear chain fiber (C) and nuclear bag fiber (B). Specialized sensory nerve endings, the annulospiral (AS) and flower spray (FS) endings, are associated with these intrafusal muscle fibers. *G,* Golgi tendon organ near junction of muscle (M) and tendon (T). Unmyelinated nerve terminal branches *(arrows)* intermingle with collagen fibers of the neurotendinous organ. (Adapted from several sources.)

bulb, Ruffini corpuscle, Merkel's disk, neuro-muscular spindle, and Golgi tendon organ.

Meissner's Tactile Corpuscles. These are located in dermal ridges, especially of the finger pads, and in other areas devoid of hair. They are formed by horizontal lamellae of epithelioid (tactile) cells and connective tissue (Fig. 2–69B).

The corpuscle is surrounded by a connective tissue capsule. Axon terminals of two or more neurons intermingle among the tactile cells.

Pacinian Corpuscles. These are present in the connective tissue underlying the skin, particularly in the hands, feet, and external genitalia. Additionally, they are present in mesenteries, capsules of joints, and the wall of the urinary bladder. They are quite large, ranging in length from 1 to 4 mm and in diameter from 0.5 to 1 mm. They are elliptic and consist of lamellae of squamoid cells arranged circumferentially about a central core (Fig. 2–69C). The central core contains the terminal of a single neuron.

Ruffini Corpuscles. The "spray endings" of Ruffini are located in walls of vascular channels, joint capsules, the dermis, and subcutaneous tissue. They are spindle-shaped, 1 mm long and 0.1 mm wide. For the most part, they are encapsulated by connective tissue (Fig. 2–69D). Naked knobbed terminals of myelinated axons form an encapsulated arborization among collagen fibers and fibrocytes (fibroblasts). They are believed to be mechanoreceptors sensitive to distortions in the collagen fiber complex as they relate to alterations caused by movement and position motion, and hence are proprioceptors. Some investigators believe that they are also heat receptors, and sensitive to touch and vibration.

Krause's End-Bulbs. These endings are located in the connective tissue of the dermis, conjunctiva of the eye, serosa, mucosa and oronasal and paraoronasal structures, ligaments, and tendons, as well as in the external genitalia as genital corpuscles. These naked terminal twigs of myelinated nerves are enveloped by fibroblasts and collagen fibrils to form a bulbous ending (Fig. 2–69E). They are believed to be mechanoreceptors.

Neuromuscular Spindles. These are sensory endings in skeletal muscle (Fig. 2–69F). They are elongated spindle-shaped structures distributed throughout the muscle parallel to the fascicle, located mostly at the belly of the muscle fiber group or near the muscle-tendon junction. They are comprised of two to twelve modified striated muscle fibers known as intrafusal fibers, encased by a connective tissue capsule. The other, or regular, muscle fibers that surround the spindle are the extrafusal fibers. Intrafusal fibers are of two types—those containing a centrally located aggregation of nuclei (nuclear bag fibers), and those containing nuclei arranged lin-

early along the fiber length (nuclear chain fibers). Nuclear bag fibers are larger in length and diameter but fewer in number than their chain counterparts. Additionally, the fibers of the nuclear bag may extend beyond the spindle into the connective tissue, investing the extrafusal fibers. The shorter nuclear chain fibers attach to the capsule or to the nuclear bag fibers inside the spindle.

The neuromuscular spindle receives sensory innervation from two distinctly diverse myelinated nerve fibers The primary fiber innervates both types of intrafusal muscle fibers by a specialized ending known as the annulospiral (primary) ending. The nerve endings wrap around th central portion of the nuclear bag and nuclear chain fibers. The secondary fibers terminate in a flower-spray (secondary) ending. This is located mostly toward the ends of the nuclear chain fibers on both sides of the annulospiral endings but may also occur at the ends of the nuclear bag fiber. Intrafusal muscle fibers of the neuromuscular spindle, a sensory receptor, are able to contract. Thus, they also receive innervation from motor nerve fibers. Both types of specialized endings measure the degree of muscle stretch. The annulospiral endings also measure the rate of muscle stretching. These stimuli are part of and therefore influence the reflex systems that counteract stretching of the muscle.

Golgi Tendon Organ. These are also known as neurotendinous organs because of their location in the tendon near the muscle-tendon junction. These too are spindle-shaped, consisting of bundles of collagen fibers encased by a connective tissue sheath (Fig. 2–69G). This receptor is associated with a myelinated nerve fiber, the naked terminals of which branch and intertwine with the bundles of collagen. These endings respond to increased tension in the tendon and, through reflex action, effect inhibition of the motor neurons that innervate skeletal muscle fibers, thereby inhibiting muscle contraction.

Other receptor types are present in the body, but these are not particularly significant to the oral histology student, except for the receptors for taste and smell.

CHAPTER 3

Development of Orofacial Structures

The male sex cell or gamete, known as a spermatozoon, fuses with the female gamete, the ovum, in a process known as fertilization. The single cell resulting from the fusion of the male and female gametes is known as the zygote, which produces a new individual. Through a series of mitotic divisions the zygote produces a berry-shaped structure, the morula, which subsequently hollows out to form a spherical blastula. The latter contains a bilaminar embryonic disc consisting of an ectodermal and an endodermal or entodermal layer. During gastrulation, in which a third germinal layer, the mesoderm, is formed, the blastula is converted into a gastrula. The germinal disc of the latter becomes trilaminar, consisting of an upper layer (ectoderm), a middle layer (mesoderm), and a lower layer (endoderm).

Continued development results in the formation of two tubular structures, the neural tube and the gut. Neurulation, or the formation of the neural tube, is a result of the upward growth of portions of the ectoderm to effect tubulation. The neural tube is the fore-runner of the brain and spinal cord. Downward folding of the lateral portions of the blastodisc eventually results in a second tube, the gut, which is lined with endoderm. This tube represents the future digestive tract, with the anterior (cephalic) end being the site of the prospective oral cavity and the posterior (caudal) end representing the future site of the anal terminal of the gastrointestinal tract.

FACE, ORAL, AND PARAORAL STRUCTURES

The Face

Approximately 1 month after fertilization, growth centers associated with the development

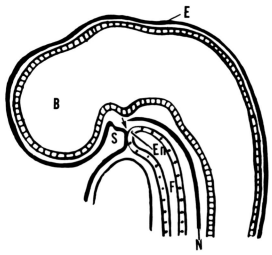

Fig. 3–1. A Diagram of prospective head and neck region in a 25-day-old embryo *(sagittal section)* showing ectoderm (E) surfacing the brain (B) and lining the future oral cavity, stomodeum, (S). Note that the ectoderm meets the endoderm (En) of the foregut (F) to form the buccopharyngeal membrane *(black arrow)* (N, notocord).

of the face, nose, palate, and jaws exhibit greatly increased mitotic activity. These centers are located around a depression known as the stomodeum, which is lined by ectodermal cells. The distalmost (caudal) portion of the stomodeum is separated from the uppermost (cephalic) segment of the primitive digestive tube, or foregut by a double membrane, the buccopharyngeal membrane, which is thus composed of endodermal cells of the foregut and ectodermal cells of the stomodeum (Fig. 3–1).

By the fourth week of development, the membrane is perforated (Fig. 3–2A). Shortly thereafter the buccopharyngeal membrane is obliterated so that the stomodeal cavity and the foregut

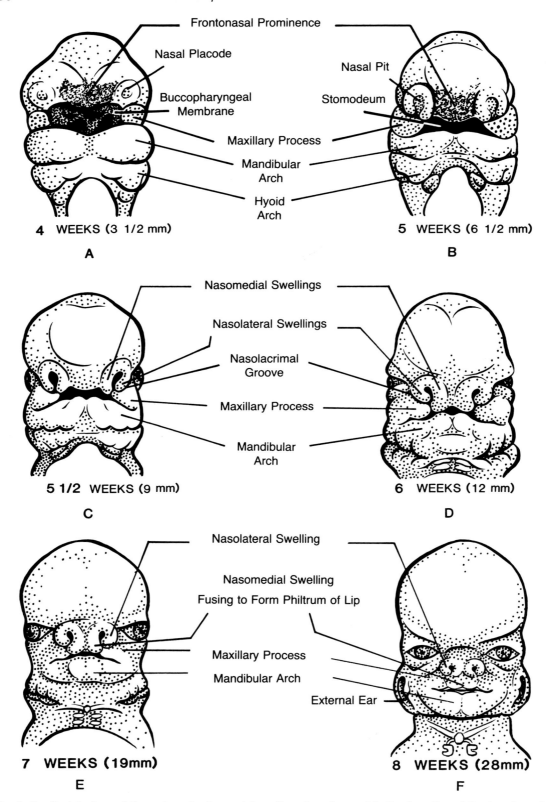

Fig. 3–2. Contributions of the various developmental swellings (prominences) in the formation of the face from *A*, 4 weeks (3.5 mm); *B*, 5 weeks (6.5 mm); *C*, 5.5 weeks (9 mm); *D*, 6 weeks (12 mm); *E*, 7 weeks (19 mm); and *F*, 8 weeks (28 mm).

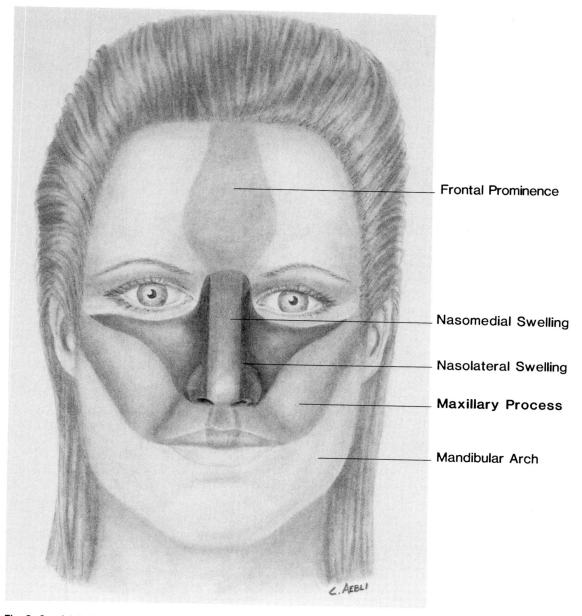

Frontal Prominence

Nasomedial Swelling

Nasolateral Swelling

Maxillary Process

Mandibular Arch

C. AEBLI

Fig. 3–3. Adult face showing contributions of the various developmental facial processes and prominence.

become continuous. At certain peristomodeal sites an influx and rapid growth of mesenchyme (derived primarily from neural crest cells) occurs, resulting in the formation of swellings, also known as prominences, processes, or epithelial thickenings (placodes). The most pronounced are the frontonasal, mandibular, and maxillary prominences (Fig. 3–2A): The frontonasal prominence is composed of frontal and nasal portions.

During the fifth week the nasolateral/lateronasal and nasomedial/medionasal swellings exhibit rapid growth around the nasal placodes, forming the nasal pits (Fig. 3–2B). The base of the frontal prominence joins the nasal contributions at the bridge of the nose so that, in the definitive structure, the nasolateral swellings produce the sides of the nose while the nasomedial/medionasal processes grow toward one another and fuse to form the middle of the nose (Fig. 3–2C and D). Furthermore, the nasomedial processes essentially produce the central portion of the upper lip (philtrum), the anterior medial portion of the maxilla, and all of the primary palate. Concomitantly, the maxillary swellings grow toward one another and join the expanding nasal processes (Fig. 3–2D and E). The growth forces of the rapidly advancing maxillary processes are so great that, in the following 2 weeks, growth of the nasal processes is limited to the area immediately under the prospective external orifices of the nose. The associated nasal and maxillary processes then fuse with one another, contributing further to the development of the nose, lips, and cheeks (Fig. 3–2E and F).

In summary, the derivatives of the frontal prominence are the forehead and the frontal bones; of the frontonasal prominence, the bridge of the nose and the nose bones; of the nasomedial process, the midsegment of the nose, philtrum (lip), and primary palate (premaxilla); of the nasolateral processes, the sides of the nose; of the maxillary process, the upper lip (except philtrum), upper cheeks, maxilla, and secondary (definitive) palate; and of the mandibular process, the mandible, lower lip, and cheeks (Fig. 3–3).

The Primary Palate

Tissue masses contributing to the facial parts also participate in the development of structures associated with the oronasal cavities. For example, the fused medionasal processes, bilaterally condensed by the encroaching maxillary processes, collectively form the intermaxillary segment (premaxilla). The latter produces three orofacial components: the philtrum (ridge) of the upper lip, the incisor-bearing segment of the maxillary arch, and the anterior portion of the palate. These structures are continuous with the nasal septum (Figs. 3–4, 3–5, and 3–6). The initial structures separating the nasal pit and oral cavity are sometimes termed the "primary palate."

The Secondary Palate

The tissue masses involved in the development of the secondary palate, which appear in the sixth week, originate from ledge-like outgrowths of the maxillary processes known as the palatine shelves or processes. Initially, primordial tissue masses are directed medially and inferiorly lateral to the developing tongue (Fig.

3–5). With continued development of the tongue and its descent into the floor of the oral cavity, the palatal shelves rise during the 8th week and grow toward one another. Between the 9th and 10th weeks of development they fuse with one another, with the primary palate, and with the nasal septum (Fig. 3–4C). With the fusion of these four segments of the palate the roof of the oral cavity, and hence the floor of the nasal cavity, are established. The nasal septum separates the right from the left nasal passage.

Nasal Chambers

As a consequence of continued growth of the nasal processes, the nasal depression (pit) becomes more deeply located and embedded in the underlying proliferating mesenchyme 6 weeks after fertilization. At first, a thin sheet of tissue known as the oronasal membrane separates the nasal pit from the developing mouth (Fig. 3–7A). With the disappearance of this membrane the two spaces (oral and nasal cavities) communicate via an opening known as the primitive choana, located immediately posterior to the primary palate (Fig. 3–7 B and C). Subsequent to the formation of the secondary palate, the paranasal tissues continue their development so that the definitive choanae, the posterior openings of the nasal fossae, come to occupy sites in the nasopharyngeal region (Fig. 3–7D).

Development, growth, movement, and merger of the various swellings or processes do not always progress synchronously or uneventfully. When developmental sequences are disturbed or interrupted, defects may occur (see below, Clinical Implications).

Paranasal Sinuses

The paranasal sinuses are bilaterally located intraosseous chambers that are identified by the names of the bones in which they are located. Hence, they are known as the ethmoidal, frontal, maxillary, and sphenoidal sinuses (Fig. 3–8). Excluding the latter, development of the sinuses is initiated between the 16th and 18th weeks of fetal life. The chambers of the sinuses are produced by an invasion and expansion of the nasal mucosa into the bone, resulting in pneumatization. Cavitation by mucosal expansion progresses slowly after birth, and is not arrested until the definitive size of the bone has been attained. The openings of the sinuses into the

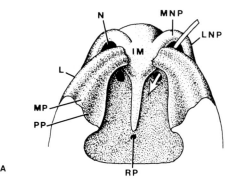

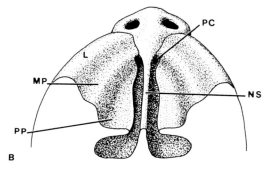

Fig. 3–4. Palatal development. *A,* 6-week-old embryo showing the relationships of the various developmental processes. Note that the oronasal cavity is open *(arrow)* and that the maxillary arch is comprised of the maxillary processes converging onto the intermaxillary segment (IM) of the fused medionasal processes (MNP). (LNP, lateronasal process; N, nares; L, lip; MP, maxillary process; PP, palatal shelves (processes); RP, Rathke's pouch). *B,* Development at 7 weeks showing the primitive posterior choana (PC) and the inferiorly oriented lateral palatal shelves (PP). (NS, nasal septum; L, lip; MP, maxillary process). *C,* Development at about 9–10 weeks showing fused primary and secondary palates (SP). The access into the nasal cavity via the anterior nares and egress by definitive posterior choanae (PC) is indicated by the arrow. Note the primary tooth germs (TG), lip (L), and incisive foramen (IF).

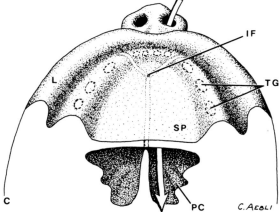

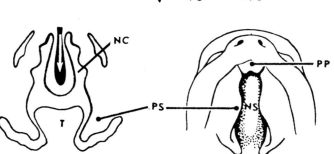

Fig. 3–5. Embryo at 6½ weeks. *A,* Palatine shelves (PS) lateral to the tongue (T), nasal chamber (NC), and nasal septum *(white arrow). B,* Prospective roof of oral cavity illustrating convergence of palatine shelves (processes) (PS), nasal septum (NS), and primary palate (PP).

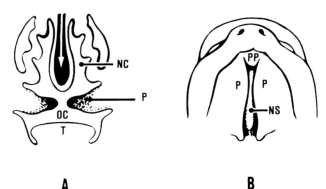

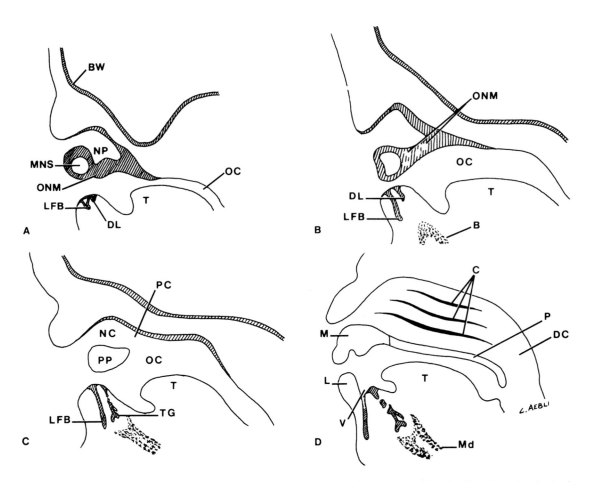

Fig. 3–6. Development of palate. *A,* Head region of about 8½-week-old embryo. Note that the tongue (T) is in a lowered position, and that the nasal septum *(white arrow)* and elevated palatal shelves (P) are converging. After these structures have fused, bilateral nasal passage (fossae), as well as a palatal roof for the oral cavity (OC), have been formed, (NC, nasal cavity). *B,* Palatal view showing fusion of the primary palate (PP) with the palatine shelves (P). Nasal septum (NS).

L. AEBLI

Fig. 3–7. Development of oral and nasal cavities observed in parasagittal sections. *A,* 6-week-old embryo showing brain wall (BW), nasal pit (NP), medial nasal swelling (MNS), oronasal membrane (ONM), oral cavity (OC), dental lamina (DL), lip furrow band (LFB), and tongue (T). *B,* Degeneration of oronasal membrane (ONM) (B, developing bone; DL, dental lamina; LFB, lip furrow band; T, tongue; OC, oral cavity). *C,* Note that the primitive oral (OC) and nasal cavities (NC) communicate via the primitive choana (PC), except in the area separated by the primary palate (PP). TG, tooth germs. T, tongue; LFB, lip furrow band). *D,* Embryo of about 10 weeks showing conchae (C), definitive choana (DC), maxilla (M), palate (P) and mandible (Md), (L, lip; V, vestibule; T, tongue).

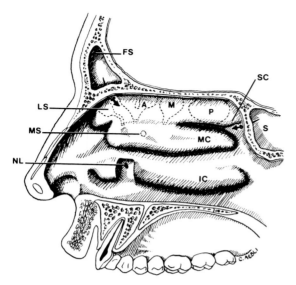

Fig. 3–8. Diagram of human adult head, (parasagittal section) showing the position of the sinuses, including those located in the ethmoid, frontal, maxillary, and sphenoid bones. Exclusive of the latter, the sinuses are formed after birth) (FS, frontal sinus; ethmoidal air cells: A—anterior, M—middle, and P—posterior; S, sphenoid sinus; conchae: SC—superior, MC—middle, and IC—inferior; NL, orifice of nasolacrimal duct; MS, orifice of maxillary sinus; LS, lacrimoethmoidal sinus). Arrows indicate communication between sinuses and nasal fossae. (Redrawn with permission from Corliss, C.E.: Patten's Human Embryology: Elements of Clinical Development. New York, McGraw-Hill, 1976.)

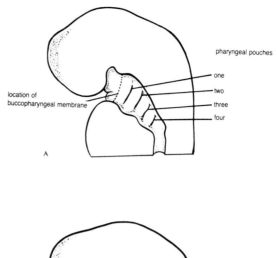

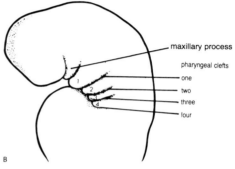

Fig. 3–9. 5-week-old embryo. *A,* Buccopharyngeal membrane and the relationships of the pharyngeal pouch openings to the lateral gut (pharyngeal) wall. *B,* Relationship of the arches (1–4) to the pharyngeal grooves/clefts (one–four). Note that arch #1 possesses a superior segment, the maxillary swelling (process). (Redrawn from Langman, J.: Medical Embryology. 3rd Ed. Baltimore, Williams & Wilkins, 1975.)

nasal passage are minute, so that certain conditions causing swelling of the mucosa, such as colds or allergies, for example, restrict the passages and produce discomfort and pain. It should be noted that the bony floor of the sinuses, especially that of the maxillary sinus, is quite thin. Thus, there is the distinct possibility that, with root tip infection or careless tooth extraction, the bone may be perforated to expose the sinuses (see Chap. 9, Paranasal Sinuses, Clinical Significance).

Pharyngeal Arches

The branchial system, also referred to as the pharyngeal system or visceral apparatus, is comprised of four parts: arches, pouches, grooves/clefts, and membranes (Fig. 3–9). Very early in development the cephalic (head) end in both humans and fish is morphologically similar. The term branchial is derived from the gill branches which develop in amphibian larvae (Gr. branchia, "gill"). According to some embryologists, in humans, such branches do not develop in the first 2 visceral pharyngeal arches. Because the

term "branchial" refers to gill branches, its use in this text is inappropriate; therefore the term pharyngeal will be used instead. The pharyngeal system is important in that its components (described later) are involved in the formation of many oral and paraoral structures. Also, many orofacial anomalies result from disturbances in the reorganizational and developmental patterns of the pharyngeal system. Late in the third and early in the fourth week, about the time that the buccopharyngeal membrane is perforated, bilateral thickenings separated by depressions appear in the cephalopharyngeal area. Toward the end of the fourth week these are distinguished as paired pharyngeal arches with intervening pharyngeal grooves. The thickenings are believed to be a result of ectodermal invaginations (clefting) and of endodermal evaginations (outpouchings) (Fig. 3–10). In their head-on path,

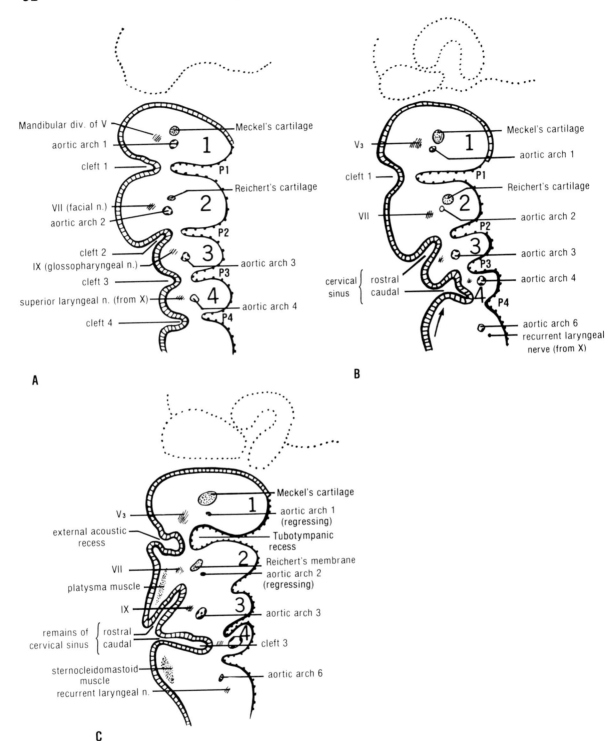

Fig. 3–10. Diagrammatic representations of development of the pharyngeal arches (1–4), clefts (grooves), and associated pouches (P1–P4). *A,* Fifth week. *B* and *C,* Within the sixth week of development. Cranial nerves, cartilages, and other structures are shown in varying formative periods. (Adapted from Corliss, C.E.: Patten's Human Embryology: Elements of Clinical Development, New York, McGraw-Hill, 1976.)

the ecto-endodermal fronts brush the mesenchyme aside, producing lateral condensations of mesenchyal tissue forming pharyngeal arches composed of ectoderm, mesenchyme, and endoderm.

The external clefts between the arches are known as pharyngeal grooves, and the internal endodermal evaginations are known as pharyngeal pouches (Figs. 3–9 and 3–10). Thus, although the arches are separated externally by clefts, internally they are separated by pouches that develop successively and in accord with the arches. The first cleft and pouch are located between the first (mandibular) and the second (hyoid) arches; the second cleft and pouch are found between the second and third arches (Fig. 3–10A). The cleft ectoderm and the pouch endoderm abut, with little or no intervening mesoderm. The site at which fusion of the ectoderm and endoderm occurs is designated as the pharyngeal membrane.

The first arch formed is the mandibular and, unlike other arches, it is comprised of two segments: a smaller superior maxillary process, and a larger inferior mandibular process. These are the primordia for the prospective upper jaw, the maxilla, and the lower jaw, the mandible, respectively. The second arch formed is the hyoid which is involved in the development of the neck and hyoid bone. Arches formed more caudally are numerically designated as the third and fourth and successive pharyngeal arches (Figs. 3–9 and 3–10). The fifth and sixth arches are rudimentary, or wanting. The external aspect of the arches is surfaced by ectoderm, while the internal aspect is lined with endoderm.

Arches, clefts, and the first through fourth pouches are well defined. Each of the first four arches contains four components: a mesodermal core surrounded by neural crest cells; a branch of the aortic arch; a cranial nerve; and a cartilaginous bar (Fig. 3–10).

Elements of the neural crest, as neuroectoderm, originate bilaterally from segments of the neural folds. These neuroectodermal cells migrate, as tissue sheets, to the developing arches. Here they constitute *ectomesenchyme* surrounding the mesoderm of the arches. The ectomesenchyme of the head region gives rise to all tissues that are usually of mesenchymal origin, such as bone, cartilage, dentin, dental papilla components (odontoblasts, fibroblasts, and some

other connective tissue elements), smooth muscle cells, and pericytes (Table 2–1). Other structures produced by the pharyngeal constituents are summarized in Table 3–1.

The cartilaginous bar of the first arch is the mandibular, or Meckel's cartilage. The cartilage bar of the second arch is known as Reichert's cartilage (Figs. 3–10 and 3–11). The cartilages contained in the other arches are not named specifically. Except for the incus and malleus bones of the ears, Meckel's cartilage is not involved in endochondral bone development in the head (Fig. 3–12). In these bones Meckel's cartilage functions as an intracartilaginous bone model. The body and ramus of the mandible, as well as the condyloid and coronoid processes (excluding the tips), are formed intramembranously. Reichert's cartilage participates in the formation of the innermost ossicles of the ears, stapes, styloid process and stylohyoid ligament, and the lesser horns and upper part of the body of the hyoid bone (Fig. 3–12). The skeletal contributions of the cartilage of the third arch include a portion of the body and the greater horns of the hyoid bone. The cartilage of the fourth arch and those of the fifth and sixth (rudimentary) arches contribute to the thyroid cartilage and to other laryngeal cartilages (e.g., cuneiform, corniculate, arytenoid, and cricoid), except the epiglottis (Fig. 3–12; Table 3–1). Innervation of specific arch structures is a product of the nerve derivation for the arch (Table 3–1).

Tongue

The tongue is the largest organ of the oral cavity, and is comprised of two parts, the body (corpus) and root (base). The body makes up the anterior two-thirds of the tongue, while the root makes up the posterior one-third. The anterior segment of the tongue is bilateral, with the line of fusion marked by a median sulcus. The V-shaped junction between the body and root bears a groove, the terminal sulcus. At the apex of the V is found the foramen cecum (Fig. 3–13C).

Structurally, the tongue is a mucosal envelope; its epithelium and connective tissue originates from the pharyngeal arch tissue. The central mass of the tongue is composed of muscle, which is mainly produced by myoblasts from the occipital somite myotomes while its connective

Table 3–1. **The Pharyngeal Arches: Their Muscular and Skeletal Derivatives and Innervation**

Arch	Nerves	Muscles	Skeletal Structures and Ligaments	
First—mandiular	Trigeminal (V)—maxillary and mandibular divisions (V₃)	Muscles of mastication (temporalis, masseter, medial and lateral pterygoids)	Mandible Malleus Incus Maxilla	Anterior ligament of malleus Spenomandibular ligament
		Mylohyoid and anteior belly of digastric Tensor veli palatini Tensor tympani		
Second—hyoid	Facial (VII)	Muscles of facial expression Stapedius Stylohyoid Posterior belly of digastric	Stapes Styloid process Stylohyoid ligament Lesser cornu of hyoid Upper part of body of the hyoid bone	
Third	Glossopharyngeal (IX)	Stylopharyngeus	Greater cornu of hyoid Lower part of body of the hyoid bone	
Fourth, fifth, and sixth (fifth and sixth arches, if present, are rudimentary)	Superior laryngeal branch of vagus and recurrent laryngeal branch of vagus (X), respectively Also pharyngeal plexus	Pharyngeal and laryngeal	Thyroid cartilage Cricoid cartilage Arytenoid cartilage Corniculate cartilage Cuneiform cartilage	

Modified from Moore, K.L.: The Developing Human: Clinically Oriented Embryology. 3rd Ed. Philadelphia, W.B. Saunders, 1982.

tissue component is formed by those from arch mesenchyme.

The body of the tongue is formed from the mandibular swellings of the first arch. Its development is initiated in the fourth week of embryonic development, and is indicated by a diamond-shaped swelling known as the median tongue bud or tuberculum impar (Fig. 3–13A). Mesechyme growing anterior and lateral to the median bud produces two additional swellings— the distal tongue buds, or lateral lingual swellings (Fig. 3–13A). During growth and devel-

opment, these fuse along the midline; they are externally delineated by the median sulcus and internally by the median septum (Fig. 3–13C). Tissue proliferation of the lateral tongue swellings overshadows that of the median tongue bud (tuberculum impar), so that the body is com-

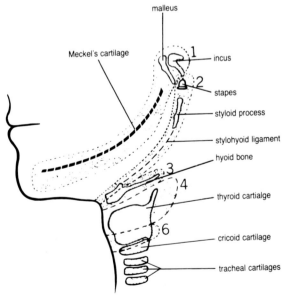

Fig. 3–12. Definitive structural derivatives of pharyngeal cartilages in arches 1,2,3,4, and 6. (Redrawn from Langman, J.: Medical Embryology. 3rd Ed. Baltimore, Williams & Wilkins, 1975.)

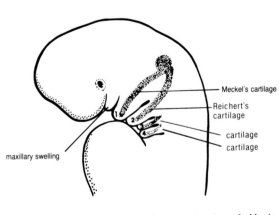

Fig. 3–11. Cartilages of the pharyngeal arches. *1,* Meckel's; *2,* Reichert's. Those of the third and fourth arches are numerically represented. (Redrawn from Langman, J.: Medical Embryology. 3rd Ed. Baltimore, Williams & Wilkins, 1975.)

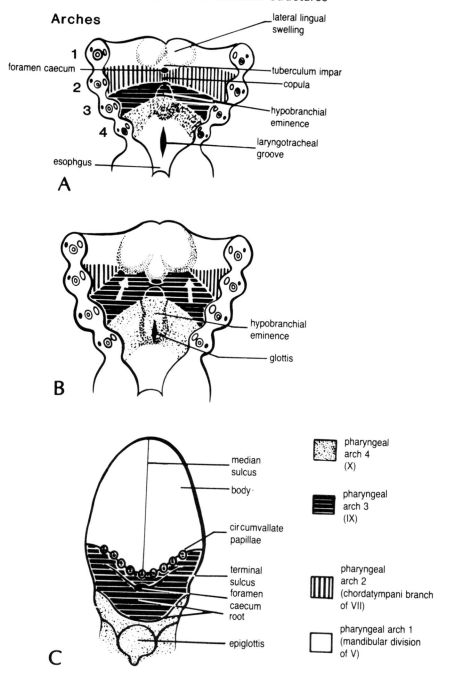

Fig. 3–13. Tongue development. *A,* Embryonic tongue development at 4 weeks showing the initial components of the prospective body (lateral lingual swellings and tuberculum impar) and that of the future root (copula and hypobranchial eminence). *B,* Fifth week of development, showing the dominant growth of the lateral lingual swellings and their encroachment on the tuberculum impar. Also note the advancing front of the hypobranchial eminence appropriating and so overshadowing the development of the copula so that in *C,* the definitive stage of development, the body of the tongue originates mostly from the lateral lingual swellings and the root from the hypobranchial eminence. The two tongue segments meet at the terminal sulcus. (White, stippled, and horizontal and vertical lines indicate the various cranial nerve relationships to the contributing primordial arches of the developing and definitive tongue.) (Redrawn from Moore, K.L.: The Developing Human, Clinically Oriented Embryology. 2nd Ed. Philadelphia, W.B. Saunders, 1977.)

posed predominantly of the anterolateral (distal-lateral) tongue segments and minimally of the median tongue bud. The mucosa of the body arises from the mandibular arch and stomodeal ectoderm, while that of the root stems from the endoderm (primitive foregut); thus, this involves pharyngeal arches 2 through 4. The mucosa arising from a specific arch is similarly innervated by the nerve of that arch.

Development of the root of the tongue involves two processes: (1) a mesenchymal proliferation known as the copula or "connector," formed by the merging ventromedial segments of the second pharyngeal arch; and (2) a hypobranchial swelling formed immediately posterior to the copula, resulting from the amalgamation of the ventromedial components of the third and part of the fourth pharyngeal arches (Figs. 3–13A and B). The remainder of the fourth pharyngeal arch produces the primordium for the epiglottis (Fig. 3–13). The various components of the tongue's root grow at different rates so that the hypobranchial contribution grows forward more rapidly, overtaking and eventually appropriating the territory of the copula. The root of the tongue is therefore almost totally of hypobranchial origin. The prospective body (composed of developing anterolateral lingual swellings and tuberculum impar) meet and merge with the growth fronts of the hypobranchial swellings at the terminal sulcus (Fig. 3–13C). The lingual epithelium of the body (see above) is of ectodermal origin, and that of the root is endodermal. The underlying connective tissue of both, however, is formed from pharyngeal arch mesenchyme.

The dorsum (dorsal surface) of the tongue bears epithelium-capped connective tissue projections known as papillae, which are of four types and shapes: filiform (thread), fungiform (mushroom), vallate (walled), and foliate (leaf-shaped). The filiform papillae are the most numerous. They are arranged diagonally in parallel, apically directed rows from the median sulcus. Intermingled with them are the fungiform papillae, which originate from the mandibular arch. The first papillae to make their appearance are the vallate (circumvallate) and foliate, early in the tenth week. They are intimately associated with and probably receive developmental impetus from the terminal branches of the glossopharyngeal nerve. The fungiform and filiform

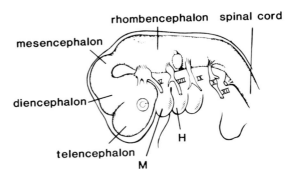

Fig. 3–14. 35-day-old embryo showing the development of primary brain vesicles and the relationship of the cranial nerves (V, VII, IX, and X) to the pharyngeal arches. Some neurons originate from ganglionic placodes. Other neurons and all supporting cells are derived from neural crest cells. (M, mandibular segment of pharyngeal arch 1; H, hyoid pharyngeal arch 2). (Corliss, C.E.: Patten's Human Embryology: Elements of Clinical Development. New York, McGraw-Hill, 1976.)

papillae develop about a week later. Taste buds, specialized sensory epithelial cells which are responsible for taste perception and are located on the fungiform, vallate, and foliate papillae, begin to develop during the eighth week, and thus presage the sites at which the papillae will appear.

The cranial nerve supply to the arches is shown in Figure 3–14 and in Table 3–1. Tongue innervation is consistent with the nerves of the pharyngeal arches contributing to tongue formation. The anterior two-thirds are innervated by the lingual nerve of the mandibular division of the fifth cranial nerve (general sensation) and by the chorda tympani of the facial nerve (taste), while the posterior third of the tongue is innervated by the glossopharyngeal nerve (IX) for general sensation and taste. The tongue region adjacent to the epiglottis, as well as the epiglottis, are innervated by the vagus nerve (X) for general sensation and taste.

Tonsils.

Three sets of tonsils are found at the entrance of the throat. Two occur singly, with one invested in the mucosa of the nasopharynx and the other in the base of tongue. The latter is the lingual tonsil and the former is the pharyngeal tonsil (adenoids). The third member of the tonsilar group is paired, and is located between the palatoglossal and palatopharyngeal folds that border the oropharyngeal isthmus. They are

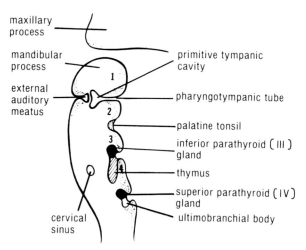

maxillary process

mandibular process

external auditory meatus

cervical sinus

primitive tympanic cavity

pharyngotympanic tube

palatine tonsil

inferior parathyroid (III) gland

thymus

superior parathyroid (IV) gland

ultimobranchial body

Fig. 3—15. Diagrammatic representation of the origins of the palatine tonsil, parathyroid glands, and thymus from pharyngeal pouches 2, 3, and 4; respectively.

known as the palatine (faucial) tonsils. Because these and other smaller patches of lymphatic tissue are circumferentially arranged around the passage from the mouth and nose to the pharynx, they are sometimes referred to collectively as the ring of Waldeyer.

Development of the tonsils is initiated in the eighth week, when the embryo is about 23 mm long. As the third month of development is approached, the palatine tonsils appear as lymphoid infiltrations in the tissue of the atrophying second pharyngeal pouch (Fig. 3–15). Concomitant with the continued formation of lymphoid tissue from the mesenchyme of the underlying pouch, ingrowths of the overlying epithelium occur as solid cords. On entering the lymphoid tissue the epithelial cords branch and, with disintegration of their core cells, fissures known as the tonsilar crypts are produced. Later in development a connective tissue capsule is produced from peripherally condensing mesenchyme. The capsule is the internal limit of the tonsilar mass. Extending from the capsule are connective tissue sheets, or septae, which divide the lymphoid mass into specific territories known as nodules. Nodular organization occurs after the sixth month of development. Crypt formation, however, is initiated earlier, at about the twelfth week. Growth of the palatine tonsils progresses slowly, appropriating more of the second pharyngeal pouch until it extends beyond its limits and bulges into the oropharyngeal

passage. Postnatally, growth continues at a leisurely pace until adolescence.

The pitted surface of the root of the tongue is due to the lingual tonsil. Development of the lingual tonsil is slow. It is initiated in the fifth month as lymphocytic accumulations in the root of the tongue (floor of the second and third pharyngeal pouches). Invasion by the epithelium occurs about the sixth month in utero and crypt formation occurs postnatally; organization of the lymphoid tissue into nodular segments may extend until the age of 5 or 6 years.

The developmental process of the pharyngeal tonsils is similar to that of other tonsilar groups, and is begun in the fourth month. Epithelial invasion begins at about the sixth month, and crypt formation occurs postnatally. Nodular organization may occur prenatally, although it may be delayed until after birth. Some embryologists believe that infiltration, which occurs in the mucosa of the roof of the second pharyngeal pouch, is a result of the local blood supply and an absence of growth tensions. Pseudocrypts are developed, which are produced by stress folds and by greatly expanded excretory ducts of mucous glands forming in the area.

Salivary Glands

Salivary glands formed anterior to the buccopharyngeal membrane arise from the ectoderm, while those produced posterior to the membrane originate from the endoderm. Thus, the smaller, or minor salivary glands of the oral cavity are ectodermally produced, while the larger, major salivary glands (exclusive of the parotids) stem from the endoderm. Irrespective of germ layer derivation, however, the developmental pattern of the salivary glands is similar; therefore one description will suffice.

Salivary glands begin as solid epithelial downgrowths into the mesenchyme. As the solid cords elongate, they branch repeatedly. At cessation of growth, their ends produce small spherical, saccular, or tubular cell masses of cuboidal or low columnar cells, called alveoli or acini, (sing., alveolus or acinus) or *tubules*, respectively (Fig. 3–16). The acinar (saccular) or tubular end-pieces of the gland form the parenchyma—which produce salivary secretions. The epithelial cords and their arborizations undergo canalization to form ducts that drain the end-pieces, or secretory units. The duct com-

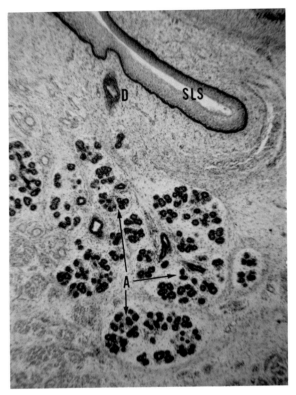

Fig. 3–16. Sublingual gland development (SLS, sublingual sulcus; D, secretory duct; A, developing alveoli. (Hematoxylin and eosin stain; ×35)

ponents are formed in the third month, and canalization occurs in the sixth month of development. During the third to sixth months, groups of alveoli or tubules (or of both) with their ducts are organized into territories known as lobules. Synthesis and release of secretions occur postnatally. Development of the minor salivary glands begins in the third month and, therefore, is initiated later than that of the major glands (parotid, fourth to sixth weeks; submandibular, sixth week; sublingual, eighth week).

CARTILAGE AND BONE FORMATION

Mesenchyme involved in the development of the head and neck structures originates from neuroectoderm (neural crest)—hence, ectomesenchyme. The cartilaginous bars of the pharyngeal arches are considered to originate from ectomesenchyme. The first arch cartilage is Meckel's (mandibular) cartilage (Figs. 3–11 and 3–12). The Reichart's cartilage belongs to the

second pharyngeal arch. The cartilages of all other arches are designated numerically.

Chondrogenesis

Cartilage development is initiated in an environment of mesenchyme. The stem cells differentiate into chondroblasts, which have the capacity for fibrillogenesis and chondromucinogenesis. These products are essential ingredients of cartilage matrices. The precursor substance of cartilage, precartilage, is similar to notochordal tissue. As the chondroblasts elaborate and become trapped in this investing matrix, they are transformed into *chondrocytes* (see Fig. 2–41). Condensation of the tissue peripheral to the cartilage bar produces a loose connective tissue sheath, the perichondrium. This is comprised of a chondrogenic layer of chondroblasts that is responsible for the deposition of cartilage matrix onto the cartilage matrix surface. The formation of cartilage on the external surface is specifically referred to as appositional growth; this is in contradistinction to that which is produced internally (within the substance of the cartilage), known as interstitial growth. Chondrogenesis, irregardless of the mechanism (appositional or interstitial), occurs in two phases: fibrillogenesis and matrix maturation. The first involves the production of the full complement of fibers and the second involves the formation of the mucoid substance, chondromucin, to complete the matrix. During matrix maturation chondroblasts are invested and develop into chondrocytes, retaining only their ability to maintain the matrix. Cartilage is avascular, and depends on the perichondrial blood vessels for accommodation of metabolic needs. Maintenance and vitality of chondrocytes and matrix require uninterrupted diffusion of metabolites through the matrix.

Osteogenesis

Bone development—osteogenesis—may occur either in an environment of mesenchyme or hyaline cartilage. In the former osteogenesis is said to be intramembranous, while in the latter it is considered to be endochondral, or intracartilaginous. The irregularly shaped bones of the face and the flat bones of the skull are produced intramembranously. These specifically include the maxilla, body and ramus of the mandible, and the nasal, lacrimal, frontal, parietal, temporal (squamous and tympanic segments), and medial pterygoid plates. All other bones are of endochondral origin.

Most of the cartilage of the mandibular arch does not contribute to osteogenic activity; rather, it develops, degenerates, and is removed. Only two small segments of the dorsal branch of Meckel's cartilage remain. These are involved in the formation of two middle ear ossicles, the incus and malleus (see Fig. 3–12 and Table 3–1). The stapes of the ear, styloid process of the temporal bone, the lesser horn (cornu) of the hyoid, and part of the body of the hyoid involve Reichert's cartilage of the second pharyngeal arch. The cartilage of the third arch participates in the production of the greater horn of the hyoid and of the remaining segment of its body.

Intramembranous Osteogenesis

This process is rapid and less involved than that of endochondral osteogenesis, and begins in the fifth week of development. The centers of ossification are initially indicated as condensations of mesenchyme with increased mitotic activity and vascularity. Mesenchymal cells differentiate into osteoblasts, which engage extensively in fibrillogenesis. This fibrillogenic period is accompanied and followed by a period of matrix maturation, during which cells and fibers are invested in ground substance to form a bone precursor, or prebone, known as osteoid. With matrix maturation, apatite crystals are seeded onto the collagen fibers during the period of calcification. Prenatal bone is elaborated in the form of bars or spicules, fusing with adjacent ones to produce a latticework (Figs. 2–45 and 2–49). The bone thus produced is primitive and temporary, consisting of spicules comprised of lacuna-containing osteocytes.

Covering the spicules are osteoblasts (see Figs. 2–45 and 2–49). Osteoblasts effect thickening of the spicules so that they may take the form of nonlamellar supporting beams, the trabeculae. Activity of the osteogenic layer lining the primitive marrow spaces (intertrabecular spaces) decreases their size. The primitive marrow spaces, which contain mesenchyme-forming primitive marrow, are consequently reduced.

By birth, the periphery of the bones of the brain case (calvarium) have become suturally joined, and the periosteum is well developed (Fig. 3–17). The sutural space is extensive, forming gaps called fontanelles at sites at which several bones meet. Such areas provide for the overlapping of bones to facilitate the birth process. Fontanelles are reduced in size as growth of the calvarium continues, so they are generally closed by early childhood. Some small spaces, however, may persist into adulthood at the sutural lines. Growth of the cranial bones results from appositional growth from the periosteum at the suture region and on the external aspect of the bone. To limit the thickness of the cranium, osteoclastic resorption occurs at the inner aspect of the bones. This remodeling is induced by the stresses caused by brain growth. The definitive architecture of flat bones is generally completed during early childhood, and includes outer and inner tables of compact bone sandwiching a layer of spongy bone, the diploe (see Fig. 2–47). This is achieved by reorganization, or remodeling of the primitive bone, marrow and marrow cavities, and involves the mechanisms of bone resorption (osteoclasia) and bone formation (osteogenesis).

Endochondral Ossification

Bone development employing endochondral ossification first involves a hyaline cartilage model produced from mesenchyme. The cartilage model widens by appositional growth from the chondrogenic layer of the perichondrium and by interstitial growth from the chondrocytes within the model. During endochondral bone formation, cartilage cells hypertrophy. These later die as a result of matrix mineralization, because diffusion of metabolites cannot occur. With death of the chondrocytes the matrix is no longer maintained, resorption occurs, and the lacunar spaces are enlarged, fusing with adjacent ones to form large, irregular, primitive marrow spaces. Osteogenic buds from the perichondrium and periosteum, consisting of capillaries and mesenchymal cells, enter the cartilage models and invade the eroded cartilage matrix spaces. Some develop into osteoblasts, and others develop into primitive marrow. Osteogenic cells line the surface of the eroding calcified cartilage, and osteoblasts differentiate and deposit bone matrix (osteoid) on the mineralized cartilage remnants. Thus, the initial formation of bone possesses degenerating calcified cartilage as its core.

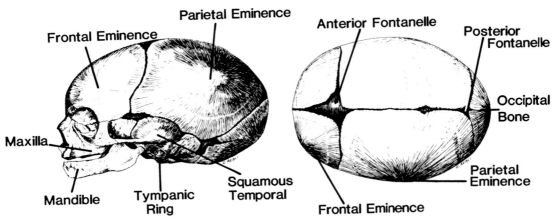

Fig. 3–17. Skull during early postnatal period showing sutures and fontanelles. *A*, Lateral aspect. *B*, Superior aspect. (Redrawn from Fitzgerald, M.J.G.: Human Embryology. Hagerstown, MD, Harper & Row, 1978.)

Endochondral ossification is involved in the formation of the long and short bones of the body. The cartilage model and resulting long bone consists of a shaft, the diaphysis, and two rounded terminals, the epiphyses (see Fig. 2–48).

In the formation of a long bone the *primary center of ossification* occurs in the diaphysis. In this area, chondroblasts of the external surface are replaced by osteoblasts. That is, the perichondrium is transformed into periosteum, and the osteogenic cells develop a layer of bone intramembranously on the surface of the cartilage model, known as the periosteal bone collar. Simultaneously, in the midsection of the diaphysis, the chondrocytes proliferate, mature, and hypertrophy. The cartilage matrix then becomes calcified by phosphate deposition by the chondrocytes. This mineralized matrix prevents diffusion of nutrients to the chondrocytes, death ensues, and the lacunae are emptied. The calcified cartilage matrix is partially eroded and becomes thinner, while the lacunar spaces enlarge, fuse, and form cavities that contain spikes of calcified cartilage.

The periosteal bone collar forms an osseous cylindric support for the central portion of the cartilage model that contains the primary center of ossification. Furthermore, the bone collar prevents the diffusion of nutrients into the cartilage model so that the sequence leading to chondroclasia is extended (chondrocyte death, matrix erosion, enlargement of primitive marrow cavity by expansion, and cavitation of the

lacunar spaces). With invasion of the area by periosteal buds as outgrowths of the periosteum, capillaries and progenitor cells (pericytes and mesenchymal cells) provide the primary center of ossification with the potential to produce connective tissue components for the formation of bone and marrow. Some of the mesenchymal cells line up along the remnants of calcified cartilage, differentiate into osteoblasts, and deposit osteoid. As a result, a latticework of bone spicules with calcified cartilage cores is formed in the diaphysis.

From the diaphyseal center of ossification bone continues to develop in bipolar (epiphyseal) directions—that is, the periosteal bone collar elongates as the perichondrium is converted to periosteum. Simultaneously, within the cartilage model, chondrocytes proliferate, mature, hypertrophy, and degenerate with matrix mineralization (Fig. 3–18). Vessels originating from periosteal buds grow toward the epiphyses en route to form plexuses, while their perithelial connective tissue produces osteogenic tissue for bone development and marrow (myeloid) tissue for blood development.

Secondary (epiphyseal) centers of ossification involve mechanisms essentially the same as those of the diaphyseal (primary) center of ossification. Vascular buds penetrate the cartilage core radially and establish the developmental sequences involved in endochondral ossification. Similarly, the perithelial connective tissue accompanying the vascular buds provides for bone and marrow development. Although car-

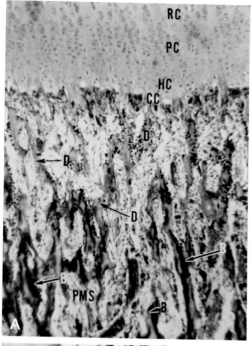

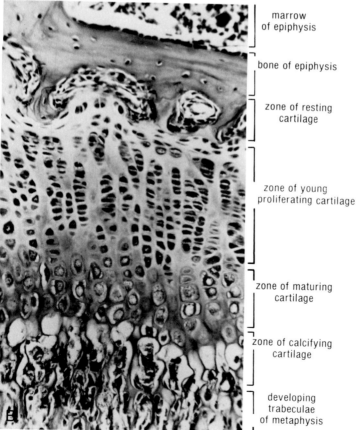

marrow
of epiphysis

bone of epiphysis

zone of resting
cartilage

zone of young
proliferating cartilage

zone of maturing
cartilage

zone of calcifying
cartilage

developing
trabeculae
of metaphysis

Fig. 3–18. Endochondral ossification. *A,* Photomicrograph showing zones of resting cartilage (RC), proliferating chondrocytes (PC), hypertrophied mature chondrocytes (HC), calcifying cartilage (CC), disintegrating cartilage (D), and developing bone (B) (PMS, primitive marrow spaces). (Hematoxylin and eosin stain; ×50.) *B,* Endochondral ossification at the epiphyseal disk (plate) showing the various zones involved in the lengthening of bone. (Ham, A.W., and Cormack, D.H.: Histology. 8th Ed. Philadelphia, J.B. Lippincott, 1979.)

tilage is retained at the articular surface, the substance of the epiphysis is transformed into trabeculae that are oriented so as to accommodate functional stresses. The nonarticular surface of the epiphysis is composed of a thin shell of compact bone. A cartilage plate, the epiphyseal disk (plate) is established for each epiphysis to separate the epiphysis from the diaphysis (Figs. 3–18B and 3–19). These form new chondrocytes and continue intracartilaginous ossification, lengthening the long bone until its definitive size has been reached. During the process of growth the epiphyseal disk consists of several layers of cartilage that exhibit different levels of activity. From the diaphysis toward the epiphysis, the layers are the following: (1) zone of bone resorption; (2) zone of osteogenesis on calcified cartilage; (3) zone of calcifying cartilage; (4) zone of chondrocyte maturation and hypertrophy; (5) zone of proliferation, or chondrocyte mitosis; and (6) zone of resting cartilage (chondrocytes).

Thus, cartilage cells proliferate, mature, hypertrophy, and degenerate, and bone is deposited onto the calcified cartilage remnants to form bony trabeculae, which are later resorbed. Lengthening of long bones is a product of chondrocyte proliferation at the epiphyseal disk. Elongation of the diaphysis continues by epiphyseal disk activity. The shaft wall is extended lengthwise by intramembranous bone formation from the osteogenic layer of the periosteum. If the activity of the periosteal collar were to progress uninterrupted, the definitive shaft would be a solid cylinder of bone, totally of bone collar origin. This is not the case, because internal osteoclastic activity not only proceeds in a bipolar direction but also peripherally, thereby resorbing the bone that is of periosteal osteoblastic origin (inner aspect of bone collar). New bone is added to the periphery by appositional growth, thus resulting in an increased diameter of the diaphysis (Fig. 3–19).

The diaphyseal segment of bone is ossified completely by birth, but the epiphyses remain partly cartilaginous. As the shaft of the bone approaches its predestined length, its composition is that of compact bone; it consists of outer circumferential and inner circumferential lamellae, with intervening haversian systems and interstitial lamellae. To arrive at this structure, intricate remodeling must occur. This involves filling in the osteoclast-induced longitudinal excavations of the periosteal bone by haversian systems, and filling out longitudinal subperiosteal trenches (Fig. 3–20). The inner endosteal and outer circumferential periosteal lamellae are formed simply by layered apposition of bone, as described earlier for the formation of the bone collar. Inasmuch as the interstitial lamellae are remnants of eroded haversian, periosteal, and endosteal units, they are significant develop-

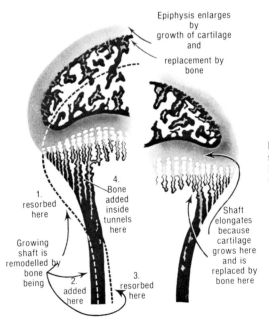

Epiphysis enlarges by growth of cartilage and replacement by bone

1. resorbed here

Growing shaft is remodelled by bone being

2. added here

3. resorbed here

4. Bone added inside tunnels here

Shaft elongates because cartilage grows here and is replaced by bone here

Fig. 3–19. Bone development at the epiphysis of a long bone, showing the interrelationship of osteoclasia and osteogenesis in bone remodeling to effect growth in width and length of diaphysis. (Ham, A.W.: J. Bone Joint Surg., *34A*:701, 1952.)

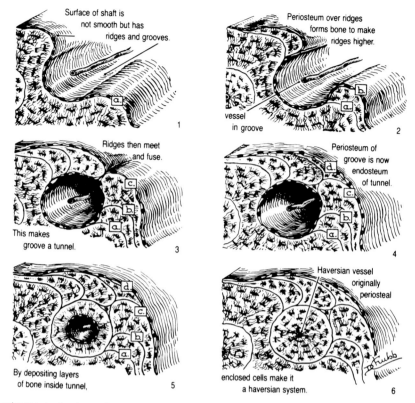

Fig. 3–20. Development of a haversian system. The longitudinal groove at the surface of the developing diaphysis is converted into a tunnel (1–4). The bone developing as a product of osteoblasts of the endosteum results in lamellae of the haversian system (5, 6). These processes add bone to the periphery of the shaft and further incorporate periosteal blood vessels into the haversian system to supply the diaphysis with a vascular supply. (Ham, A.W., and Cormack, D.H.: Histology. 8th Ed. Philadelphia, J.B. Lippincott, 1979.)

mentally. Areas in which bone resorption is followed by bone deposition are separated by *cementing lines*. These stain more intensely, are collagen poor, and canaliculi-free.

Epiphyseal disks continue their activity until mitotic activity ceases and the cartilage has been replaced by bone. The sites formerly occupied by the disks are marked by an epiphyseal line.

Endochondral ossification of short bones does not include epiphyseal or secondary centers, as for long bones. Rather, epiphyseal cartilage is replaced by bipolar extensions of the diaphyseal center.

Bone of adults, and especially that of growing children, is transient, continually being remodeled by erosion and replacement to produce a different internal architecture able to accommodate new forces and stresses. Remodeling of bone begins immdiately following the deposition of bone on the calcified cartilage spikes in developing long bones, and its progresses throughout life.

The size and shape of bones, as well as the environments (loose connective tissue versus cartilage) required for osteogenic activity, are genetically determined, as are the duration and sequences of osteogenic actvity. Hormones and vitamins variously contribute to the development and maintenance of bone.

CLINICAL IMPLICATIONS

Developmental defects of the face and paraoral apparatus, including the lips, jaws, and palates (hard and soft), result from the inability of the primordial centers (processes) to develop, grow, migrate, and fuse. The failure of these developmental sequences to progress normally may be due to genetic defects or to teratogens, including alcohol, drugs, hormones, vitamins, and radiation.

Cleft Lip

The occurrence of lip clefting is about 1:900 in the white and 1:1500 in black populations. The

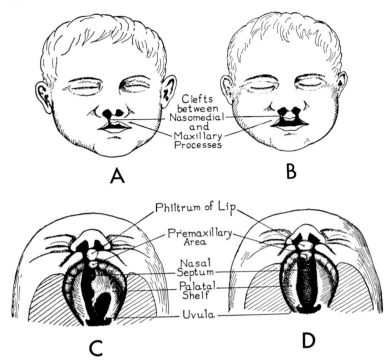

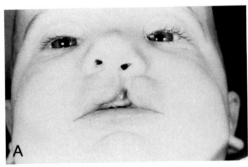

Fig. 3–21. Cleft lip and palate. *A,* Unilateral cleft lip. *B,* Bilateral cleft lip. *C,* Unilateral cleft lip and palate. *D,* Bilateral cleft lip and palate. (Patten, B.M.: Human Embryology. 2nd Ed. New York, Blakiston, 1968.)

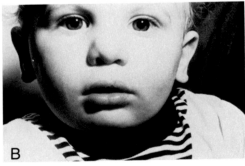

Fig. 3–22. *A,* Child with unilateral cleft lip. *B,* Same child after reconstructive surgery. (Courtesy of H. Wilhelmsen, D.D.S., M.D.)

simplest form involves either the right or left side, lateral to the median line of the upper lip, where the medionasal (nasomedial) processes fail to join and fuse with the maxillary lateral nasal processes (Figs. 3–21A and 3–22). Clefting occurs in the second month of development, and is found with equal frequency on the right or left side. Less often, however, neither the right nor the left medionasal processes fuses with their lateral nasal and maxillary complements, resulting in bilateral lip clefting (Fig. 3–23A). If only the lip is involved, only the externalmost segment of the processes fails to fuse—that is, there is only superficial involvement.

Cleft Lip and Jaw

With more extensive failure of fusion of the processes, more distal involvement of the oral structures occurs. Accordingly, the cleft will be deeper and will extend varying distances distally through the maxillary arch into the hard and soft palates (Fig. 3–21C and D). If the palatal shelves, which are medial extensions of the maxillary processes, fail to meet and fuse with one another and with the descending nasal septum, not only will lip clefting occur but, in addition, the oral and nasal cavities will communicate directly so that the primitive condition of the nasobuccal cavity is retained (Fig. 3–21C and D).

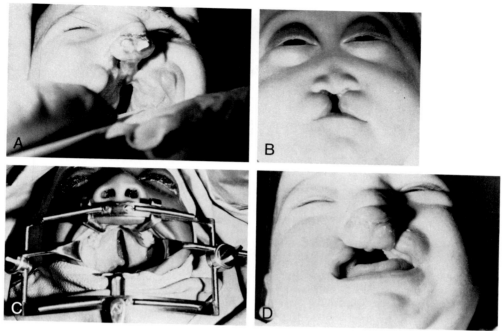

Fig. 3–23. Developmental anomalies resulting from fusion failures of primordial processes associated with the face and oral cavities. *A,* Bilateral cleft lip and palate. *B,* Median cleft of upper lip resulting from bilateral fusion failure of the nasomedial processes. *C,* Incomplete cleft palate. *D,* Unilateral cleft lip. (Courtesy of H. Wilhelmsen, D.D.S., M.D.)

Clefting of the secondary palate is caused by the failure of the palatal processes to fuse. This developmental anomaly may occur independently of lip and arch clefting (Fig. 3–23C). Other less common abnormalities result from the failure of fusions of these primordial processes, including such conditions as the median cleft of the upper lip and arhinia (Fig 3–23B), as well as cleft of the lower jaw, and oblique facial clefts. In ad-dition to agenesis of the premaxilla, arhine-cephale involves the anterior segment of the brain and skull. Arhinia involves only the nose. These may occur in various combinations (Fig. 3–23A and B). An oblique facial cleft occurs when the entire maxillary process fails to merge with the lateral nasal process. Currently, so-phisticated plastic surgery techniques can repair most developmental defects (Fig. 3–22).

CHAPTER 4

Development of Dental and Paradental Structures

The mouth is divided into two regions, the oral vestibule and the oral cavity proper. The entrance to the oral cavity is known as the vestibule, which is limited exteriorly by the lips and cheeks and interiorly by the gingivae (gums) of the maxillary (upper) and mandibular (lower) arches. The oral cavity proper is bound anteriorly by the lingual gingivae of the arches, superiorly by the palates (hard and soft), and inferiorly by the floor, or sublingual sulcus. In addition to the tongue, which is the largest organ, the mouth contains tonsils, salivary glands (major and minor), and teeth, with their attachment apparatus. This chapter is concerned only with the development of teeth and their anchoring structures.

Teeth are comprised of two parts, crowns and roots. The latter are anchored in their sockets (alveoli) by a regularly arranged dense fibrous connective tissue, the periodontal ligament. The core of the tooth is comprised of a mucoid-like connective tissue, the dental pulp, encased in a mineralized connective tissue, the dentin. The dentin of the crown is covered by enamel, the hardest calcified tissue of the body. Dentin of roots, radicular dentin, is covered by cementum, a calcified connective tissue similar to bone. Root(s) and crown join at the cervix, or neck, of the tooth. Externally the junctional area is identified as the cementoenamel junction (CEJ). The grinding surface of the crown bears irregular conical elevations, called cusps. The chisel-shaped surfaces of the anteriormost or incisor teeth are called incisal edges.

Throughout life, humans grow two sets of teeth: deciduous or primary (baby or milk) dentition, and succedaneous or permanent dentition. The normal complement for the former is

20, 10 per arch. The total number of permanent teeth is 32, or 16 per arch. Teeth differ dimensionally and morphologically from the anterior (mesial) segments to the posterior (distal) segments of the arches. Arches for primary dentition bear 10 teeth each. In a quadrant, which is half of an arch, primary dentition consists of central incisors (mesialmost on the arch), distally followed by a lateral incisor, canine, first molar, and second molar. For permanent dentition each arch contains 16 teeth, and each quadrant has 8 teeth—a central incisor, lateral incisor, canine (cuspid), first premolar, second premolar, first molar, second molar, and third molar. In some people third molars are not produced; in others they may not grow into the oral cavity, and are designated as impacted.

The discussion that follows is concerned with the induction processes involved in odontogenesis, development of dental tissues (enamel, dentin, dental pulp, cementum), and with the tooth's anchoring soft and hard tissues (periodontal ligament, alveolar process, attachment epithelium).

INDUCTION

The process by which products of specific cell groups/layers direct the developmental course(s) of neighboring cell groups or layers during embryogenesis is known as induction. Cell products affecting the developmental path are known as inductors or organizers. Inductor substances are also known as evocators, and their capacity to stimulate tissue to react is called competency. The precise nature of the evocators or inducing substances is unknown, but the material is probably a small protein molecule using a larger molecule as the vehicle for its transmission from the inducing to the responder tissue. Several generalizations have been made regarding induction, among which are that period and duration of induction in a tissue are precise and that species specificity is not evident.

The first or primary organizers are exemplified by the notochord, primitive streak, and paraxial mesoderm. After the primary organizers have established the basic architectural plan of the embryo, they are succeeded by secondary organizers. In oral histology, secondary organizers are of special importance in tooth development (odontogenesis) and in alveolar process development. For example, there is no question that inductive interactions occur between the stomodeal ectoderm and the underlying mesenchyme, resulting in the initiation, growth, and elongation of the dental lamina. Similarly, inductive activity occurs between the dental lamina and the mesenchyme, resulting in the formation of the deciduous and succedaneous tooth germs. Likewise, reciprocal inductive interaction occurs between the inner enamel epithelium and the dental papilla of the tooth germs. That is, the inner enamel epithelium influences the development of the initial segment of dentin, and this dentinogenic activity influences the preameloblasts to mature to ameloblasts.

It has also been suggested that the stratum intermedium of the tooth organ engages in inductive activity because synthesis of enamel normally does not occur in the absence of this layer. Inductive interaction probably occurs between the tooth organ and the developing bone, because the alveolar process or sockets of roots do not develop in the absence of the dental germs. Inductive mechanisms are also present in other tissues of the developing oral and paraoral tissues, but they are not of particular significance in this discussion.

LAMINAE

The initial appearance of the primordial tissues involved in the formation of the oral cavities, arches, and dental anlagen occurs between 6 and 6½ weeks of embryonic development with the formation of two basic laminae, vestibular and dental. Laminae are sheets or cords of stomodeal ectoderm that initiate development at about the same time. Although their paths differ, both grow into the underlying mesenchyme. The vestibular lamina is responsible for the formation of the oral vestibule. The dental lamina sequentially produces and/or results in the formation of four other laminae: the lateral, successional, parent, and rudimentary laminae. With the possible exception of the latter, all participate actively in odontogenic processes.

Vestibular Lamina

This lamina is also referred to as the lip furrow band. The path of growth of the lip furrow band into the mesenchyme is superficial, extending rapidly from its mesial site of origin distally to circumscribe the limits of the oral vestibule. Thereafter, until the definitive size of the vestibule has been attained, the vestibular lamina grows slowly and synchronously with the development of other orofacial structures. The vestibular lamina is characterized by distinct features—that is, it is a widening epithelial band or sheet, whose core (central) cells disintegrate to produce cavitation (Fig. 4–1). The cavity thus formed is the vestibule of the mouth. The epithelium that is retained contributes to the mucosal lining for the vestibule. Later, further epithelial downgrowths into the connective tissue result in the formation of the minor salivary glands (see page 97). The ectoderm lining the external wall of the vestibule forms the skin for the cheeks and the vermillion border of the lips. Thus, the lip furrow band not only forms the oral vestibule but, in its developmental path, participates with the dental lamina in defining the maxillary and mandibular arches into which the dental laminae will proceed in odontogenesis (tooth development) (Fig. 4–1C and D).

Dental Lamina

Development of the dental lamina is marked by pronounced proliferation of the basal cell layer of the oral ectoderm so that the epithelium is thickened in the area of increased mitotic activity (Figs. 4–1A and 4–2). With continued proliferation of the basal cells, an epithelial sheet is formed that invades the underlying mesenchyme. The developmental path of the dental lamina is arcuate; it progresses bilaterally from its most mesial site to circumscribe the most distal regions of the maxillary and mandibular dental arches (Fig. 4–3).

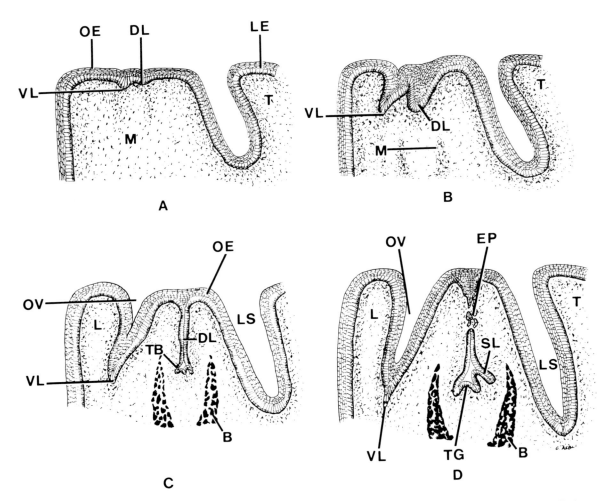

Fig. 4–1. Schematic representation of development of laminae and dental germs. *A,* Initiation of dental (DL) and vestibular (VL) laminae, (T, tongue; OE, oral epithelium; LE, lingual epithelium; and M, mesenchyme). *B,* Further development of dental (DL) and vestibular (VL) laminae, and condensation of mesenchyme (M), presaging bone development (T, tongue). *C,* Progression of the vestibular lamina (VL) to define the lip (L) and dental arch (prospective alveolar process), bordered by the oral vestibule (OV) and by the lingual sulcus (LS). Note that the dental lamina (DL) has elongated and bears the tooth organ as a primary tooth bud (TB). Note further that intramembranous bone development (B) has progressed further toward the oral cavity, circumscribing the labiolingual aspect of the dental (alveolar) arch (OE, oral epithelium). *D,* In a later stage, note that the lip (L), vestibule (OV), and membrane bone (B) have developed further. Disorganization of the original dental lamina results in epithelial pearls (EP). The tooth organ (TG) has developed into the cap stage, and formation of the succedaneous laminae (SL) is initiated, (LS, lingual sulcus; T, tongue; VL, vestibular laminae).

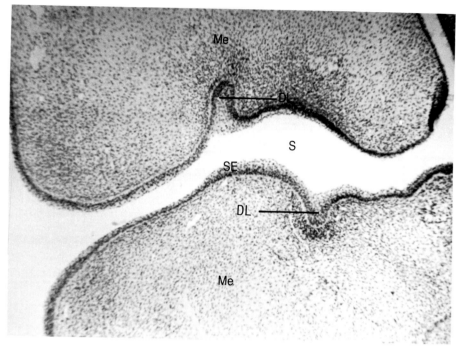

Fig. 4–2. Photomicrograph showing stomodeal cavity (S) lined by ectoderm (SE). The dental lamina (DL) is growing into mesenchyme (Me). (Hematoxylin and eosin stain; ×50)

Lateral Lamina

With the formation of the anlagen (tooth germs or organs) for primary teeth on the labial (facial) aspect of the dental laminae, some of the dental lamina is extended labially as an epithelial bridge connecting the differentiating tooth organ with the dental lamina. This lateral extension of the dental lamina is called the lateral lamina (Fig. 4–3). Occasionally, growth forces of mesenchyme against the lateral lamina produce a cavitation between the lamina and the dental organ, called the enamel niche (Fig. 4–4). Neither the lateral lamina nor the enamel niche, however, is functionally important.

Successional Lamina

After the formation of the primary tooth germs, the free terminal or tip of the dental lamina continues to grow into the mesenchyme. By the fourth month of development, the extension of the dental lamina is called the successional lamina because it is responsible for the production of tooth germs that succeed the 20 exfoliated primary teeth (Figs. 4–1, 4–3, 4–4, and 4–5). Thus, the primary dentition is replaced by permanent dentition, tooth for tooth—that is, incisor for incisor and cuspid for cuspid. The primary molars are replaced by permanent bicuspids.

Parent Dental Lamina

The dental lamina initiated during the sixth week of embryonic development provides tooth germs for the 20 primary teeth (10 maxillary, 10 mandibular). Distal growth of the laminae in the dental arches does not cease with the production of the primary tooth germs; rather, they continue to grow distally. These laminae are now designated as the parent dental lamina. By the end of the fourth month in utero, the parent laminae produce tooth germs for the first permanent molars. Tooth germs for the second and third permanent molars are formed by the parent dental lamina after birth. The second molar germs develop in infants at about 9 months of age, and the third molars are initiated in 4-year-olds. The parent dental lamina, therefore, provides dental primordia for all teeth lacking primary predecessors.

Rudimentary Lamina

Laminae associated with odontogenesis begin to atrophy and/or disintegrate rapidly after the

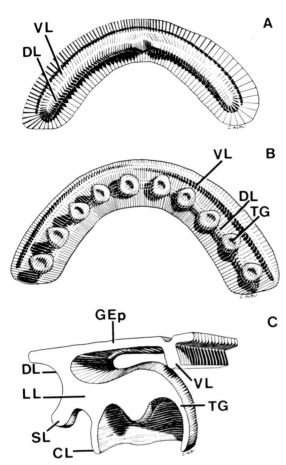

Fig. 4–3. Schematic representation of laminae and tooth germs. *A,* Dental lamina (DL) and vestibular lamina (VL). *B,* Tooth (anlagen) germs (TG) in late bud stage of development. These represent the germs for the ten primary teeth. Note their relationship to the dental lamina (DL) and the vestibular lamina (VL). *C,* Orobuccal section showing vestibular lamina (VL) and the relationship of the tooth germ (TG) in the bell stage to the lateral lamina (LL), succedaneous lamina (SL), and original dental lamina (DL), (GEp, prospective gingival epithelium; and CL, cervical loop).

establishment of the tooth germ. These processes occur initially in the lamina joined to the epithelium—the orodental epithelial junction. From this area, disorganization of the dental laminae progresses toward the developing tooth germ. Some cells of the laminae persist and tend to aggregate, and sometimes proliferate as keratinized epithelial whorls or arranged as nests (nidi) known as *epithelial pearls* (formerly misnamed and interpreted as glands of Serres) (Figs. 4–5 and 4–6). Epithelial pearls may develop into tooth germs, producing extra teeth known as supernumerary teeth, tooth-like tumors, or cyst linings.

STAGES OF ODONTOGENESIS

The development of teeth and their acquisition of functional competency require three interrelated and interdependent stages: development of the crown (coronal): development of the root (radicular); and eruption. Coronal odontogenesis (tooth formation) involves amelogenesis (enamel formation) and dentinogenesis (dentin formation). Enamel formation is the primary activity of a four-layered enamel organ. The latter is produced by the dental laminae as a tooth germ that passes through three cytomorphologic periods: (1) bud-stage (initiation); (2) cap-stage (growth and differentiation); and (3) bell stage (histodifferentiated and functionally competent four-layered enamel organ). The enamel organ participates in matrix synthesis and matrix maturation. Dentin formation occurs concomitantly with amelogenesis. Three periods are required in dentinogenesis: fibrillogenesis, matrix maturation, and matrix mineralization.

Root odontogenesis consists of dentinogenesis and cementogenesis (cementum formation). With root development, which involves its longitudinal growth, the crown is pushed through the subgingival connective tissue to arrive in the oral cavity and to meet its antagonist of the opposing arch. Growth of the crown from an intraosseous to an intraoral location is called tooth eruption. Thus, radicular odontogenesis and tooth eruption are concomitant activities involving dentinogenesis and cementogenesis. Dentinogenesis does not cease with the formation of the crown; rather, it continues to develop radicular dentin. The latter stimulates cementogenesis. Cementum, the external covering of the root, interfaces with the periodontal tissues. With the alveolus formed by the alveolar bony process and the anchoring activity of the periodontal tissue, the root, is secured by the cementum. The alveolar process, periodontal ligament, and cementum, along with other tissues (attachment epithelium), constitute the dental attachment apparatus.

In the following discussion maturation of the prospective periodontium and endodontium are included, because their development is both in-

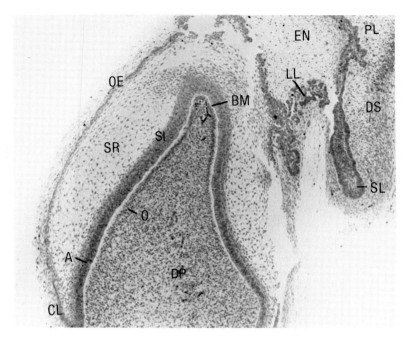

Fig. 4–4. Tooth germ in the bell stage of development showing the enamel niche (EN), lateral lamina (LL), parent lamina (PL), and successional lamina (SL) surrounded by connective tissue of the dental sac (DS). The enamel organ is comprised of the outer enamel epithelium (OE), stellate reticulum (SR), stratum intermedium (SI), inner enamel epithelium or prospective layer of ameloblasts (A), cervical loop (CL), basement membrane (BM), preodontoblasts (O), and dental papilla or future dental pulp (DP). Araldite-embedded and toluidine blue stain; ×4)

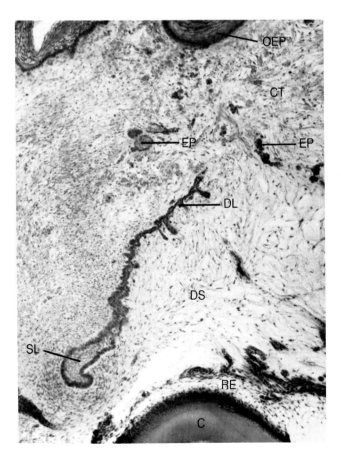

Fig. 4–5. Tooth organ showing mineralized enamel of crown (C), and overlying reduced enamel epithelium (RE). The parent dental lamina (DL) in the subepithelial connective tissue (CT) has atrophied into epithelial pearls (EP) (SL, successional lamina; DS, dental (sac/follicle). (Hematoxylin and eosin stain; ×50)

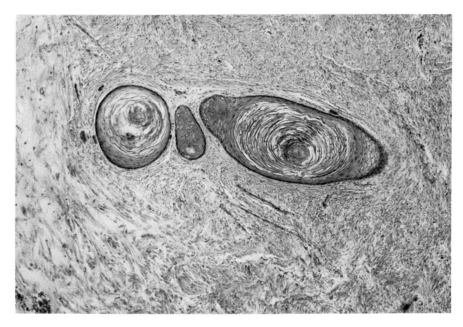

Fig. 4–6.   Epithelial pearls in the prospective gingival connective tissue. Note the whorl-like appearance of the epithelial cells. (Hematoxylin and eosin stain; ×120)

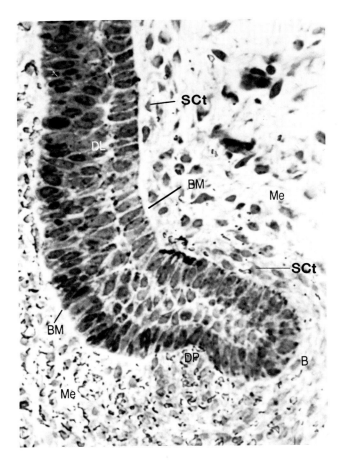

Fig. 4–7.   Dental lamina (DL) and its lateral extension, the dental organ, in the late bud or early cap stage (B) is invested in mesenchyme (Me). Note the sheath-like arrangement of connective tissue (SCt) abutting the dental lamina and the dental organ. The cell concentration at the inferior concavity of the dental organ is the prospective dental papilla (DP). Note that the cells of the core of the tooth germ are polygonal and the surface cells are low columnar (BM, basement membrane). (Epon-embedded, toluidine blue stain; ×400)

trinsically and integrally involved in odontogenesis.

Coronal Formation

The crown develops through the processes of amelogenesis and dentinogenesis.

Enamel Organ Development

Three stages are involved in the formation of the enamel organ.

Bud Stage. Subsequent to the formation of the dental lamina, ten tooth (dental) germs, or buds, are produced on each arch. Tooth buds are proliferations of the dental laminae that represent the presumptive primary teeth (Figs. 4–1 and 4–3). The mandibular buds are the first to appear (seventh week), and the maxillary buds follow a few days later. By the eighth week the two sets of ten buds have been produced. The surface cells of the bud abut the basal lamina. Initially, they are low columnar cells, and the deeper cells are polygonal with small intercellular spaces (Fig. 4–7).

Cap Stage. The cells of the bud increase in number rapidly, thereby enlarging the tooth organ. During growth the base of the bud becomes indented and the space is occupied by mesenchyme. With continued growth the resulting concavity forms a conical core of mesenchyme, called the dental papilla. With growth of the dental germ and its dental papilla, the cells are rearranged so that the dental organ changes morphologically from bud shape to cap shape (Fig. 4–8). Cells of the cap exhibit pronounced differences in size, shape, and location, so that four areas/sites may be distinguished: (1) a single layer of cuboidal cells covering the external surface of the cap, called outer enamel epithelium; (2) a single layer of low columnar cells lining the dental papilla, called the inner enamel epithelium; (3) several layers of polygonal cells overlying the latter, called the stratum intermedium; and (4) multilayered polymorphous core cells of the cap, the prospective stellate reticulum. Although the cells are closely packed at sites one, two, and three, those comprising the core of the cap possess lengthening cell processes so that the intercellular spaces become progressively enlarged (Fig. 4–12).

As the cap develops, a local surge of mitotic activity at the third site results in a temporary swelling known as the enamel (Ahren's) knot (Figs. 4–9 and 4–10). The rapidly dividing cells overflow into the core of the cap to form a cell

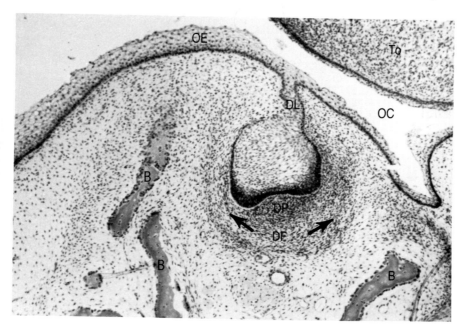

Fig. 4–8. Section of the mandibular arch showing the tongue (To) in the oral cavity (OC), oral epithelium (OE), and the dental lamina (DL). The tooth organ in a cap stage of development is surrounded by mesenchyme. The inferior border of the dental organ exhibits a dense cell concentration of the dental papilla (DP). The dental follicle (DF) surrounds the tooth organ (B, spicules of membrane bone). (Hematoxylin and eosin stain; ×50)

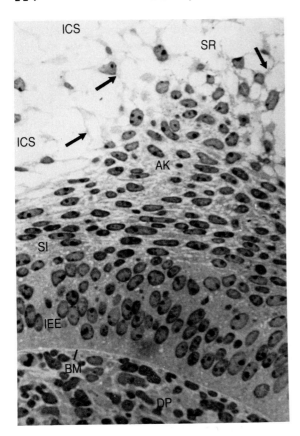

Fig. 4–9. Segment of the tooth organ in the late cap or very early bell stage of development. The section is taken at the level of the crest of the dental papilla (DP); visualized are the inner enamel epithelium (IEE) and stratum intermedium (SI). Note the relationship of the enamel (Ahren's) knot (AK) to the stellate reticulum (SR) and to the stratum intermedium. Large intercellular spaces (ICS) formed by long branching processes *(arrows)* are seen in the stellate reticulum (BM, basement membrane). (Epon-embedded, toluidine blue stain; ×400)

cord, the enamel cord, spiraling to the outer enamel epithelium. Here, a slight epithelial depression is produced, called the enamel navel. Within a few days the cap is enlarged and transformed into a bell-shaped structure. It is during early bell stage that the enamel knot and cord disappear. There are two hypothetical functions of the enamel knot: (1) it causes partial delineation of the vestibular and lingual segments of the dental papilla; and (2) it provides cells for the enlarging and maturing enamel organ.

During the cap stage the mesenchyme of the dental papilla exhibits greatly accelerated mitotic activity, so that cell density is greatly increased (Fig. 4–8). The cells produced are mostly mesenchymal cells and fibroblasts, the

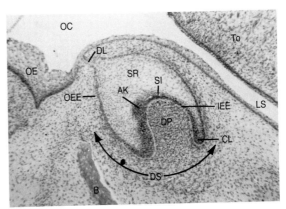

Fig. 4–10. Section of tooth organ in the bell stage of development showing the cervical loop (CL), dental papilla (DP), inner enamel epithelium (IEE), stratum inermedium (SI), stellate reticulum (SR), and outer enamel epithelium (OEE). Note the relationship of the tooth organ to the dental lamina (DL) and paraoral structures (To, tongue; LS, lingual sulcus; OC, oral cavity; OE, oral epithelium; B, bone spicule; DS, dental sac; AK, Ahren's knot). (Hematoxylin and eosin stain; ×50)

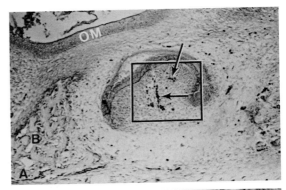

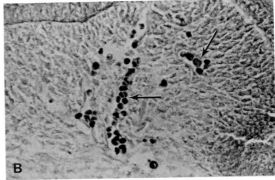

Fig. 4–11. Tooth organ in early bell stage of development showing blood vessels, *(arrows)* of the dental papilla. *A,* Bone spicules (B) and oral mucosa (OM). *B,* Dental papilla enlarged to show developing blood vessels with cells *(arrows).* (Erythrocytic zinc stain; dithizone technique) (Falkler, W.A., Jr., and Provenza, D.V.: Detection of developing vessels in tooth primordia using the dithizone reaction and the fast green staining method of Van Gieson. J. Balt. Coll. Dent. Surg., *27:*80, 1973.)

processes of which in later stages tend to increase in length to create larger intercellular spaces, thereby enlarging the dental papilla. The intercellular spaces contain reticular and collagenous fibers, nerves, and blood vessels. There is strong evidence indicating in situ development of blood vessels (Fig. 4–11). These are initiated in the substance of the papilla and grow toward its apex, where enamel is deposited first.

The dental follicle, or dental sac, is the connective tissue that invests the developing tooth organ. It will form the periodontal membrane and ligament (Figs. 4–8 and 4–10). As an immature tissue, its cell and fiber compositions are similar to those of the dental papilla. During the appositional stage of amelogenesis, the dental follicle matures into a loose fibrous connective tissue; in the eruptive stage, this becomes more densely fibrous, and may be referred to as the periodontal membrane. As a definitive tissue, when the tooth has assumed its occlusive position, the membranous features are lost and the tissue becomes the periodontal ligament. Thus, in its development to the periodontal ligament, the tissue of the dental sac progresses from mesenchyme to loose fibrous connective tissue to dense fibrous connective tissue irregularly arranged to its definitive stage, dense fibrous connective tissue regularly arranged (ligamentous).

Bell Stage. In this stage the enamel organ completes morphodifferentiation, through cell multiplication, growth, differentiation, and rearrangement. The cap is transformed into a bell-shaped structure with four definite layers (Fig. 4–10). The outer enamel epithelium, which in the early bell stage is cuboidal, later becomes squamoid (Fig. 4–12). The transition in cell shape, proceeding from the crest of the bell to the rim (cervical loop), is synchronous with appositional activity. The cervical loop region of the tooth organ is bilaminar, comprised of inner and outer enamel epithelial layers (Figs. 4–13 and 4–14). The inner enamel epithelium (preameloblasts) are low columnar cells containing centrally located nuclei, poorly developed endoplasmic reticulum, and Golgi apparatus (Fig. 4–14). The latter is located toward the proximal end of the cell, between the nucleus and the terminal bar. In the more differentiated preameloblast, as observed in the early bell stage, the cell is more columnar. The nucleus is more basally located and the Golgi system has mi-

grated toward the more distal cell segment (Fig. 4–14, stage B). In the bell stage, organelle populations hypertrophy in anticipation of amelogenesis. The cells lengthen and eventually attain maximum dimensions of 4 μm in diameter and 80 μm in length (Fig. 4–14, stage C). The crest cells, located at the prospective incisal edge or cusp tip, are the first to differentiate and engage in amelogenesis. Thereafter cell maturation and enamel synthesis progress toward and over the slope, so that the amelogenic period is shortened toward the cervical loop (Figs. 4–14 and 4–15). Accordingly, the thickest enamel is at the cusp tip or incisal edge, and the thinnest enamel is at the cervix (neck) of the tooth.

Overlying the inner enamel epithelium is the stratum intermedium. It is a multicellular layer, with round to flat components. Their processes are short and blunt, and are joined to those of adjacent cells by desmosomes (Fig. 4–16). Large intercellular spaces containing numerous microvilli distinguish this layer from adjacent ones. The cytoplasm is organelle-rich.

In the late cap stage the intercellular spaces of the core cells increase in size, especially those near the stratum intermedium (Fig. 4–8). By the appositional stage the intercellular spaces have become enlarged throughout the core; this area of the enamel organ is referred to as the stellate reticulum (Figs. 4–12 and 4–15). The increased size of the intercellular spaces results from lengthening of the cell processes concomitant with the production of large amounts of mucoid substance (acid mucopolysaccharides) released into the cell interspaces. Contacts between processes are maintained by desmosomes (Fig. 4–12). Some oral histologists believe that the increase in the volume of the stellate reticulum provides space for the tooth's coronal tissues.

The dental papilla similarly increases in volume during the bell stage of development. Although this may be partially a result of cell division, it is also probably due an increase in the size of the intercellular spaces. The cells of the dental papilla include mesenchymal cells, fibroblasts, and preodontoblasts. The latter originate from the mesenchymal cells or from fibroblasts located at the periphery of the dental papilla. Their differentiation into odontoblasts occurs first at the apex of the dental papilla and proceeds over the slope of the papillae in harmony

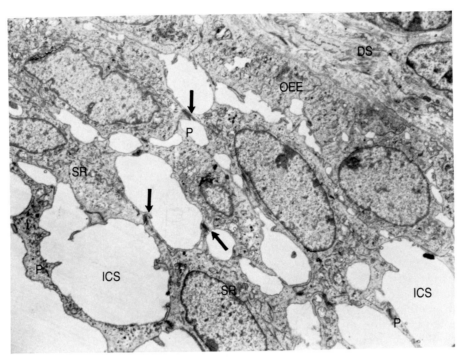

Fig. 4—12. Electron micrograph of section of dental sac (DS), outer enamel epithelium (OEE), and stellate reticulum (SR). Note that the intercellular spaces (ICS) are formed by cell bodies and their lengthening cell processes (P). Arrows point to desmosomes. (Epon-embedded, uranyl acetate and lead citrate stain; ×3000)

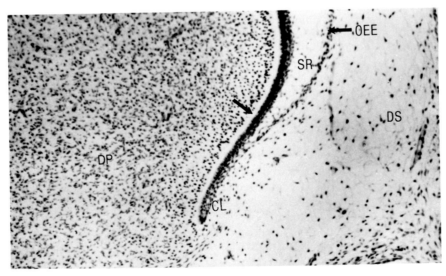

Fig. 4—13. Segment of tooth organ near the cervical loop (CL), showing its relationship to the outer enamel epithelium (OEE) and stellate reticulum (SR). The basement membrane *(arrow)* is pronounced, and is bordered in this area by the inner enamel epithelium and the preodontoblasts (DP, dental papilla; DS, dental sac). (Hematoxylin and eosin stain; ×80)

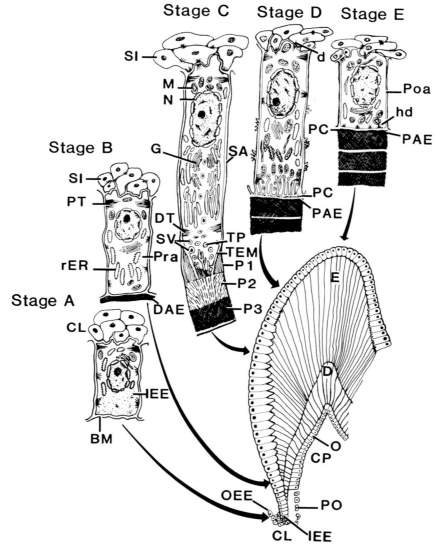

Fig. 4–14. Diagrammatic representation of the life cycle of an ameloblast. In stage A, there are some cytologic changes of the inner enamel epithelium (IEE), beginning as the least developed cells at the cervical loop (CL), as differentiating preameloblast (Pra) approaching the slope of the crown in stage B, as secretory ameloblasts engaged in production of enamel prisms in stage C, as secretory ameloblasts (SA) depositing the surface aprismatic enamel and primary cuticle in stage D, and as protective reduced enamel epithelium or postameloblasts in stage E. The low columnar inner enamel epithelial cells of the cervical loop (CL) region in stage A are the last to mature, while those approaching the slope and those along the slope of the crown are progressively taller columnar cells with hypertrophying organelle populations anticipating or engaged in synthesis. The low columnar postameloblasts (Poa; reduced enamel epithlium) having completed their secretory functions, serve in tooth eruption and protection of the crown. Note that the activity of the ameloblasts in amelogenesis and of the odontoblasts (O) in dentinogenesis are synchronous. Subsequent to the formation of the aprismatic enamel in stage B, the mature ameloblasts reach maximum height and their cytoplasmic structures, which participate in enamel synthesis, are intracellularly stratified. The phases of amelogenesis in stage C are as follows: phase 1 (P_1), formation of Tomes' process (TP); phase 2 (P_2), formation of the prism template (TEM); and phase 3 (P_3), filling and mineralization of the matrix (contained in the template). Note that the stratum intermedium (SI) overlies the proximal terminal of the ameloblasts (stages B, C, D, and E). At the cervical loop (CL), however, the IEE is continuous with the outer enamel epithelial (OEE) cells. The stratum intermedium and postameloblasts remain united via desmosomes (d), forming the primary attachment (junctional) epithlium joined to the tooth by the primary cuticle (PC) and hemidesmosomes (hd) shown in stages D and E (PT, proximal terminal bar; DT, distal terminal bar; DAE, distal aprismatic enamel; rER, rough endoplasmic reticulum; G, Golgi complex; M, mitochondria; N, nucleus; SV, secretory vesicles; E, enamel; D, dentin; BM, basement membrane; PAE, proximal aprismatic enamel; CP, pulp chamber; Po, preodontoblasts). (Adapted from Ten Cate, A.R.: Oral Histology: Development, Structure, and Function. 2nd Ed. St. Louis, C.V. Mosby, 1985.)

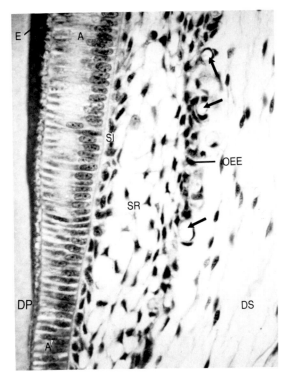

Fig. 4–15. Appositional stage of odontogenesis showing the periphery of the dental papilla (DP), enamel matrix (E), ameloblasts (A), stratum intermedium (SI), stellate reticulum (SR), outer enamel epithelium (OEE), dental sac (DS), and blood vessels *(arrows)*. Note that the enamel-synthesizing ameloblasts (A) are taller than the differentiating ones (A'). Note also the polarization of the nucleus in the ameloblast. (Hematoxylin and eosin stain; ×320)

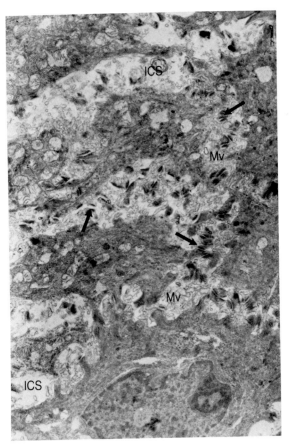

Fig. 4–16. Electron micrograph of stratum intermedium showing intercellular spaces (ICS) packed with microvilli (Mv). Desmosomal connections *(arrows)* are numerous. (×5,000)

with the maturation of the inner enamel epithelium and preameloblasts. Thus, differentiation of the preodontoblasts and preameloblasts occurs synchronously toward the cervical loop in anticipation of the activities associated with the appositional stage of development (dentinogenesis and amelogenesis).

Additionally, there is further evidence of in situ production of blood vessels. The intrapapillary vascular network awaits the arrival of and linkage with the blood vessels accompanying nerves from sources outside the developing primitive pulp—that is, from the projected radicular supply, which originates from the fundus of the socket.

The four-layered enamel organ is invested with loose fibrous connective tissue, in which are located fibroblasts and mesenchymal cells. The latter are found mostly near the blood vessels of the dental sac (prospective periodontal

membrane or ligament) (Fig. 4–15). Fibroblasts, the predominant cell type, are oriented generally parallel to the surface of the tooth organ and define the boundary of the dental sac. The intercellular spaces contain collagen fibrils. Some of the peripherally located fibers become intercalated with those external to the dental sac. These tend to provide stability and strength to the tooth organ.

Appositional Stage (Amelogenesis)

Synthesis and secretion of the products of amelogenesis are involved in this stage of development. Enamel formation is presaged by definite changes in the enamel organ components. The cells of the outer enamel epithelium at the crest become squamoid, and their continuity is disrupted by gaps. These provide spaces for the invasion of connective tissue elements (fibers, cells, vascular buds) from the dental sac (Fig. 4–17). As the blood vessels of the dental follicle advance through the discontinuities, the mucoid intercellular material of the stellate reticulium is withdrawn so that the cells tend to collapse on one another. Most of the cells are destroyed or removed, but a few persist. These are reorganized into cell islands, epithelial pearls, similar to those of the rudimentary lamina described above. Extension of the vascular channels into the stellate reticulum brings the substances essential for metabolism and enamel matrix synthesis closer to the principal cells involved in amelogenesis (stratum intermedium, ameloblasts). Except for an increase in the number of microvilli in the intercellular spaces, the cells of the stratum intermedium are not cytologically different from those of the bell stage.

During the appositional stage, ameloblasts acquire their maximum height and their organelles increase in number and become polarized (Figs. 4–14, 4–15, and 4–18). That is, the nucleus occupies the proximal third of the cell, the part adjacent to the stratum intermedium. The Golgi complex and endoplasmic reticulum occupy most of the middle third of the cell. The distal or dental papilla-facing third is filled with rough endoplasmic reticulum and secretory vesicles

(Figs. 4–14, 4–18, and 4–19) Mitochondria are located throughout the cell, although most are concentrated at the proximal third of the cell. Terminal bars are found at both the proximal and distal ends of the ameloblast. Tomes' processes (described below) extend enamelward from the distal terminal bar.

Enamel formation begins immediately after the initiation of dentinogenesis. The first layer of dentin is juxtaposed by a thin layer of enamel, which is aprismatic (Fig. 4–14). Having deposited the "foundation" layer of enamel the maturing ameloblasts retreat, complete differentiation, and begin to deposit organic enamel matrix as protein *amelogin*. This occurs in daily increments averaging 4 μm to form enamel prisms. This occurs in three phases: (1) formation of Tomes' process; (2) formation of the enamel prism template; and (3) matrix completion (maturation) and mineralization for the enamel prism (P1, P2, and P3 in stage C Fig. 4–14).

Phase 1. After the elaboration of the aprismatic layer, matrix is secreted into the lateral intercellular spaces of the ameloblast from the distal terminal bar forward. The laterally deposited matrix compresses the distal tip of the ameloblast, forming a cone-shaped protoplasmic terminal, the Tomes' process, which measures

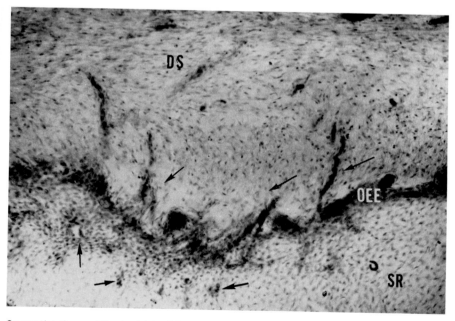

Fig. 4–17. Connective tissue of the dental (sac) follicle (DS), outer enamel epithelium (OEE), and stellate reticulum (SR) in the late bell stage of development. Note that the outer enamel epithelium is thrown into folds, and that blood vessels *(arrows)* have advanced onto and have invaded the stellate reticulum. (Hematoxylin and eosin stain; × 80)

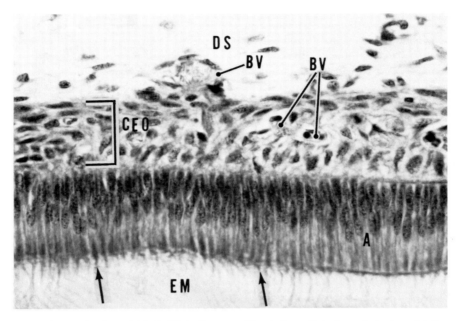

Fig. 4–18. Photomicrograph of ameloblasts (A) in the appositional stage. Note the polarized nuclei, Tomes' processes *(arrows)*, enamel matrix (EM), and the condensed three layers of the enamel organ (CEO). Blood vessels (BV) are numerous in th dental sac (DS), and enter the papillary ridges formed by the folded epithelium of the tooth organ to be in closer proximity to the cells participating in the enamel-forming activities. (Hematoxylin and eosin stain; ×320)

about 4 μm in diameter and length (Figs. 4–14, 4–18, and 4–19).

Phase 2. During this period, secretory ameloblasts and the overlying cells of the stratum intermedium retreat. In doing so, a honeycomb-shaped depression is produced, the enamel prism template (Fig. 4–14C).

Phase 3. This period is characterized by enamel matrix synthesis and secretion filling the depression. Seeding the matrix lattice with apatite crystals occurs simultaneously with or immediately following matrix secretion. The lattice pattern (arrangement) and seeding of crystals occur with specific orientation, probably under the influence of the plasmalemma of the Tomes' process (Figs. 4–14 and 4–19).

The three phases of amelogenesis are repeated every 24 hours, so that about 4 μm of enamel matrix are deposited daily. Thus, ameloblasts produce enamel rods (prisms) composed of 4 μm-thick additions. More than one ameloblast may participate in the production of an enamel rod, so that each rod is not the exclusive territory of a given ameloblast. Furthermore, although the ultimate number of daily increments is equal to the number of days of activity, such is not the case for the externalmost prism segments. Surface enamel prisms are not di-

mensionally the same as those of the enamel substance or those near the dentinoenamel junction. Rather, they are thinner and wider, because they must accommodate the greater surface area of the crown.

Enamel synthesis does not end abruptly but diminishes gradually, forming progressively more shallow but wider increments (Fig. 4–14). Ameloblasts producing incisal edges or cusps may produce rods consisting of hundreds of increments. The ameloblasts of the cervical area, however, which are active for shorter periods (i.e., a few days or weeks), produce short rods of only a few increments. The final finishing layer of enamel, only a few microns thick, is aprismatic. Hence, the outermost and innermost layers of enamel are aprismatic at the surface and at the dentinoenamel junction (Fig. 4–14). It is believed that enamel matrix mineralization may, in fact, be initiated in the mantle dentin, proceeding bidirectionally from this area (toward the enamel and dentin).

In the process of mineralization, the crystals formed initially are needle-shaped, measuring about 3 nm in thickness and 30 nm in width. Over 1200 crystals are crowded into a square micrometer. As the crystal growth progresses to maturity by accretion, some fuse, eventually be-

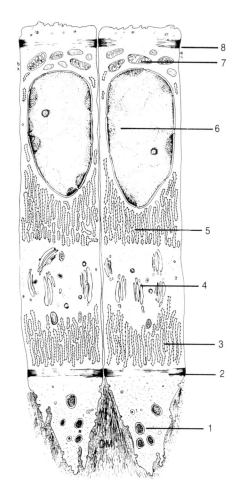

A

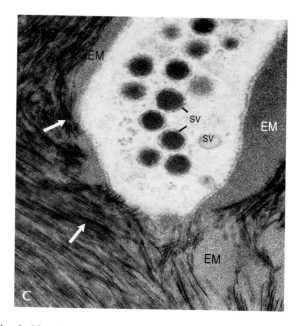

C

Fig. 4–19. Ameloblast engaged in enamel synthesis. *A,* Diagram showing cytoplasmic zones of secretory ameloblast: Tomes' process (1); distal terminal bar (2); distal endoplasmic reticulum (3); Golgi complex (4); supranuclear endoplasmic reticulum (5); nucleus (6); mitochondria (7); and proximal terminal bar (8). *B,* Distal region of a secretory ameloblast showing Tomes' process (TP), intercellular enamel matrix (EM), calcifying enamel matrix *(white arrow)* and secretory granules *(black arrows),* terminal web (TW), and endoplasmic reticulum (DER). (×4,200) *C,* Distal terminal Tomes' process abutting enamel matrix (EM) and calcifying enamel matrix *(arrows).* Note the secretory vesicles (sv) in Tomes' process. (×36,000)

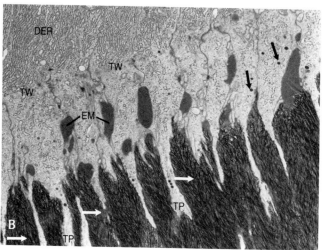

B

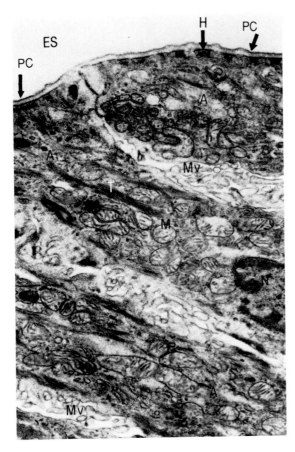

Fig. 4–20. Electron micrograph showing segments of postameloblasts (A) elaborating the primary cuticle (PC). Note that the intercellular spaces are packed with microvilli (Mv), (N, nucleus, M, mitochondria, T, tonofilaments; ES, enamel space (ES); H, hemidesmosomes). (×10,000) (Provenza, D.V., and Sisca, R.F.: Fine structures of monkey (Macaca mulatta) reduced enamel epithelium. J. Periodontol., *41*:313, 1970.)

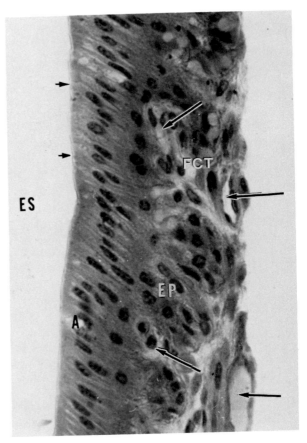

Fig. 4–21. Section of reduced enamel epithelium. The postameloblasts (A) elaborate the primary cuticle *(arrowheads)*. Cells of other epithelial layers have been organized into epithelial (pegs) ridges (EP) separated by connective tissue ridges of the dental follicle (FCT). Contained in the latter are blood vessels *(arrows)* and various cell types (ES, enamel space). (Araldite-embedded, toluidine blue stain; ×120) (Provenza, D.V., and Sisca, R.F.: Fine structure features of monkey (Macaca mulatta) reduced enamel epithelium. J. Periodontol., *41*:313, 1970.)

coming hexagons 65 nm in width, 30 nm in thickness and a couple of micrometers in length (see Fig. 5–5). Space for crystal growth is provided by the removal of some organic components by the secretory ameloblasts. This function of the ameloblast has been confirmed by electron microscopy and tracer studies. In mature enamel, the number of crystals per micrometer is about halved and the organic component is identified as enamelin.

After the predestined amount of enamel has been deposited, the ameloblasts produce a nonmineralizing thin organic membrane, the primary cuticle (Figs. 4–14 and 4–20). Thereafter the ameloblasts become shortened and, except for tonofibril bundles coursing through the cell,

the organelle populations are greatly decreased (Figs. 4–14 and 4–20). These, now the postameloblasts, are attached to the enamel cuticle by hemidesmosomes (Figs. 4–14, stages D and E; and 4–2). Postameloblasts with their overlying cells, which are mostly stratum intermedium components comprise the reduced enamel epithelium (Fig. 4–20). The latter provides protection for the crown during tooth eruption. If breaks occur in the reduced enamel epithelium, components of the dental sac pass through the openings and arrive at the enamel surface, where they produce a matrix that calcifies on the enamel. These deposits are known as coronal cementum and afibrillar cementum, because the collagen of this matrix does not bear the characteristic 64-nm periodicity of mature collagen. Afibrillar cementum occurs mostly in the cervical area of the crown.

Eventually the reduced enamel epithelium fuses with the oral epithelium (see Fig. 4–34). Over the tooth tip(s) both epithelia thicken, atrophy, and are sloughed off, providing a path for tooth eruption. The remaining reduced enamel epithelium covering the crown forms a kind of collar, which adheres to the tooth surface. With the emergence of the tooth into the oral cavity, the reduced enamel epithelium is known as the primary attachment (junctional) epithelium (see Figs. 4–34, 8–39, and 8–41). This is later desquamated and replaced by a germinal layer from the gingival epithelium to form the permanent attachment (junctional) epithelium, known as the secondary junctional epithelium (see Fig. 4–35). Some scientists believe that the moribund reduced enamel epithelium, which is the first attachment epithelium, is not desquamated. Rather, it is thought that it degenerates and condenses to become part of the primary enamel cuticle. This is subsequently removed by the abrasive action of mastication or brushing.

Dentinogenesis (Dentin Formation)

The changes in the components of the dental papilla that lead to the establishment of a dentinogenic layer will be described here in terms of the developmental stages of the enamel organ with which they are inextricably associated.

Mantle Dentin. Fibroblasts and collagenous elements are separated from the dental lamina by a basement membrane and its basal lamina. These are continuations of those of the arch epithelium. Early in the bud stage, the cells and fibrils surrounding the dental lamina and bud are sheath-like in arrangement (Fig. 4–7). The initiation of a dental papilla is indicated first by the formation of a concavity on the inferior surface of the bud (Fig. 4–7). With tooth germ enlargement and development of the cap stage, the papilla is also enlarged and becomes more deeply embedded in the tooth organ (Fig. 4–8).

With the continued growth of the enamel organ to form a four-layered bell-shaped structure, not only has the volume of the dental papilla increased but conspicuous changes are also noted in the apex of the structure, as follows.

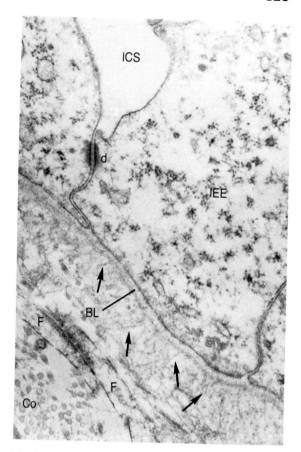

Fig. 4–22. Electron micrograph of a segment of the dental papilla at the prospective dentinoenamel junction, (IEE, distal cell segments of the inner enamel epithelium; BL, basal lamina; aperiodic fibrils *(arrows)*; Co, collagen fibrils; d, desmosome; F, filopodia; ICS, intercellular spaces). (Araldite-embedded; × 15,000) (Provenza, D.V., Fischlschweiger, W., and Sisca, R.F.: Fibres in human dental papillae. Arch. Oral Biol., *12*:1533, 1967, Pergamon Press, Ltd.)

The fibroblasts and/or mesenchymal cells adjacent to the preameloblasts are separated from them by a space up to 20 μm occupied by the basement membrane. Fine aperiodic filaments appear, oriented at right angles to the basal lamina (Fig. 4–22). The filaments probably are type IV collagen and are probably produced by cells of the prospective inner enamel epithelium. The most peripheral cells of the dental papilla become oriented more or less perpendicular to the future dentinoenamel junction. Differentiation of these cells as preodontoblasts is indicated by increases in organelle populations. The cell processes are lengthened and extend through the basement membrane to the basal lamina.

The interface between the papilla and the enamel epithelium assumes an irregular profile because of the formation of blunt distal processes of the ameloblasts. Furthermore, the space between the lengthening preodontoblast processes becomes filled with type I collagen so that, by the time the processes arrive at the basal lamina, many of the collagen fibrils have been aggregated into bundles. These tend to fan out and become aligned perpendicular to the basal lamina or become intercalated between the terminals of the ameloblast and odontoblast processes. These collagen fibril bundles of diameters of 0.1 μm to 0.2 μm are known as (von) Korff's fibers (Figs. 4–23 and 4–24). They are the dominant fiber for the first 20 μm of dentin formed, the mantle dentin. This dentin differs from that formed thereafter (circumpulpal dentin) in that it is composed of three fibril types: aperiodic filaments, collagen fibrils, and collagen fibers (fibril bundle aggregates, the Korff's fibers).

Until recently, it was believed that the Korff's fibers of mantle dentin were argyrophilic fibrils aggregated into bundles en route between the odontoblasts to the prospective dentinoenamel junction where they became intercalated between the scallops of the junction. Based on electron microscopic investigations, it has been shown that the fibrils are not present in the intercellular spaces of the prospective odontoblasts. Thus, it has been proposed that these type I collagen fibrils are produced by and during the differentiation of odontoblasts from the ectomesenchymal cells peripherally located in the dental papilla. The silver-staining intercellular components, observed by optical microscopy and formerly interpreted as reticular fibers, are in fact artifact and not reticular/argyrophilic (collagen type III) fibrils. The stainable material is probably proteoglycans of the ground substance produced initially by ectomesenchymal cells of the early dental papilla and later by differentiating odontoblasts. Thus, the Korff's fibers of mantle dentin observed at the distal terminals of the odontoblast processes and oriented perpendicularly to the dentinoenamel junction are fibril-bundle aggregates of type I collagen produced by (pre)odontoblasts.

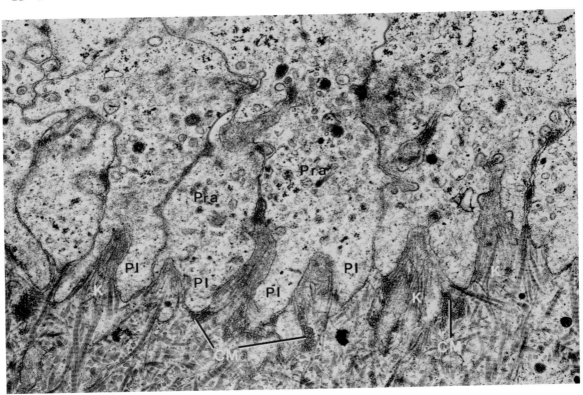

Fig. 4–23. Mantle dentinogenesis in progress at the distal terminals of a preameloblast (Pra) with Korff's fibers (K) between the terminals of the preameloblasts (Pl) and coarse matrix deposits (CM) in mantle dentin. (×19,000)

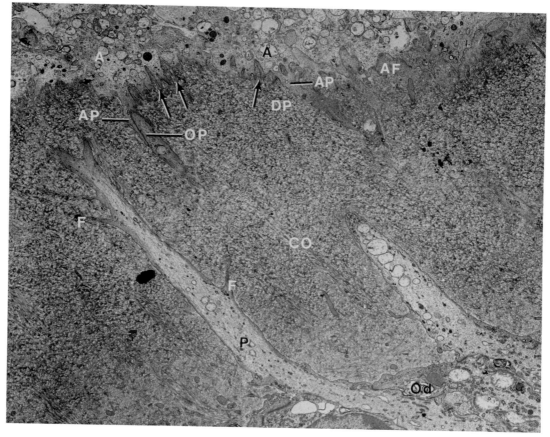

Fig. 4–24. Electron micrograph of mantle, through the layer of aperiodic filaments (AF), dentinogenesis showing major odontoblast processes (P) with filopodia (F). The ameloblast process (AP) projects through the layer of aperiodic filament (AF), arriving in the mantle dentin to become intimately associated with the terminal of the odontoblast process (OP) (Od, odontoblast; Korff's fibers *(arrows)*; A, ameloblasts; and Co, collagen) (Uranyl acetate and lead citrate stain; ×5,100) (Sisca, R.F., and Provenza, D.V.: Initial dentin formation in human deciduous teeth. Calcif. Tissue Res., 9:1, 1972.)

As soon as the mantle dentin space has acquired its full complement of collagen, more ground substance is produced obscuring the fibrils of the area. This period of matrix maturation, produces what is known as predentin (dentinoid). Subsequently, predentin intervenes between the mineralized dentin and the distalmost surface of the odontoblast cell body. Predentin that becomes mineralized is replaced so that it is a persisting layer, present for the life of the pulp.

Mineralization is the terminal step leading to dentin as a definitive tissue. The process involves seeding the predentin matrix with apatite crystals. These begin as tiny spherules that grow by accretion and eventually fuse with their neighbors until a uniform calcifying front has been formed. All matrix components are mineralized except the odontoblast processes. These become trapped in canaliculi, in dentin known as tubules (Fig. 4–25). Just as a layer of nonmineralized matrix (predentin) lies between calcified dentin and the odontoblast cell body, a microspace-containing matrix is found between the process and the peritubular dentin (Fig. 4–25).

With completion of mantle dentinogenesis, ameloblasts juxtaposed to the odontoblasts complete differentiation and deposit aprismatic enamel on the dentin. This dentinoenamel junction is comprised of aprismatic enamel and mantle dentin.

Odontoblasts. The progenitor cells of odontoblasts are the ectomesenchymal cells located at the periphery of the dental papilla. The first indication of differentiation is a change in cell

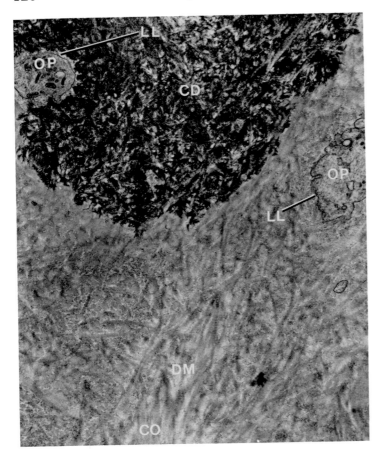

Fig. 4–25. Area of dentin formation showing calcifying dentin (CD) surrounding an odontoblastic process (OP). The latter is separated from the calcified dentin by a thin organic membrane, the lamina limitans and/or a layer of predentin (LL). The dentin matrix (DM) has not as yet attained its full complement of ground substance; thus, the collagen fibrils (Co) are not masked and show striations. (Epon-embedded, uranyl acetate and lead citrate stain; × 12,000)

shape from stellate to low columnar as a preodontoblast. Additionally, nuclei become polarized and occupy the proximal (pulpward) cell segment. This location is retained. With further development, the distal processes lengthen and extend toward the prospective dentinoenamel junction. The preodontoblast cell bodies lengthen gradually, becoming tall columnar to accommodate the increased number of organelle populations. As a fully differentiated (secretory) odontoblast the cell consists of a distal process, which is confluent with the terminal web. The layers of organelles from the terminal web basalward include the zone of distal endoplasmic reticulum, zone of the Golgi complex and mitochondria, zone of supranuclear endoplasmic reticulum, nuclear zone, and infranuclear zone (Fig. 4–26). Various techniques involving histochemistry, radiolabeled proline, and electron microscopy have provided evidence that procollagen elements are synthesized in the endoplasmic reticulum, processed and packaged in the Golgi apparatus, and secreted from the distal

cell surface into the intercellular space, where they are further organized into collagen fibrils.

From the distolateral surfaces basally, odontoblasts are joined by a variety of junctional complexes including the zonulae occludens and adherens, macula adherens (desmosomes), and nexus (gap junctions). The latter are especially interesting since they have been shown to join neighboring odontoblasts' cell bodies and processes to form a syncitium. Gap junctions located between odontoblasts and the subodontoblastic structural milieu, including nerves, have been observed, which may be functionally significant.

Circumpulpal Dentin. Circumpulpal dentin is deposited on mantle dentin. This tissue subtype, produced hereafter, differs from mantle dentin principally in the size of the fibrils in the matrix. These fibrils are much smaller (about 50 nm in diameter), but they exhibit the 64-nm banding of type I collagen. Further differences in circumpulpal dentin involve aperiodic filaments and fibril bundles (Korff's fibers) which are absent. Additionally, the collagen fibrils

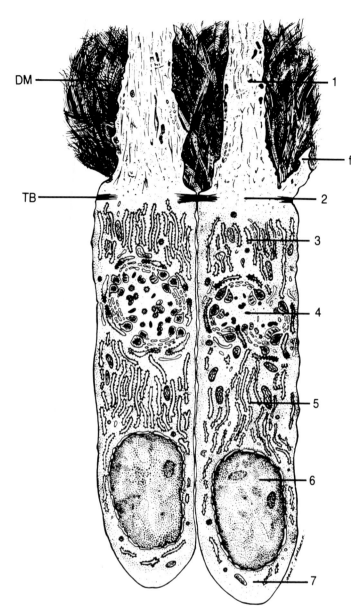

DM

TB

1

f

2

3

4

5

6

7

Fig. 4–26. Secretory odontoblast showing various cytoplasmic zones: process (1); terminal web (2); distal endoplasmic reticulum, (3); Golgi complex (4); supranuclear endoplasmic reticulum (5); nuclear zone (6); infranuclear zone (7), (DM) dentin matrix; f, filopodia; and TB, terminal bar).

form a dense meshwork, destined to be oriented somewhat perpendicular to the prospective dental tubules. Further, the first ground substance of mantle dentin, hence that of the earliest stage of the dental papilla development, is a product of mesenchymal cells and later of differentiating and mature odontoblasts. The ground substance of circumpulpal dentin, on the other hand, is produced solely by odontoblasts.

The mineralization process is the same in both mantle and circumpulpal dentin. The dentin constituting the walls of the tubules—the peritubular dentin—is more highly calcified than that located between the tubules (intertubular

dentin) (see Chapter 6, Dentin). Separating the mineralized dentin and the odontoblast process is an organic membrane (mostly glycosaminoglycans) known as the *lamina limitans*. It should be noted that all oral histologists are not in agreement regarding the presence of this membrane. Some are of the opinion that only the plasmalemma of the odontoblasts occupies the area. Others believe the space is occupied by an organic substance.

Root Formation

With cessation of amelogenesis, the crown is fully formed and root development is initiated.

The latter results in growth of the tooth toward the oral cavity, a process known as tooth eruption. The soft connective tissue forming the core of the root, the radicular pulp, is encased by two hard tissues: dentin which is a continuation of its coronal counterpart, and cementum, which is a newly produced tissue that forms the external root surface. Three relationships may ocur between cementum and enamel at their junction: most often, cementum overlaps enamel; less often, cementum simply abuts enamel; and least often, cementum and enamel do not meet, leaving a space that exposes the underlying dentin. Dentin forms the greater part of the root tissues, except at the apices of older teeth, in which the tips may be composed exclusively of cementum.

Root Sheath and Diaphragm

As the differentiating preameloblasts approach the cervical loop, the cells of the inner and outer enamel epithelia at the cervical area engage in mitotic activity, resulting in the elongation of the loop. This lengthening bilaminar extension of the loop is known as Hertwig's epithelial root sheath (Fig. 4–27; also see Fig. 4–33). It defines the limits of the pulp and determines the number, size (thickness and length), and shape of the roots.

The number of roots is determined by changes in sheath morphology. For single-rooted teeth, a single cone-shaped sheath is formed. For others, a cone-shaped sheath must be provided for each root; these are joined at the prospective furcation site of the root. Sheaths for multirooted teeth are produced simultaneously. The cone-shaped template is formed by medial ingrowths of the root sheath as bilaminar epithelial processes (Fig. 4–28). In birooted teeth, two processes are formed that advance and fuse at the prospective site of root bifurcation. Trirooted teeth require three converging processes, and quadrirooted teeth require four processes. Thus, for all except single-rooted teeth, initial growth of the root sheath is bidirectional, longitudinal, and medial (Fig. 4–28).

At the tip (apex) of each growing root the sheath is inclined sharply medially (inward ± 45°), forming what has been designated as an epithelial diaphragm (see Fig. 4–42). The sheath and its diaphragm remain active until the root has been fully formed. Thereafter, they disappear or form cell rests (of Malassez) (Fig. 4–29).

Root Dentin

Dentinogenesis is uninterrupted from the crown throughout the length of the root. The process is the same for both crown and root,

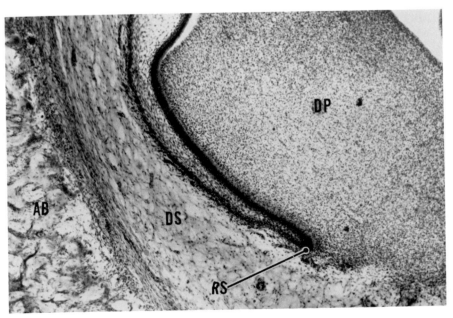

Fig. 4–27. Photomicrograph showing a segment of the enamel organ at the level of the developing root sheath (RS), (DS, dentral sac; DP, dental papilla; AB, developing alveolar bone). (Hematoxylin and eosin stain; × 50)

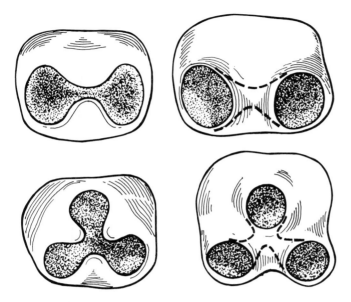

Fig. 4–28. Diagrammatic representation of root sheath activity in the formation of two and three roots. Two epithelial processes of the root sheath grow medially toward one another and fuse. The growth pattern for three roots is made by the formation of three epithelial processes, which grow (medially) toward the midline and fuse.

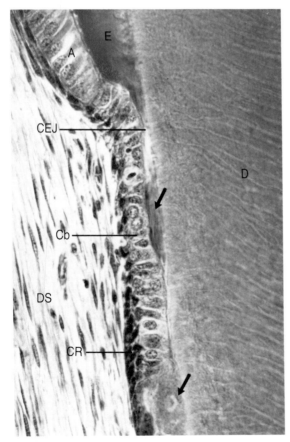

Fig. 4–29. Developing root at the cementoenamel junction (CEJ), showing postameloblasts (A), enamel (E), cementoblasts (Cb), dental sac (DS), and cementum matrix *(arrows),* (D, dentin; CR, cord of cell rests of Malassez). (Hematoxylin and eosin stain; ×320)

except for three differences: (1) radicular dentin matrix is deposited against the root sheath, instead of ameloblasts; (2) in the root, the path of the dentinal tubules is different; and (3) in the root, dentin is covered by cementum.

Cementum

The root sheath separates the odontoblastic layer lining the prospective root canal from the dental follicle (future periodontal membrane or ligament). With deposition and mineralization of root dentin matrix, shrinkage occurs, so that the calcified tissue pulls away from the root sheath and disrupts its continuity. These breaks provide openings through which collagen fibers, and cementoblast progenitor cells (mesenchymal cells or fibroblasts) from the dental follicle (prospective periodontal ligament), enter and become aligned against the dentin. These connective tissue elements (e.g., cells, fibers, ground substance) succeed in isolating the detached root sheath. Although most of the latter disintegrate and undergo autolysis, others persist as cell cords or islands known as epithelial rests of Malassez (Fig. 4–29; also see Figs. 4–31 and 8–4). It is possible, although rare, for root sheath fragments to cling to the dentin. In these instances, the inner enamel epithelial component may differentiate into ameloblasts and form aberrant islands of enamel on the root surface known as enamel pearls (Fig. 4–30). It is also possible for premature breaks to occur in the root sheath, resulting in confluency of the connective tissue of the dental follicle and that of

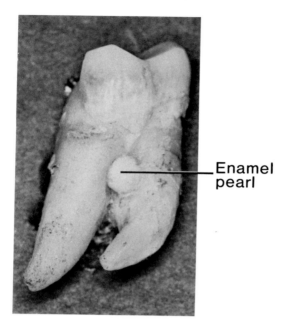

Enamel pearl

Fig. 4–30. Enamel pearl on root.

The cementum first formed, (hence, the oldest) is located more toward the cervix. This cementum does not contain cementocytes and is thus known as acellular cementum. Acellular cementum is produced during periods in which cementogenic activity is sufficiently slow to provide time for the comentoblasts to retreat toward the periodontal connective tissue. Although acellular cementum is limited mostly to the upper third of the root, it may be found in any location in which matrix production and calcification occurs at a leisurely pace. On the other hand, as the tooth approaches the oral cavity, matrix synthesis occurs rapidly and the cementoblasts are trapped in the calcifying matrix. This cementum is appropriately known as cellular cementum (Fig. 4–31).

Alveolar Process

The alveolar bone forming the root sockets represents processes or extensions of the body of the mandible and maxilla. The mandible is the second bone of the body to begin development (Figs. 4–8 and 4–10). This occurs early in the seventh week of fetal growth, and it is marked by an increase in mitotic activity of the mesenchymal cells that border Meckel's cartilage. As in the case of other bones of the face, the mandible is formed by intramembranous osteogenic activity. Labyrinths of spicules are produced, which follow paths predetemined by the growth of Meckel's cartilage. The latter, however, does not form a cartilage model, so it is not involved in the osteogenic process that forms the mandible and maxilla, and their alveolar processes.

The maxilla, the third bone of the body to develop, begins development in the seventh week of fetal growth. Three intramembranous centers of ossification are initiated, one for each maxillary process and the third for the intermaxillary (premaxilla) segment. As the centers enlarge and grow toward one another, they eventually fuse to form a continuous arch.

As the tooth organs (Figs. 4–8 and 4–10) and subsequently the definitive dental tissues are developed, the bone spicules formed near them are incorporated into the body of the maxilla or mandible. If tooth germs are absent, bone for the jaws would continue to develop until the bone masses for the body of the mandible and maxilla had been produced. Such is not the case

the pulp. When this occurs, dentinogenic and cementogenic activities pass over the connective tissue of the break. With subsequent dentinogenic and cementogenic activity, a canal is formed from the periodontal tissue to the pulp, known as an accessory canal.

Differentiating cementoblasts that line up against the dentin form a distinct layer, which engages in cementogenesis. This also involves three phases: (1) fibril formation; (2) matrix maturation by the influx of ground substance; and (3) mineralization. The second and third phases occur in rapid succession or concurrently. The matrix fibrils or intrinsic fibrils are produced by the cementoblast and are oriented parallel or at a slight angle to the dentin surface. The investing matrix produced by cementoblasts obscures the fibrils and, with maturation of the tissue, is referred to as precementum or cementoid (see Fig. 8–5). As in the case of dentin a layer of cementoid persists, which separates the calcified matrix from the cementoblastic layer. Collagen, in the form of Sharpey's fibers from the periodontal tissue, are inserted in the matrix so that, with calcification, the collagen bundles are anchored in the matrix (Fig. 4–31; also see Fig. 8–5). These fiber bundles form the principal fiber groups of the periodontal ligament; they serve to anchor the root(s) into their sockets.

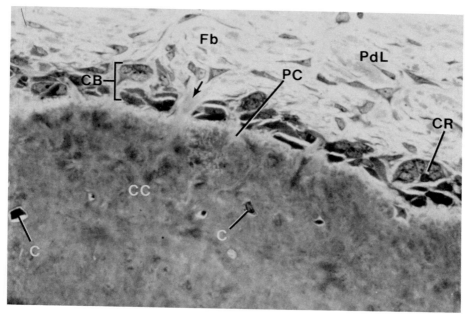

Fig. 4–31. Cellular cementum (CC) showing insertion of Sharpey's collagen bundles *(arrow)*, cementocytes (C), cementoblasts (CB), and fibroblasts (Fb) in the periodontal ligament (PdL). Cell rests of Malassez (CR), cementoid precementum (PC). (Hematoxylin and eosin stain, ×330)

for the alveolar processes, which are in fact extensions of the body of the mandible and maxilla. Thus, in cases of complete anodontia (absence of dentition), alveolar processes are not formed. It follows, therefore, that the stimulus for the development of alveolar processes is provided by the development and growth of tooth organs.

Bone formed during the development of the crown is incorporated into the body of the jaw bones. With root development, osteogenic activity associated with the formation of the sockets (alveoli) for elongating roots is added to the body of the mandible and maxilla as a process or extension, called the alveolar process. Because osteogenic actvity of the alveolar process is stimulated by root growth, bone development ceases when the tooth acquires its occlusal (functional) position in the oral cavity. Furthermore, if a tooth is extracted, the bone of the process— hence, the bone accommodating the root—will be resorbed. Development of the alveolar process involves the formation of supporting beams of bone, trabeculae. Later, as the alveolar process thickens, tables of compact bone are deposited, sandwiching the trabecular core. These three regions of the alveolar process are given special names. The central core of trabeculae, or diploe, is called the spongiosa. The bony plate of compact bone lining the alveolus is the cribriform plate (L. cribrum, sieve), while that forming the vestibular and lingual surfaces of the alveolar processes is the cortical plate (Fig. 4–32).

Numerous collagen fibers (Sharpey's) originating from the periodontal ligament are inserted as extrinsic fibers into the cribriform plate or the cementum. These form the principal fiber bundles of the periodontal ligament. They develop fully and exhibit preferred orientation after the teeth meet their antagonists and acquire functional competency with masticatory and other forces. Additionally, the numerous openings on the cribriform plate, which present a sieve-like appearance, provide sites through which vascular, lymphatic, and nerve elements pass to and from the periodontal tissue, especially the middle third.

Periodontal Ligament

The dense connective tissue in which the fibrous elements surrounding the tooth are definitively arranged in accord with their special functions is known as the periodontal ligament (Fig. 8–25). The regular arrangement of the type I collagen fiber bundles accounts for its title of "ligament."

There are four stages of development:

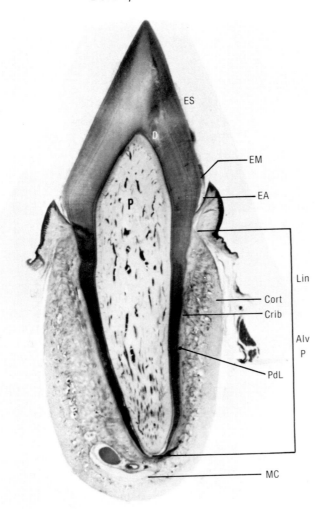

Fig. 4–32. Incisor and supporting tissues visualized by buccolingual section. Shown are enamel spaces (ES), enamel matrix (EM), dentin (D), epithelial attachment (EA), pulp (P), periodontal ligament (PdL), and cortical (Cort) and cribriform (Crib) plates of the alveolar process (Alv P), (MC, mandibular canal; Lin, lingual aspect). Note that the spongiosa is missing in this incisor, and that the lingual cortical plate is thicker than the vestibular.

1. It originates as mesenchyme, which surrounds the early tooth primordium.
2. It is subsequently transformed into a loose connective tissue, the dental follicle (L. follis, bag), or sac, surrounding the developing tooth (Fig. 4–33).
3. Later, with root development, the dental follicle matures into a dense fibrous connective tissue, the periodontal membrane.
4. Finally, with occlusal forces, it becomes a dense fibrous connective tissue in which the fibers are regularly arranged, the periodontal ligament.

With progression through each stage, the tissue becomes denser by the acquisition of more fibers; the cell population is correspondingly decreased. Prior to the acquisition of the ligamentous character of the tissue, the collagen fiber groups are inserted into the cementum, while the others are inserted into the cribriform plate of the developing alveolar process. These are identified as Sharpey's fibers.

During development of the periodontium, the extrinsic pericemental Sharpey's fibers course irregularly through the periodontal space; fiber groups disassociate from one bundle to join neighboring ones, eventually meeting the Sharpey's fibers of the cribriform plate. Because this organizational feature of the principal collagen fiber bundles is most apparent in the more central region of the periodontal membrane, the term "intermediate plexus" has been applied to this area. Whether this plexus is an anatomic feature or simply an artifact is uncertain. Present research has suggested that the plexus does not contribute to tooth emergence.

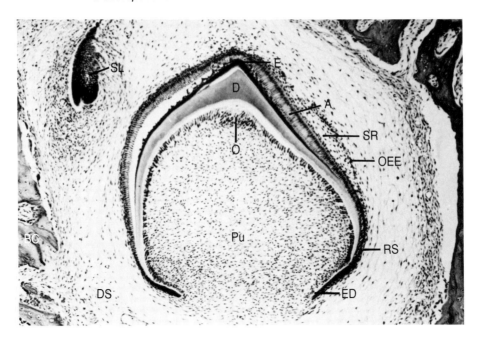

Fig. 4–33. Primary tooth in the appositonal stage of development sharing its bony crypt (BC) with the permanent tooth/successional lamina (SL), (E, enamel; D, dentin; A, ameloblasts; O, odontoblasts; Pu, dental papilla; DS, dental sac; RS, developing root sheath; SR, stellate reticulum; OEE, outer enamel epitheium; ED, epithelial diaphragm). (Hematoxylin and eosin stain, ×50)

When the erupting tooth meets its antagonist of the opposite arch and is in full function, islands of loose connective tissue remain in the periodontal ligament among the principal fiber groups. This loose connective tissue is known as interstitial (space) tissue (see Figs. 8–21, 8–28B, 8–30, and 8–35). The principal fiber groups and the interstitial tissue serve specific functions (see Chapter 8).

Other fibers reported in the periodontal tissue include elastic fibers, which are associated with blood vessels, and oxytalan fibers, which are possible precursors of elastic fibers. Special stains are required for their resolution. Type III collagen and fibronectin filaments are also reported to be components of the tissue. Fibronectin is a glycoprotein related filament. These filaments appear to be implicated in cells (mostly fibroblasts) and are concerned with movement, extension, and adhesion to certain tissue substrates. The ground substance in which the cellular and intercellular elements of the periodontal tissue exist is comprised of proteoglycans and glycoproteins.

Tooth Eruption

Tooth eruption involves two activities of synthesis, which have been described earlier: root

dentinogenesis and cementogenesis. There are about a dozen interrelated factors suspected, but these have not been convincingly proven to contribute to tooth eruption. Other than root dentinogenesis and cementogenesis, additional factors implicated in tooth eruption include the following: (1) osteogenic activity involved in the growth and lengthening of the alveolar process; (2) lytic actvity by the reduced enamel epithelium; (3) action of the epithelial diaphragm and its associated cushioned hammock ligament; (4) vascular and tissue fluid pressures, both intra-dental and peridental; (5) volumetric increase in dental pulp by mitotic activity; (6) collagen synthesis and fibroblast contraction; (7) tensional and growth forces transmitted alveolodentally by collagen fiber bundles; (8) hormonal influences, mainly those of the thyroid and pituitary; and (9) pressures exerted by facial muscles, especially the masticatory group. While bone is eroded over the crown it is deposited at the fundus of the socket. This, in conjunction with the cushioning action of the hammock ligament and fluid pressures at the base of the root, appear to facilitate tooth eruption. Results of in vivo experiments, however, tend to suggest that tissue growth (e.g., of bone, cementum, dentin, pulp) and tensional forces involving the collagen bundles and fibroblasts, as well as periapical vascular pressures, are primary eruption factors.

The relationship of root elongation and tooth eruption through dentinogenic and cementogenic activity is evident. In a more simplistic concept for eruption, it has been suggested that growth forces of the tooth push the connective tissue aside, condensing and incorporating it into the periodontal tissue (Fig. 4–34). On the other hand, a more intricate concept involves lysis and removal of the connective tissue overlying the advancing front of the growing tooth, possibly by the reduced enamel epithelium. After removal of the connective tissue, the reduced enamel epithelium covering the tooth and the approximating gingival epithelium proliferate, thicken, and fuse. The core cells subsequently disintegrate, resulting in the formation of a passageway extending through the depth of the fused epithelia overlying the emerging tooth tip(s). Additionally, there is evidence indicating that the growth forces condensing the connective tissue intervening between the reduced enamel epithelium and the oral epithelium occlude blood vessels, thereby effecting tissue necrosis. This is manifested clinically as "sore spots." The thin layer of epithelium is sloughed off, providing an epithelium lined opening for tooth emergence (Figs. 4–34 and 4–35).

The reduced enamel and oral (gingival) epithelia are structurally united by the desmosomal connections and persist to participate in development of the dentogingival epithelial junction. It is noteworthy that the noncellular organic layer, which consists of a basal lamina-hemidesmosome complex, is known as the epithelial attachment. This is in contradistinction to the cellular layer, which consists initially of the reduced enamel epithelium exclusively, and later totally of oral epithelium-derived components; these form the primary junctional or attachment epithelium and secondary junctional (attachment) epithelium, respectively. Thus, the reduced enamel epithelium constituting the first attachment is referred to specifically as the primary epithelial (junctional) attachment (Fig. 4–36A and B).

The remnant of the enamel organ attached to and forming an epithelial cuff on the coronal surface is the final, or attachment, stage of the postameloblasts. As more and more of the crown enters the oral cavity, the superior segment of the attachment epithelium becomes detached, forming a shallow trough between the gingiva and the enamel. This trough, which surrounds the crown, is called the gingival sulcus. In time

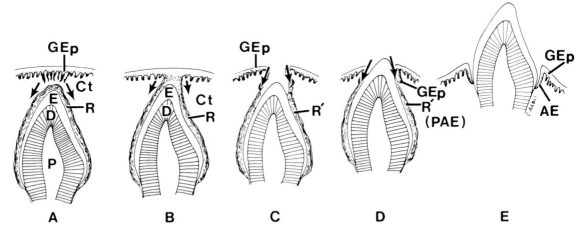

Fig. 4–34. Diagrammatic representation of tooth emergence showing the relationships among the prospective gingival epithelium (GEp), connective tissue (Ct), and the reduced enamel epithelium (R). The simplistic concept provides for the lateral condensation of the connective tissue around the emerging tooth *(black arrows)*, which is subsequently fused with the periodontal ligament *(A–E)*. *B*, The epithelia of the gingiva and tooth organ proliferate and fuse. The core cells of the fused epithelia disintegrate to form an epithelial-lined opening (C) for the emergence of the tooth. *D*, The primary attachment epithelium (R') is composed exclusively of the reduced enamel epithelium, and covers most of the crown. Note the relationship of gingivally derived epithelium (GEp) and its migration path, which replaces the primary attachment epithelium (PAE). *E*, The primary attachment epithelium (R'/PAE) is replaced by the components of the oral epithelium. Also note that the attachment epithelium (AE) is located totally on the crown (E, enamel; D, dentin; P, pulp chamber).

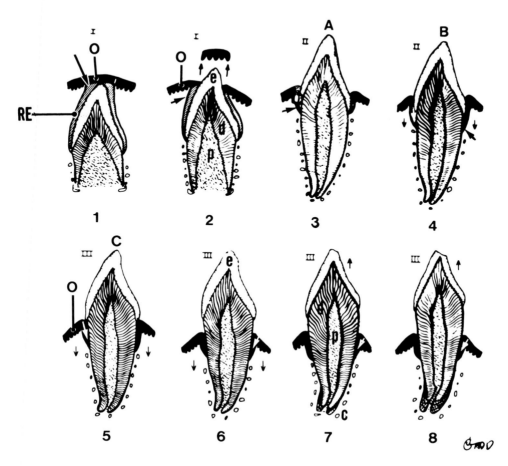

Fig. 4–35. Diagram of tooth emergence (I, II) and progressive development (I, II) and relocation (I, II, III) of the attachment epithelium from a total crown position (1–3) to a crown-root position (4), and finally to an exclusive root location (5–8). Drawings 1 and 2 show the attachment epithelium in stage I. Note that the oral epithelium (O) has fused with the reduced enamel epithelium (RE), as indicated by the arrow. Drawing 2 shows the sloughed epithelium *(vertical arrows)* above the emerging tooth tip. Over the slope of the crown, the reduced animal epithelium *(horizontal arrow)* has fused with the oral epithelium (O), forming the attachment epithelium. Note that the attachment epithelium consists only of the reduced enamel epitelium (primary attachment or junctional epithelium), and is located entirely on the anatomic crown, (e, enamel; d, dentin; p, dental pulp).

Drawings 3 and 4 show stage II, the transitional stage, in which the attachment epithelium is composed minimally of the primary junctional epithelial components *(black arrowheads)* located at the base of the tooth's cervix *(small black arrows)*; the greater segment of the attachment epithelium is composed of epithelial components derived from the gingiva—hence, secondary attachment epithelium *(large white arrows)*. Note that in drawings 2 and 3 the attachment epithelium is located exclusively on the crown, while in drawing 4 it is located on both the enamel and cementum.

Drawings 5, 6, 7, and 8 show stage III of attachment epithelium development, in which the structure consists totally of components derived from the oral epithelium, and thus is the definitive or secondary attachment epithelium *(large white arrows)*.

In these drawings note the progressive recession, thereby increasing the size of the clinical crown. Note that further attrition of the incisal edge is accompanied by an increase of apical cementum.

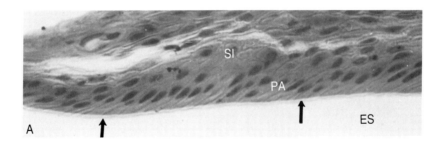

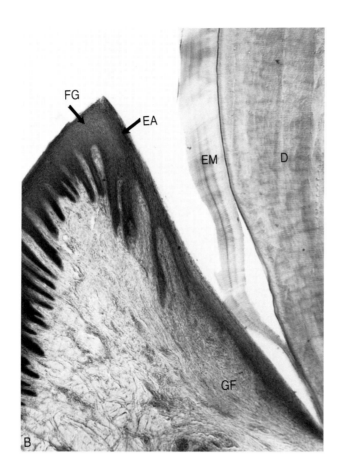

Fig. 4–36. *A,* Reduced enamel epithelium overlying the crest of an erupting crown. The enamel space (ES) is produced by decalcification. The primary attachment epithelium, consisting of the reduced enamel epithelium (postamelolasts, PA), and stratum intermedium (SI) is connected to the enamel via the epithelial attachment *(arrows)* comprised of the basal lamina/hemidesmosome complex. (Toluidine blue stain; × 120) *B,* Erupting tooth showing epithelial cuff and its associated primary attachment epithelium (EA) and free gingiva (FG), (GF, gingival fibers; EM, enamel matrix; D, dentin). (Hematoxylin and eosin stain; × 50) (*A* from Provenza, D.V., and Sisca, R.F.: Fine structure features of monkey (Macaca mutatta) reduced enamel epithelium. J. Periodontol., *41*:313, 1970.)

the primary attachment epithelium is replaced by the epithelium originating from the gingival epithelium. This then is the secondary attachment epithelium; it is comprised exclusively of elements originating from the gingival epithelium. (Its migratory replacement path is shown in Figures 8–38 and 8–41.) It has been estimated that renewal of the secondary attachment epithelium occurs in less than a week, consistent with the renewal period of gingival epithelium.

It should be noted that, although some investigators believe that the reduced enamel epithelium, as the original attachment epithelium, is sloughed off, others believe that it degenerates and is compressed to become part of the primary enamel cuticle, which is removed with the abrasive action of mastication or brushing. Additionally, although the attachment epithelium initially covers the entire crown, the epithelium shifts to more cervical positions with continued eruption. It may remain in this location or shift apically to occupy cementoenamel locations, or it may even migrate to occupy an exclusive cementum location, in which case recession has occurred. (Figs. 4–34 and 4–35; also see Fig. 8–41). The attachment epithelium forms a protective barrier, sealing off the oral cavity from the periodontal ligament.

REPLACEMENT OF DECIDUOUS TEETH

Humans are diphyodonts, that is, during life they possess two sets of teeth. The first set of 20, 10 each for the maxillary and mandibular arches, is called the primary (milk, deciduous, or baby) teeth, and all are shed by the twelfth year. They are replaced by a secondary set of 32, 16 for each arch (10 successors plus 6 molars). These are referred to as the secondary (permanent, or adult) teeth.

Developmental Stages of Permanent Teeth

The primordial tissue for the permanent teeth develops lingually to the primary tooth germs; it is simply an extension of the original dental lamina (Fig. 4–4). The developmental history for permanent teeth is the same as that for primary teeth. Both sets of teeth are formed simultaneously, but are obviously in different developmental stages (Table 4–1). The development and growth of primary teeth are essential because they influence arch growth and provide stimuli for the formation of arch space for the secondary teeth, which are larger and more numerous. Premature loss of primary teeth, especially cuspids and molars, may result in malocclusion or malpositioning of the secondary teeth. Development of the arches is generally sufficient to accommodate the increased number and size of the secondary teeth. If arch development is not in accord with the size of the developing teeth, crowding or diastemata (spaces between the teeth) may occur. Finally, primary teeth are required for speech articulation.

Chronology of Primary and Secondary Dentition

The chronology of human dentition given in Table 4–1 shows that each quadrant of primary dentition bears five teeth. Starting at the midline they are the central incisor, lateral incisor, cuspid, and two molars (first and second). Up to the time of the emergence of the first tooth, which is at about the seventh or eighth month of life, all but the second and third permanent molars are present in the maxillary and mandibular arches in varying developmental stages from the lamina to tooth eruption.

Permanent tooth organs continue developing long after the primary teeth have entered the oral cavity and met their antagonists. Between the ages of 7 and 11 years, the primary teeth are replaced by the permanent successors. It is during this 4-year period that the mouth is characterized by mixed dentition (Fig. 4–37).

Differences between Primary and Permanent Teeth and their Adnexa

There are a number of different features between the primary and secondary dentition.
1. Primary teeth are fewer and smaller.
2. The crowns of primary teeth are whiter, rounder, and wider in regard to the relationship between the mesiodistal diameter and the cervico-occlusal height.
3. The cusps of primary teeth are taller and more angular, and the occlusal fossae are deeper.
4. The vestibular aspect of the cervical ridges of primary teeth, especially of the first molars, is pronounced.
5. The vestibular and lingual faces of primary

Table 4–1. Chronology of the Human Dentition

Deciduous Tooth	Hard Tissue Formation Begins Fertilization Age in utero, Weeks	Amount of Enamel Formed at Birth	Enamel Completed (Mo. after Birth)	Eruption (Mean age in Mo., ± 1.S.D.)	Root Completed (yr.)
Maxillary					
Central incisor	14	(13–16)	Five-sixths	1½	10 (8–12)	1½
Lateral incisor	16	(14⅔–16½)	Two-thirds	2½	11 (9–13)	2
Canine	17	(15–18)	One-third	9	19 (16–22)	3¾
First molar	15½	(14½–17)	Cusps united; occlusal completely calcified plus a half to three-fourths crown height	6	16 (13–19) boys (14–18) girls	2½
Second molar	19	(16–23½)	Cusps united; occlusal incompletely calcified; calcified tissue covers a fifth to a fourth crown height	11	29 (25–33)	3
Mandibular					
Central incisor	14	(13–16)	Three-fifths	2½	8 (6–10)	1½
Lateral incisor	16	(14⅔–)	Three-fifths	3	13 (10–16)	1½
Canine	17	(16–)	One-third	9	20 (17–23)	3¾
First molar	15½	(14½–17)	Cusps united; occlusal completely calcified	5½	16 (14–18)	2¼
Second molar	18	(17–19½)	Cusps united; occlusal incompletely calcified	10	27 (23–31) boys (24–30) girls	3

From Lunt, R.C., and Law, D.B.: A review of the chronology of calcification of deciduous teeth. J. Am. Dent. Assoc., 89:878, 1974; based on Logan and Kronfeld, slightly modified by McCall and Schour, suggested by Lunt and Law, for the calcification and eruption of the primary dentition. Copyright by the American Dental Association. Reprinted by permission.

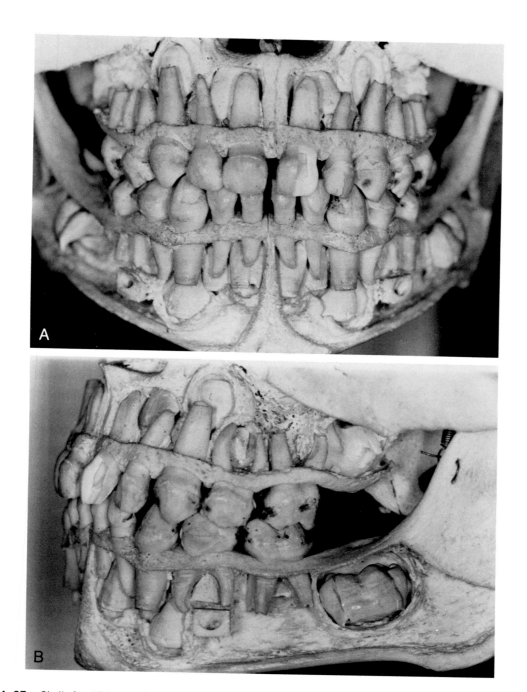

Fig. 4–37. Skull of a child approximately 6 years old, showing mixed dentition. *A,* Frontal view. *B,* Lateral view. (Courtesy of M. Davidson.)

molars at the cervical bulge are less rounded.

6. The cervices of primary molars are more constricted.

7. The cementoenamel junction of primary teeth is abrupt, forming a definite ridge.

8. The enamel thickness of primary teeth is not only less, but is rather uniform over the coronal surface (about 1 mm).

9. The cervical enamel rods of primary teeth are inclined incisally and occlusally.

10. The dentin of primary teeth is reduced in quantity and is more homogeneous and lighter in color.

11. The pulps of primary teeth are correspondingly larger, and conform to the profile of the dentinoenamel junction more exactly.

12. The pulp horns of primary teeth are longer and more slender.

13. The roots of primary teeth are longer and narrower, and their canals are generally wider.

14. The roots of primary teeth are very flared.

15. The apical foramina of primary teeth are larger.

16. The cementum of primary teeth is shallow, and the amount of cellular cementum is greatly reduced.

Furthermore, the lamina dura of the alveolar process is thicker, the bony (trabecular) trajectories are fewer, and the intervening marrow spaces are correspondingly larger. Finally, the periodontal ligament of primary teeth is not as well developed as that of permanent teeth.

Eruption of Permanent Teeth

During early stages of development, both the primary and permanent tooth organs are accommodated in a single crypt and share the same dental sac tissue (Fig. 4–33). Later, however, their growth movements are such that the permanent tooth organs become more deeply situated. With bone growth, the two dental germs are separated and occupy separate crypts. Eventually the primary teeth emerge and the permanent tooth organs assume positions of eruption, which for the premolars is below the root bifurcations of the primary molars (Fig. 4–38), and for the incisors and cuspids is below and usually lingual to the roots of their primary predecessors (Figs. 4–39 and 4–40). In most cases, the permanent teeth erupt to occupy the sites

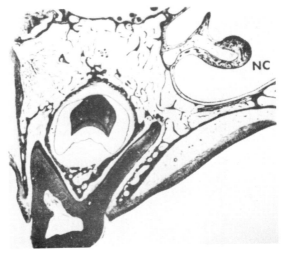

Fig. 4–38. Photomicrograph showing the relationship of the permanent maxillary premolar germ to the primary molar. Note the flared roots of the primary molar (NC, nasal cavity). (Hematoxylin and eosin stain; × 6) (Churchill, H.R.: Meyer's Normal Histology and Histogenesis of the Human Teeth and Associated Parts. Philadelphia, J.B. Lippincott, 1935.)

of the primary teeth. As the primary tooth emerges to take its functional position in the oral cavity, the germ of the permanent tooth grows in its crypt and, except for a narrow corridor, called the gubernacular canal, is occupied by connective tissue known as the "gubernacular cord." This cord is the connective tissue link between the crypt connective tissue and the mucous membrane (Figs. 4–39 and 4–40). It has been suggested that the gubernacular cord provides the directional path for eruption of the permanent tooth.

The first indication of eruption is the removal of the ceiling of the bony crypt. This results in the merging of the fundal connective tissues of the alveolus and those of the crypt. With progressive growth of the permanent tooth, the crown of the latter encroaches onto the root and compresses the intervening soft tissue. This causes root resorption (Figs. 4–40 and 4–41). In fact, the compressive forces of the emerging permanent tooth cause resorption of all the hard tissues in its path—the alveolar bone, cementum, and dentin of the primary tooth. In some instances even enamel may be resorbed.

The mechanisms involved in the removal of the hard tissues are the same as those described for bone (osteoclasia). Odontoclasia, or erosion of the dental hard tissues (cementum and dentin),

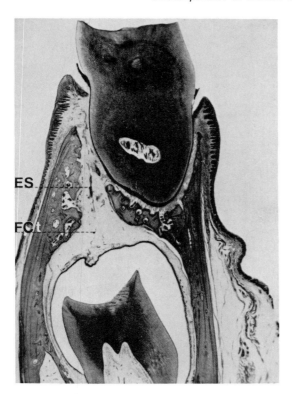

Fig. 4–39. Crypt for the developing permanent molar and its surrounding follicular (dental sac) connective tissue (FCt) are separated from the periodontal ligament of the primary tooth by a bony ceiling, which accommodates a narrow passage, the gubernacular canal. This contains connective tissue, blood vessels, nerves, and the remains of the successional lamina (ES). (Hematoxylin and eosin stain; ×8) (Churchill, H.R.: Meyer's Normal Histology and Histogenesis of the Human Teeth and Associated Parts. Philadelphia, J.B. Lippincott, 1935.)

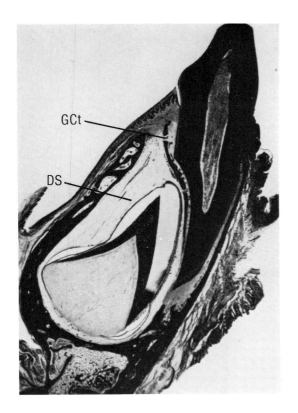

Fig. 4–40. Photomicrograph showing the lingual relationship of the developing permanent incisor to its primary predecessor. Erosion of the crypt ceiling and root of the primary incisor is shown. Note the thickening of the bone at the fundus of the socket of the developing permanent tooth and the gubernacular canal (GCt), which becomes confluent with the subgingival connective tissue, (DS, dental sac). (Hematoxylin and eosin stain; ×8) (Churchill, H.R.: Meyer's Normal Histology and Histogenesis of the Human Teeth and Associated Parts. Philadelphia, J.B. Lippincott, 1935.)

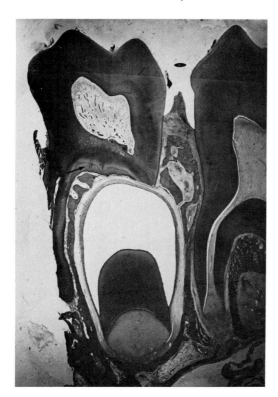

Fig. 4–41. Root resorption induced by development of the permanent successor. Temporary repair in the bifurcations is indicated by a black arrow. (Hematoxylin and eosin stain; ×5) (Courtesy of M.S. Aisenberg).

Fig. 4–42. Exfoliation of the primary tooth. Most of the root has been resorbed, and the primary tooth (DC) is being retained only by its pulpal connection to the subjacent connective tissue. Bone apposition at the base of the socket (FTB) for the permanent successor is illustrated. Note that the permanent tooth appropriates all the space of the primary predecessor and much more adjacent tissue to construct its own socket. The epithelial root sheath separates the connective tissue of the pulp from that of the prospective fundal periodontal ligament, (ES, enamel space; D, dentin; P, pulp; FTB, fundal trabecular bone). (Hematoxylin and eosin stain; ×5) (Courtesy of M.S. Aisenberg.)

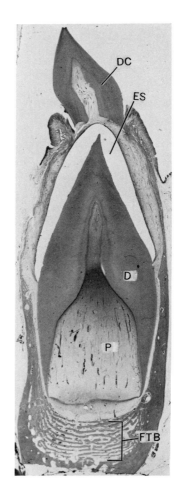

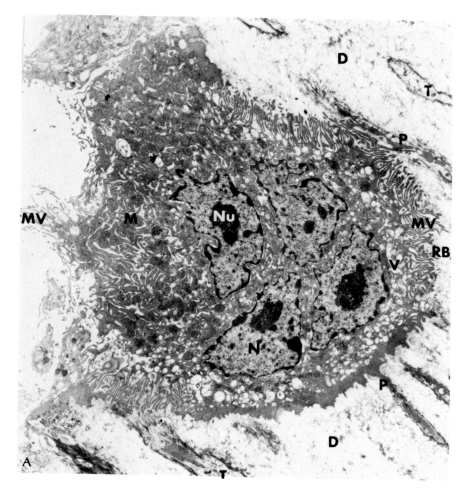

Fig. 4–43. Odontoclasts. *A,* Odontoclast in Howship's lacuna in eroding dentin (D). The cell contains four nuclei (N). The ruffled border (RB) and vacuoles (V) are abundant. Dentinal tubules (T) are sectioned tangentially. Note cytoplasmic process (P) in tubule (Nu, nucleoli; MV, microvilli). (Original magnification × 7000)

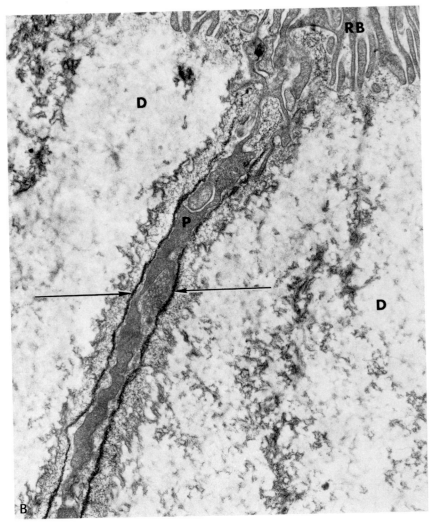

Fig. 4–43 Continued. *B,* Dentinal tubule *(arrows)* containing cytoplasmic processes (P) of the odontoclast. The process originates from the ruffled border (RB) portion of the odontoclast (D, dentin matrix). (Original magnification ×30,000.) (Freilich, L.S.: Ultrastructure and acid phosphatase cytochemistry of odontoclasts; effect of parathyroid extract. J. Dent. Res., *50:*1047, 1971.)

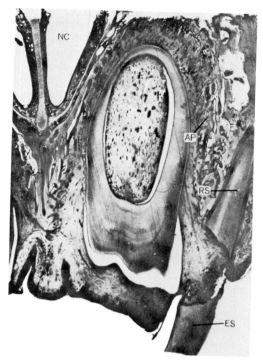

Fig. 4–44. Photomicrograph showing root remnants (RS) of primary tooth being incorporated by bone apposition into the alveolar process (AP) (NC, nasal cavity; ES, exfoliating root of primary tooth). (Hematoxylin and eosin stain; × 10) (Courtesy of M.S. Aisenberg.)

involves cells and (Howship's) lacunae (Figs. 4–41, 4–42, and 4–43). These hard tissue destroyers, variously known as odontoclasts, cementoclasts, and dentinoclasts, are cytomorphologically similar to the osteoclasts (Fig. 4–43).

The fate of the tooth organ tissues and the mechanisms engaged in the eruption of the permanent teeth are similar to those for primary teeth; these have been discussed earlier in this chapter.

Eruption-Induced Tissue Changes

In addition to resorption of bone, cementum, and dentin, tooth emergence also affects the dental and paradental soft tissues, such as the periodontal ligament, dental pulp, and the attachment epithelium of the primary teeth.

The periodontal tissue becomes disorganized, and the ligament's features disappear. That is, the tissue is no longer densely fibrous with its collagenous elements and principal fiber bundles definitely arranged. Disorganization is initiated in the fundus of the socket, so the first to disappear are the apical fiber bundles, the oblique groups, and finally the horizontal bundles and those of the alveolar crest. Some of the blood vessels are compressed and even rup-

tured, which tends to hasten bone, cementum, and dentin resorption. The loose connective tissue that results from resorption may be incorporated later in the formation of the periodontal ligament for the permanent dentition.

The dental pulp is least affected. The blood and nerve supplies are maintained sufficiently to sustain dentinogenic actvity of the coronal odontoblasts during exfoliation. In fact, primary teeth, although weakened by disorganization of the attachment apparatus tissues (i.e., the alveolar process, periodontal ligament, and cementum), are not shed because of the strong structural bonds existing between the pulp of the primary tooth and its underlying soft tissue (Fig. 4–42).

Exfoliation

In time, the socket becomes so shallow that its attachment apparatus is functionally inadequate to stabilize the tooth, especially when subjected to masticatory forces. Thus, the tooth is shed. In some cases, the tooth may become so loose and so annoying that it can be removed by the child. Following exfoliation, the area heals until the crown of the permanent successor penetrates the tissue to emerge into the oral cavity.

CLINICAL IMPLICATIONS

Bone or tooth fragments, or both, that are not resorbed may work themselves to the surface, where they are freed. Primary molars possess widely flared roots under which the permanent premolar germs develop. Incomplete resorption may result in retention of large root segments (Fig. 4–44). These may fuse with the bone of the alveolar process, or may remain free. In either case they are potential producers of cysts.

In rare instances primary teeth are not shed during the normal period. Hereditary and other causes (e.g., rickets or endocrine disturbances) may be responsible for this condition. Retention of the primary teeth should not be allowed, because this could cause the eruption path of the successional teeth to be in a lingual or vestibular direction, thereby resulting in malocclusion.

CHAPTER 5

Enamel

PROPERTIES

The anatomic crown of teeth is covered by an acellular hard tissue tissue, known as enamel, which is the hardest tissue of the body. The matrix synthesized by ameloblasts is entirely organic. The enamel matrix forms a complex in which the principal component is amelogin protein. Mineralization involves hydroxyapatite (calcium phosphate) crystals that are seeded within the enamel protein. As the crystals increase in size they appropriate more and more of the matrix space, until the final composition of definitive fresh enamel is 97% mineral, 0.3% organic, and 2.7% water by weight. It is the mineral content of enamel that makes it vulnerable to acid attacks, and leads to dental caries. Enamel is pearl white to gray in color. It is also translucent, a property that increases proportionately with mineralization. Enamel is so brittle (more than bone, cementum, and dentin) that were it not for the resiliency of the underlying dentin, it would collapse under masticatory forces.

Thickness. Enamel thickness varies with the shape of the tooth and its location on the crown. The thickest enamel is found at the crest of the cusps, or incisal edges (up to 2.5 mm). It becomes thinner over the coronal slope with minimal thickness (less than 100 μm) at the cervix and along the fissures and pits in the multirooted teeth. Color intensity tends to increase toward the cervix, because the enamel is progressively thinner and thus reveals the deeper yellow hues of the underlying dentin. Enamel of the cusps is thicker than that of the incisal edge. Additionally, cuspal enamel in multicuspid teeth is thicker than that in bicuspids.

Cementoenamel Junction. Three relationships may exist between the cementum and enamel at their junctional sites: (1) cementum may overlap enamel; (2) the ends of cementum and enamel may abut; or (3) the ends may be separated by a space, leaving the underlying dentin exposed (Fig. 5–1). With age and gingival recession, the dentin may become sensitive to thermal, chemical, and mechanical stimulation, as for example in brushing and periodontal scaling. The frequency with which the various cementoenamel relationships occur is about 60% for the first, 30% for the second, and 10% for the third. Scanning electron microscopy has revealed that the cementoenamel junction on a given tooth may exhibit all three conditions.

STRUCTURAL COMPONENTS

Transmission and scanning electron microscopy have been used to study replicas of thin undecalcified enamel. These techniques are particularly useful for investigating the crystalline features of enamel. Optical microscopy often involves acid-etched ground sections. These are cut with diamonds, ground to thicknesses of less than 50 μm, and polished to remove scratches (artifacts).

Most morphologic characterizations of enamel are based on optical microscopic observations of specimens sectioned specifically along certain anatomic planes. These planes are the sagittal, (longitudinal section along a buccolingual path), transverse horizontal section, and facial, longitudinal section along a mesiodistal path, tangent to the facial surface of the tooth (Fig. 5–2). Meaningful interpretations must take the following into account: sectional planes, optical interference resulting from crystalline structure, directional paths, and orientation of rods.

Electron microscopic examination of the dentinoenamel junction shows that enamel and den-

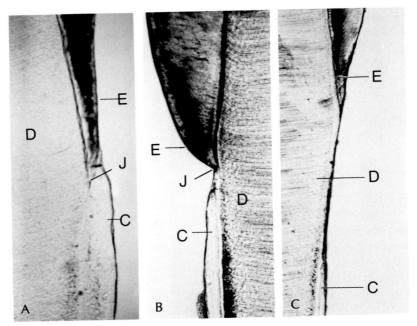

Fig. 5–1. Cementoenamel junction relationships. *A,* Cementum (C) overlaps enamel (E) at the junctional site (J), (D, dentin). *B,* Simple contact between enamel (E) and cementum (C) at the junction (J). *C,* Enamel (E) and cementum (C) do not meet, leaving the dentin (D) exposed. (Ground section; ×50)

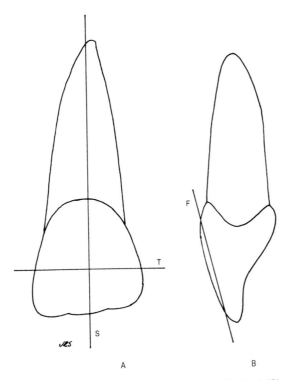

Fig. 5–2. Sectional planes of a tooth. *A,* Sagittal (S), transverse (T). *B,* Facial (F).

tin, and obviously their crystalline substructures, intermingle. This condition is made possible by the absence of a tissue barrier between the two tissues.

The dentinoenamel junctions of teeth sectioned longitudinally reveal that their gross profiles simulate the external outline of the crown. Microscopic examination, however, shows that the dentinoenamel junction is scalloped (see Fig. 5–16A). The concavities of this junction are about 70 μm in diameter, and are directed toward the external surface.

Enamel Rods (Prisms)

Except for its innermost and outermost layers, which are aprismatic (rodless), enamel is composed of structural entities called rods (prisms), which have associated interrod (interprismatic) and sheath components. Rods are somewhat cylindric and radiate uninterrupted from the aprismatic layer of the dentinoenamel junction to the aprismatic layer of the external surface. Rods in transverse section form a mosaic of inverted flask-shaped units, with each comprised of a bulb and neck segment. This arrangement of the enamel rods has also been referred to as the "arcade" or "keyhole" configuration. The principal mass of the rod is occupied by the bulb, while the neck of the flask or tail of the keyhole forms the interrod (interprismatic) territory (Figs. 5–3 and 5–4A and B). Some investigators interpret the bulb-shaped structure as the enamel rod viewed transversely, bearing a definite relationship to the interrod enamel. The marginal area of the bulb is known as the sheath. Its organic content is greater, so it is less mineralized than the rest of the rod (Fig. 5–3).

The crowns of incisors are composed of over 8½ million rods, while those of the molars may number over 12¼ million. The rod is narrowest at its point of origin, at the dentinoenamel junction, and it widens gradually toward the surface. Here its width is about double that at the dentinoenamel junction, and the width of rod increment may be wider than it is thick. The average width of a rod is 4 μm.

Mineralization occurs immediately following enamel matrix synthesis and secretion by the ameloblasts, and is initiated by the deposition of ribbon-like apatite crystals within the matrix space. Crystals grow by accretion (see page 151) and, when viewed transversely, appear hexagon-shaped (Fig. 5–5). They are stacked to form long ribbons. Electron microscopic examination of rod segments indicates that microspaces are found between the crystals (Figs. 5–3 and 5–5). The intercrystal spaces are fewer and smaller as

rod mineralization increases (Fig. 5–5). Crystals exhibit preferred orientation in enamel rods, which probably reflects the molecular arrangement of the constituents of organic matrix (amelogen/proteins and polypeptides). In general, ribbons of crystals in the peripheral area of the bulb are observed in transverse section because they are oriented parallel to the rod length (Fig. 5–4). From this area toward the neck segment the crystals fan out so that in the neck (interprismatic) territory they are viewed longitudinally and are oriented perpendicular to the adjacent bulb margin (Fig. 5–4). The deflection in crystal alignment and its association with the arrangement of the crystals of adjacent rods produce a herringbone pattern (Fig. 5–4).

Enamel mineralization is divided arbitrarily into two stages; primary and secondary maturation. Enamel at the dentinoenamel junction is the first to calcify and the first to become fully mineralized. Mineralization is initiated at the incisal or cusp tip. Initial or primary calcification, which accounts for up to 30% of the definitive mineral content, occurs very rapidly, almost simultaneously with matrix deposition. The maturation period, during which the rate of calcification decelerates is also known as the secondary stage. Secondary mineralization proceeds in paths established for matrix deposition in amelogenesis. Enamel acquires its full com-

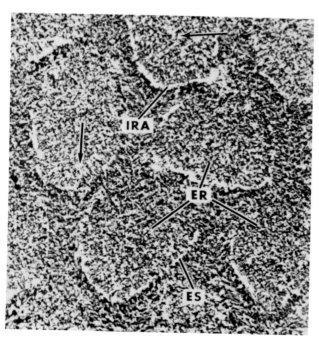

Fig. 5–3. Replica of etched ground section. Enamel rods (ER) are visualized in transverse section. The light areas indicated by arrows are intercrystalline spaces *(arrows)*. The light perimeter outlining the enamel rod is the enamel sheath (ES) (IRA, interrod area). (Enlarged from ×4,000) (Courtesy of D.B. Scott.)

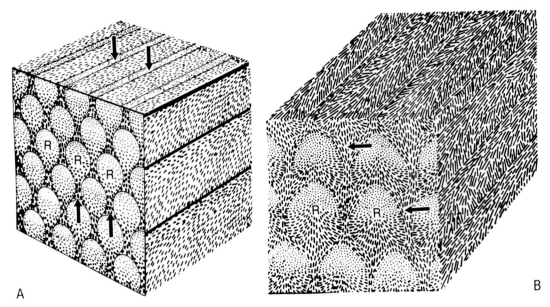

Fig. 5–4. Diagrammatic representations of two groups of enamel rods. *A*, Three-dimensional drawing of a group of arcade/key-shaped rods showing the interrelationship of the enamel rods (R) to the interrod (interprismatic) areas *(arrows).* The interrod areas are extended regions of the rod. *B*, Round or polygonal enamel rods (R) showing the relationship to the interrod area *(arrows).* Note the orientation of the apatite crystals in various regions of the rod, as viewed longitudinally and transversely.

plement of mineral by the time the crown emerges into the oral cavity.

Striations. Enamel rods are comprised of numerous incremental units that represent the daily deposition of enamel matrix. Cross striations appear along the length of the prisms, separating the incremental units. These are especially prominent in slightly acid-etched ground sections (Fig. 5–6). Striations therefore represent the boundaries of increments. They are revealed by acid etching because the striated areas are less mineralized, and thus are less affected by acid than the rod substance. Scanning electron microscopic observations show that the striations exhibit periodic alterations in prism width, reflected as alternating constrictions and bulges.

Sheaths. The peripheral area of the prism (bulbous segment), except for the interprismatic area (neck segment), exhibits a microspace of increased organic and therefore decreased mineral content. This marginal zone is called the rod sheath (Fig. 5–3). These areas are especially prominent in the bulb segment. When the organic content is greater the mineral content is decreased, and the crystal arrangement is dif-

ferent. The neck segment of the rod has no sheath. The crystal configuration in the neck, or interprismatic area, is longitudinal (Fig. 5–4).

Arrays of Enamel Rods. Cleavage of enamel tends to occur in planes paralleling the path of the rods. One of the most frequent operations in general dentistry involves removal of diseased (carious) enamel. In this procedure, diseased enamel, decalcified by bacteria, is cut away. Some of the surrounding "sound" (noncarious) enamel is also removed. A cavity of appropriate size and shape is made by burrs, chisels, and other instruments. The prepared cavity accommodates the restorative material. In cutting sound enamel, one must consider the path and arrangement of the enamel rods from the dentinoenamel junction to the surface of enamel. In cavosurface (marginal) preparations, cleaving the unsupported enamel by hand instrumentation is required (see below, Repair).

Paths. The path of the enamel rods from the dentinoenamel junction is initially straight. Many, however, change course shortly after their departure from the dentinoenamel junction (Fig. 5–7). Some may swerve to the right, while others move to the left. Beyond the inner

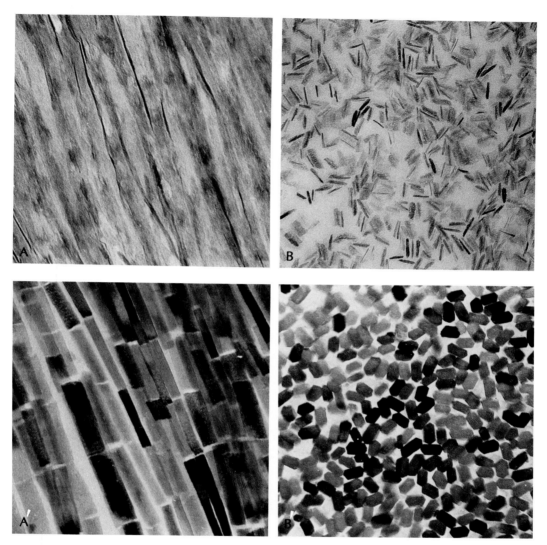

Fig. 5–5. Electron micrographs of enamel, showing developing crystals viewed longitudinally *(A)* and transversely *(A')*, with mature crystals in longitudinal *(B)* and transverse *(B')* views. Note differences in crystal dimensions and shape initially and with maturation. The diminution in the size of the intercrystalline space with maturation is shown. (×160,000) Courtesy of M.U. Nylen.)

third of the enamel, the rods tend to return to their original paths and follow a more direct course to the surface. In some areas, particularly below the cusp surface, the rod paths are tortuous, resulting in areas of "gnarled" enamel (Fig. 5–8). Differences in enamel rod paths are believed to provide strength and stability to enamel under masticatory stresses.

Arrangement. Enamel rods are arranged in planes to resist masticatory forces. In describing their orientation, the internal (dentin aspect) or external surfaces are used.

All rods, except the cervical ones in perma-

nent teeth, are oriented at right angles to the dentinoenamel junction. The cervical rods of permanent teeth are inclined toward the gingiva.

The more accurate reference is the free surface to which the rods are perpendicular (Fig. 5–9). Accordingly, rods of incisal edges, cusps, ridges, and other regions of the crown form right angles with tangents drawn to the tooth surface (Fig. 5–9). The cervical rods of primary teeth, however, are arranged parallel to the incisal or occlusal surface.

Hunter-Schreger Bands. When longitudinal

Fig. 5–6. Acid-etched ground section of enamel, showing rods coursing from the lower right to the upper left of this photomicrograph, (striation, *black arrows*). (×320)

ground sections are viewed with reflected light, enamel areas in which differences in rod paths occur exhibit an optical phenomenon. Such sections will show light (parazones) and dark (diazones) bands, which correspond to the swerving paths of the rods of the inner third of the enamel. The light and dark zones are known collectively as Hunter-Schreger bands (Fig. 5–10A). The parazones are believed to be longitudinally sectioned rods, and the diazones transversely sectioned ones (Fig. 5–10B). Other than an optical phenomenon resulting from the plane of sectioning, other less likely causes for the production of the Hunter-Schreger bands include differences in the degree of calcification and organic content and differences in permeability.

Stripes of Retzius. Longitudinal and transverse sections of the crown may reveal rod bundles of varying widths and color (brown) intensities. These are known as lines or stripes (striae)

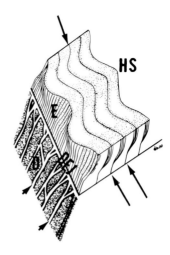

Fig. 5–7. Diagram showing the interrelationship of the enamel (E) and dentin (D). Note the straight and wavy paths of the enamel rods from the dentinoenamel junction (DEJ), producing tufts *(arrows)* and Hunter-Schreger bands (HS). The dense black line of the DEJ represents aprismatic (rodless) enamel. Although the dentinal tubules (arrowheads) are shown to extend to the dentinoenamel junction, they may actually extend into the enamel for a short distance.

of Retzius, and they represent disturbed periods of growth. In longitudinal section the stripes of Retzius form arcs over cusps and incisal edges (Fig. 5–11). Arcs that are not contained completely in the substance of enamel terminate on the surface of the crown as shallow depressions circling the tooth. The trough of the waves or depressions are referred to as perikymata (Gr. peri, around + kyma, wave) (Fig. 5–11). Because these are the external (surface) manifestations of disturbed growth periods, they have also been called imbrication lines of Pickerill. In transverse section the striae form concentric circles about the dentin that resemble the annual growth rings of trees (Fig. 5–12). A few narrow Retzius lines are of normal occurrence. When present in increased numbers, however, particularly as broad bands, they may indicate periods of metabolic disturbances or disorders that occurred during amelogenesis. Retzius lines contain fewer apatite crystals, so they are areas of increased organic and decreased mineral content (hypocalcified areas). Among the more likely explanations for the lines of Retzius are differences in the organic-to-inorganic ratio, disturbances in the mineralization cycle, local changes in rod path, and lag in enamel matrix production or maturation.

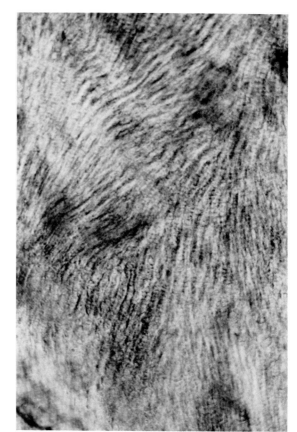

Fig. 5–8. Gnarled enamel produced by the irregular paths of the rods. (Ground section; ×320)

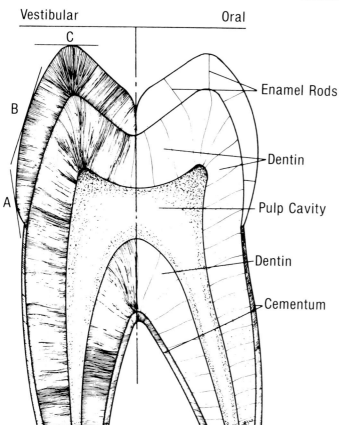

Vestibular   Oral

C

B

A

Enamel Rods

Dentin

Pulp Cavity

Dentin

Cementum

Fig. 5–9. Diagram showing differences in the orientation of the enamel rods over the coronal surface. The vestibulo-oral aspect of the permanent molar illustrates the paths of enamel rods and dentinal tubules. The right side shows the orientation of enamel rods. Note that the more cervical rods are gingivally inclined. The left side shows that the rods course perpendicular to the tangent. Also note the orientational path of the dentinal tubules. (Courtesy of J. Bianco, Jr., D.D.S.)

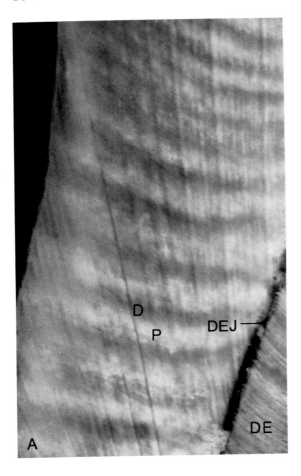

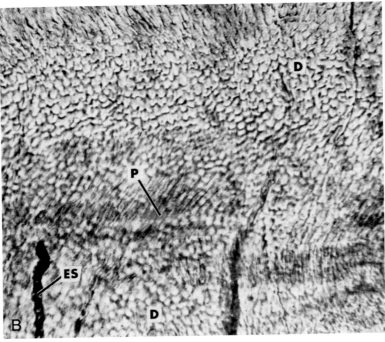

Fig. 5–10. Hunter-Schreger bands. The bands are visualized with reflected light. *A,* Hunter-Schreger bands show dark diazones (D) and light parazones (P) (DEJ, dentinoenamel junction; (De, dentin). (Ground section; ×50) *B,* The wavy course of the enamel rods, as they depart from the dentinoenamel junction, results in one area of their length being sectioned longitudinally, with adjacent areas being sectioned transversely. In human enamel, parazones are rods cut longitudinally and diazones are rods cut transversely (ES, enamel spindle). (Longitudinal ground section; enlarged from ×430)

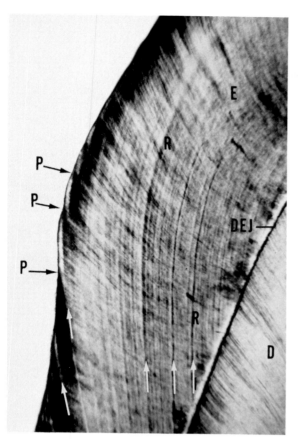

Fig. 5–11. Crown viewed in longitudinal section showing enamel (E), dentin (D), dentinoenamel junction (DEJ), and lines of Retzius (R) reflected as arcs *(white arrows),* terminating as shallow depressions on the tooth surface as perikymata (P). (Ground section; ×32)

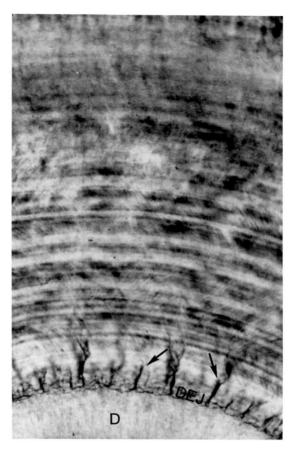

Fig. 5–12. Crown segment viewed transversely, showing dentin (D), scalloped dentinoenamel junction (DEJ), enamel spindle *(arrows),* and lines of Retzius concentrically arranged in enamel. (Ground section; ×50)

Neonatal Line. Enamel produced during embryonic development contains only a few stripes of Retzius. Because of this it is thought to be of superior quality to the enamel produced after birth. Embryonic enamel contains fewer striae because of the fetus's salubrious uterine environment. At birth, when the newborn must assume a more "free-living" existence, the trauma is registered in the enamel as an exaggerated stripe of Retzius called the neonatal line (Fig. 5–13). Once the infant becomes adjusted to the new environment and way of life, there are no further additions to the width of the neonatal line.

THE EXTERNAL SURFACE

Structures identified on the surface and outer segment of enamel include aprismatic enamel,

primary dental cuticle, secondary coatings, perikymata, lamellae, pits, and fissures.

Aprismatic Enamel. A surface layer of aprismatic (rodless) enamel about 30 μm thick is found in all primary teeth and in most permanent teeth. The aprismatic layer is most prominent in the cervical enamel, and may be absent in the cusps and incisal edges. The crystals in this type of enamel are more closely packed in parallel array, and are oriented at right angles to the striae of Retzius. Thus, aprismatic enamel is more mineralized than the underlying rods.

Dental Cuticle and Secondary Coatings. The final activity of the ameloblast is that of elaborating and secreting an organic (noncalcifying) layer up to 1 μm thick (see Fig. 4–20). This structure is known as the enamel cuticle, primary (dental) cuticle, Nasmyth's membrane or, more recently, the epithelial attachment. The

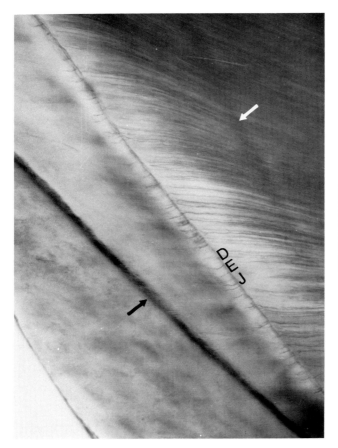

Fig. 5—13. Primary tooth segment viewed longitudinally, showing neonatal line in dentin *(white arrows)*. juxtaposed by dentin formed during embryonic development and by that formed postnatally. The neonatal line in enamel *(black arrow)* is bordered dentinward by prenatal enamel and surfaceward by postnatal enamel (DEJ, dentinoenamel junction). (Ground section; ×80)

enamel cuticle envelops the entire crown. Subjacent to the cuticle is a 5-nm thick layer of delicate apatite crystals heterogeneously dispersed among somewhat larger plate-like ones. These become confluent with the crystals of the surface layer of enamel. The cuticle and its associated crystals are abraded by the grinding and frictional forces of mastication. Consequently, the dental cuticle is worn away soon after eruption; some believe this occurs within a matter of hours. In the more sheltered areas, such as the neck of the tooth, the cuticle may persist for longer periods.

Replacing the primary cuticle are surface coatings secondarily acquired. These are of three types: (1) those derived from salivary components, designated as salivary pellicles; (2) those derived from bacteria, known as dental plaque; and (3) calculus, or calcified dental plaque.

Perikymata. Tooth surfaces, particularly those that have not been exposed to the abrasive forces of chewing for long periods, appear wavy, consisting of crests and troughs (perikymata).

These overlay the ends of rod groups or striae of Retzius. The perikymata of the neck of the crown are more numerous (30/mm) and are more distinct than those of the cusps or incisal edges. In the latter perikymata may be totally absent, often fewer than 10/mm. In profile the perikymata appear at the interface of overlapping shingles (Fig. 5–11). With age, the enamel surface becomes worn and the perikymata become less distinct, especially on the facial and lingual surfaces of the crown. In the more sheltered regions, such as the proximal surfaces, particularly of the molars, wear of the perikymata occurs less rapidly. Some teeth may not bear perikymata.

Pits and Fissures. Enamel development begins at the tips of the prospective cusps and incisal edges, and proceeds cervically. Eventually, adjacent cusps meet at their bases and fuse (Fig. 5–14). The fusion areas form developmental or segmental lines (grooves). The grooves or lines are found at specific locations on the buccal, lingual, and occlusal surfaces of the posterior teeth. They are named according to the

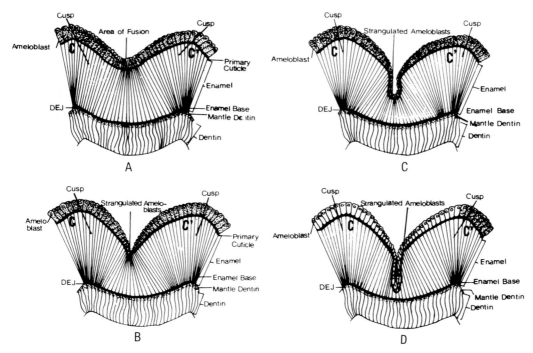

Fig. 5–14.   Diagram of groove or pit development showing the relationship of adjacent growth centers (C–C'). *A,* Growth centers located in close proximity. Note that the slope is slight, forming a shallow depression with fusion of its adjacent counterpart. The thickness of the enamel at the fusion site, under these conditions, is not much thinner than the adjacent areas of the cusp (e.g., intermamelon grooves located labially on anterior teeth). *B,* The growth centers are further apart, and the slope is steeper, so the cusps are higher and the depressions deeper. As the ameloblasts retreat during amelogenesis their bases meet, thereby preventing further development. Thus, the enamel is thinner under these conditions. *C* and *D,* The slopes of the cusps are acutely angled so that the duration of amelogenesis is sharply curtailed, resulting in pits or fissures bordered by short enamel rods, (DEJ, dentinoenamel junction).

portions of the crown they connect. To identify the location of a development groove (DG), certain landmarks should be established on the occlusal aspects of the crown of the premolar (bicuspid) and molar teeth (Fig. 5–15). These locations may be summarized as follows:

1. Developmental grooves of the bicuspid (Fig. 5–15A)
   a. Central (CDG)
   b. Distobuccal (DBDG)
   c. Distolingual (DLG)
   d. Mesiobuccal (MBDG)
   e. Mesiolingual (MLG)
2. Developmental grooves of the maxillary molar (Fig. 5–15B)
   a. Buccal (BDG)
   b. Central (CDG)
   c. Distolingual (DLG)
3. Developmental grooves for the mandibular molar (Fig. 5–15C)
   a. Central (CDG)
   b. Distobuccal (DBDG)

   c. Lingual (LDG)
   d. Mesiobuccal (MBDG)

Fissures are deep clefts produced in multicuspid teeth in association with developmental lines. They represent those longitudinal defects that result from acute angulation of the slopes of the enamel organ segments forming the cusps (Fig 5–14C and D). Ameloblasts of the steep slopes grow toward one another so that, at the bases of the cusp, they become so compressed that further growth is impossible. Not only are fissures produced by the lack of fusion, but the enamel at these sites is much thinner. Fissures that course from mesial to distal are more commonly found along the central segmental (developmental) groove of posterior teeth (Fig. 5–15 A and C), with the exception of those that have a transverse ridge. They frequently occur along the buccal (BDG) and distolingual (DLG) developmental lines of the maxillary molars (Fig. 5–15B). In the mandibular molars they are

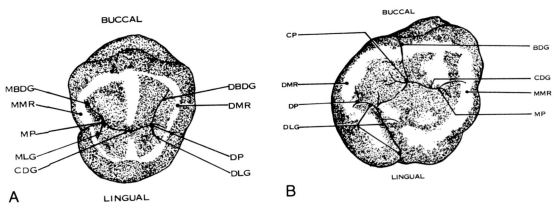

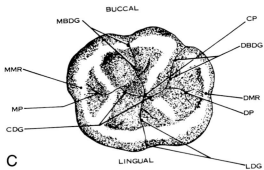

Fig. 5–15. Anatomic landmarks of occlusal surfaces. *A*, Maxillary left bicuspid. *B*, Maxillary right first molar. *C*, Mandibular right first molar. *Developmental grooves:* buccal developmental groove (BDG); central developmental groove (CDG); distobuccal developmental groove (DBDG); distolingual groove (DLG); lingual developmental groove (LDG); mesiobuccal developmental groove (MBDG); mesiolingual groove (MLG). *Pits:* central pit (CP); distal pit (DP); mesial pit (MP). *Ridges:* distal marginal ridge (DMR); mesial marginal ridge (MMR).

associated with the mesiobuccal (MBDG), distobuccal (DBDG), and lingual (LDG) segmental grooves (Fig. 5–15C).

Pits are enamel depressions that are often so small that they may escape detection by the dental probe. They can be found at the ends of developmental lines or points at which two or more segmental grooves intersect. They are more commonly associated with developmental grooves in posterior teeth (Fig. 5–15B and C). These are identified as the mesial and distal pits of bicuspids (Fig. 5–15A), and as the mesial, distal, and central pits of the maxillary and mandibular molars (Fig. 5–15B,C).

Fissures and pits are potential sites of caries invasion because food may be ground into them, and/or because of the inaccessibility of the area for cleaning. Because pits may be so small that they may be overlooked, many dentists restore or seal them prophylactically.

Lamellae. Lamellae are of three types. Type A represents longitudinal enamel segments in which the mineral content is less than the organic component (Fig. 5–16A). Type B lamellae are longitudinal cracks that contain cellular debris, probably the remnants of the enamel organ

(Fig. 5–16B). Because types A and B are produced during the terminal stages of development, they may be classified as developmental enamel lamellae. Type C lamellae are also longitudinal cracks, but they are produced after the tooth has erupted. These may have debris ground into the crevices, but the debris need not be cellular.

Lamellae are most frequently found in the cervical enamel. In addition to pits and fissures, they may be associated with the developmental grooves in molars and premolars. Lamellae originate at the surface and proceed to and even beyond the dentinoenamel junction for short distances. It has been suggested that lamellae are related to structures called tufts (see below) which are located at the dentinoenamel junction (Fig. 5–16A). The organic material in the cracks may become mineralized. In such cases the organic material serves as the matrix for crystal deposition and growth, and the mechanism is not unlike that employed in amelogenesis.

Cervical Enamel. In addition to differences in the arrangement of enamel rod groups, the number of perikymata, and the prevalence of lamellae, other differences may be found in

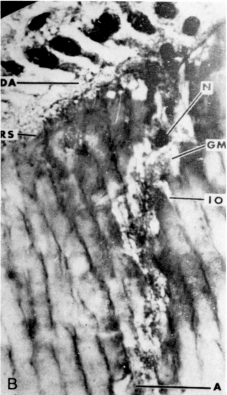

Fig. 5–16. Lamellae. *A,* Ground section of tooth showing dentin (D), scalloped dentinoenamel junction (DEJ), enamel tufts (T), and lamellae (L). In this section the lamellae do not extend beyond the DEJ. (×50) *B,* Lamella of permanent molar. The surface of the enamel rods (RS) is covered by the distal ends of ameloblasts (DA). The lamella contains cell fragments of ameloblast (A), granular material (GM), nuclei (N), and interrod opening (IO). (Ehrlich's hematoxylin and eosin stain; ×510) (Hodson, J: An investigation into the microscopic structure of the common forms of enamel lamellae with special reference to their origin and contents. Oral Surg. Oral Med., Oral Pathol., 6:305, 1953.)

cervical enamel. These involve the degree of mineralization and the thickness of the daily increments of enamel. Many investigators have shown that the less mineralized (hypomineralized) and, therefore, softer areas occur more frequently in the cervical region. Hypomineralized areas contain more organic material. The organic content may be such that, even after decalcification, enough is left to be stained for optical microscopic examination (Fig. 4–36B). It has been estimated that hardness differences of up to 40% may exist between the cervical and more occlusal enamel.

The daily increments of enamel at the surface may be much smaller than those more deeply situated. This is particularly true of those segments comprising the perikymata. Consequently, the lines of Retzius are much closer to each other near the surface of the crown than near the dentinoenamel junction. As indicated earlier, surface enamel may not show daily rod structure—thus the term "aprismatic enamel." On the other hand, the outer surface of the crown may be composed of very distinct prisms that slant sharply to form a smoother surface. Although the preceding features may apply to enamel located anywhere on the crown, they are more relevant to a discussion of the more cervically located enamel.

THE INTERNAL SURFACE

The inner third of the enamel varies from that of the remainder in regard to hardness, rod path, and structure.

Dentinoenamel Junction. If one relates the contour of the tooth surface to that of the dentinoenamel junction, differences in enamel rod length that occur from the cusp to the cervix may thus be explained. It is obvious that the longest rods—and the thickest enamel—are found over areas subjected to the greatest forces.

Although the macroscopic profile of the dentinoenamel junction in longitudinal section is somewhat S-shaped, the microscopic contour is scalloped (Figs. 5–7, 5–16A, and 5–17). The diameter of each concavity of the scallop, which is directed enamelward, may house a bundle containing as many as 20 enamel rods.

Foundation (Aprismatic) Enamel. The enamel forming the foundation for the rod group (the enamel immediately adjacent to the dentin) does not clearly exhibit rods with their sheaths and interrod structures. Accordingly, this enamel is aprismatic. Test results have indicated that this enamel is more highly calcified than that of the rods. At certain intervals, however, corresponding to those at which enamel tufts are present, the enamel is less calcified.

At the ultrastructural level enamel and dentin are seen to interdigitate, so that the apatite crystals of each intermingle. This is possible because barriers between the two hard tissues do not exist.

Tufts. Ground transverse sections of enamel demonstrate structures that have the appearance of clumps of grass, called enamel tufts. They begin at the dentinoenamel junction and may extend to the inner third of enamel, while in the cervical area they may extend to the surface (Figs. 5–6, 5–16A, and 5–17). The tuft appearance is actually an optical effect.

Most scientists believe that tufts are less calcified (hypomineralized) enamel rods, therefore containing more organic material than the neighboring prisms. Whether all the components of the tufts are poorly calcified is not certain. Because the enamel rods begin near the dentinoenamel junction and assume a straight course for a short distance, rows of them viewed on end convey the impression of a common origin. With deviation in the directional paths of the rods (right or left) the "tuft" effect is produced (Figs. 5–7 and 5–18). The poorly calcified enamel rods of the tufts, which are brownish in color, contrast sharply against the whiter background of the more calcified rods.

Other concepts related to the nature of the enamel tufts have been proposed but have not been widely accepted. For example, some oral histologists believe that tufts are crackless lamellae types. Others are of the opinion that they represent a type of organic membrane originating at the dentinoenamel junction. As the membrane proceeds into the enamel, it breaks up into numerous fibril-like structures of differing diameters and lengths, which are scattered between the enamel rods.

Spindles. Longitudinal ground sections of enamel exhibit irregular club-shaped structures in the adjacent area of the dentinoenamel junction called enamel spindles (Fig. 5–19). Spindles are minute blind canaliculi packed with air and debris during the cutting and polishing processes

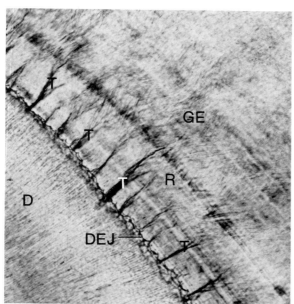

Fig. 5–17. Crown segment in transverse section showing dentin (D), scalloped dentinoenamel junction (DEJ), enamel tufts (T), and lines of Retzius (R), (GE, area of gnarled enamel; E, enamel). (Ground section; × 50)

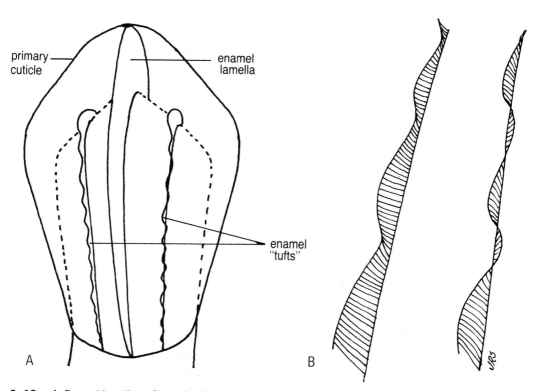

Fig. 5–18. *A,* Enamel lamella, tufts, and primary cuticle possess insoluble organic residues that remain after prolonged demineralizing procedures. *B,* Two enamel tufts. Fearnhead, R.: The proceedings of the Third International Symposium on Tooth Enamel. J. Dent. Res., 58:914, 1979.)

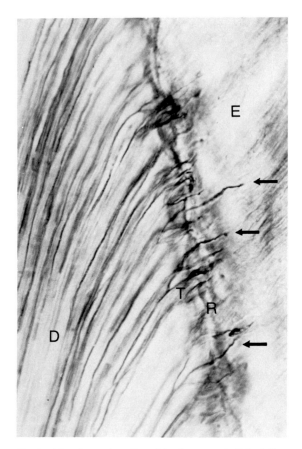

Fig. 5–19. Ground section of tooth segment. Orientation of enamel rods (R) and scalloped dentinoenamel junction are plainly visible (E, enamel; D, dentin; enamel spindles *(arrows)*; T, branching dentinal tubules). (×320).

of slide preparation. At one time they housed the terminals of the odontoblast processes. In sections in which the bases of the spindles and dentinal tubules lie in the same plane, one can see that the two are continuous. The orientation of the spindles is not necessarily the same as that of the rod paths. For example, at the cusp tips, the orientation of the spindle is similar to that of the rods. Along the cusp slope, however, the spindles angle toward the cervix.

The precise mechanism involved in the location of odontoblastic terminals in the enamel territory, which causes tufts, is not fully understood. Some investigators have observed odontoblastic processes extending into the ameloblastic layer and, based on this, they suspect that the ends of the odontoblasts are trapped in the matrix with the formation of enamel. Others have noted that the processes of the ameloblasts

may project well into the dentin matrix (Fig. 4–24). In this instance enamel matrix is deposited in the territory of dentin surrounding the odontoblastic processes, which cannot retreat. There are no fixed boundaries for enamel and dentin. Thus, territorial invasions by both ameloblasts and odontoblasts processes can occur. In the one case, the limits of the enamel are extended into the dentin after the ameloblasts retreat. In the other case, the processes grow into the ameloblastic layer and remain even after enamel has been formed and the ameloblast retreats. Other views have been proposed, but their acceptance has been so limited that they do not warrant discussion.

AGING AND REPAIR OF ENAMEL

Over time enamel undergoes various degenerative changes, thus necessitating restorative procedures.

Wearing Changes

The changes occurring during the life of the tooth are primarily those of "wear" which result from masticatory forces. The primary cuticle, which is nonmineralized, is the first to disappear because of wear. This occurs shortly after the tooth enters the oral cavity. On the other hand perikymata, which are depressions between the ridges, wear more slowly because the surface enamel is mineralized. The cusps and incisal edges become more blunt over the years because of wear (Fig. 5–20). The less shallow anterior teeth suffer more than the molars and premolars from the abrading action. Similarly, the vestibular and lingual aspects of teeth wear more than the mesial and distal surfaces. The amount of wear also varies with diet and hardness of teeth. For example, teeth of those whose diets contain mainly soft foods will be less affected. More calcified rod groups withstand frictional and other forces better than hypocalcified enamel.

Chemical changes occurring with age are not clearly understood. Most investigators agree that age tends to reduce enamel permeability, and enamel may therefore become less susceptible to caries attack. Color changes also occur— that is, teeth appear less white.

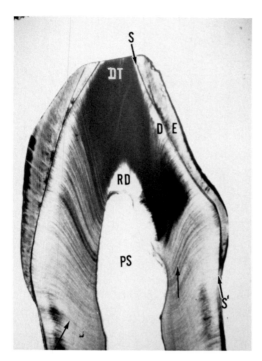

Fig. 5–20. Primary incisor showing worn incisal edge and mass of dead tracts (DT), beneath which is an area of sclerotic reparative dentin (RD). Note the S-shaped curvature of the dentinoenamel junction over the slope, and the cervix of the crown from point S to S′. Also note the S-shaped curvature of the dentinal tubules *(arrows)* from the dentinoenamel junction to the pulp chamber (PS) (D, dentin; E, enamel). (Ground section; ×8)

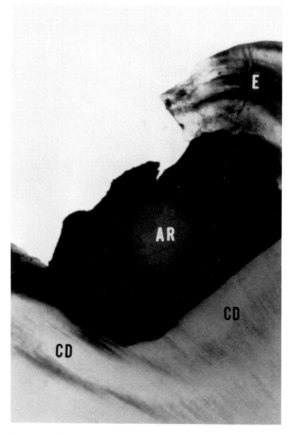

Fig. 5–21. Ground section of crown with amalgam restoration (AR). Note carious dentin (CD) surrounding the margin of the prepared cavity (E, enamel).

Repair

Only ameloblasts produce enamel. If they are destroyed they cannot be replaced, because they neither engage in mitotic self-perpetuation nor possess permanent mother (stem) cells. Having completed enamel formation, the cells are incapable of further amelogenic activity; thus, biologic repair or replacement of enamel is impossible. Accordingly, diseased, fractured, or otherwise damaged enamel can only be repaired through operative procedures. Repair is effected by preparing a cavity and filling it with an appropriate restorative material, such as are amalgam (silver-mercury alloy), gold, or acrylic (plastic). In preparing a cavity, knowledge of the orientation of the enamel prisms is of prime importance, because enamel is not elastic but very brittle. It depends on a base of sound dentin for its elasticity and support. Rods that are not supported by dentin give way quickly under masticatory pressures. Breaks or improperly sealed areas around the fillings result in "leaks," which act as new avenues for caries invasion (Fig. 5–21).

Clinical studies have shown that prepared cavities reduce the original strength of dental tissues by about 50%. Amalgam restorations, which are most effective in building contact points, do not succeed in restoring the strength lost in cavity preparation. Acrylic restorations that are bonded to the accommodating hard tissues do not necessitate undercutting the dental tissue. Consequently, fractures of restored teeth are less likely to occur with such methods.

Cavities are not all of the same shape; rather, they are prepared in very definite ways depending on the location of the decay lesion. The various types of prepared cavities and directional paths of the rod groups involved are illustrated in Figure 5–22. Some of the noncarious enamel and dentin is removed with the diseased ma-

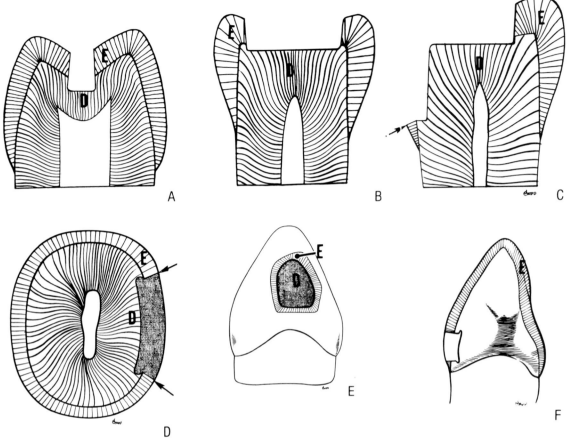

Fig. 5–22. Prepared cavities illustrating direction of enamel rods (E). Note that all enamel rods remaining after preparation of the cavity are supported by dentin (D). *A*, Class I preparation of premolar in longitudinal section *(buccolingual aspect)*. *B*, Class I preparation showing mesiodistal aspect. *C*, Class II preparation on posterior teeth only. Note the bevel *(arrow)*. This is made to obtain the full length of the enamel rods. *D*, Class II preparation in transverse section. Note that the enamel rods are maintained in full length *(arrow)*. *E*, Class III preparation showing inclination of enamel rods and proximal outline. *F*, Class V preparation of cuspid. Not illustrated are class IV preparations which are made for anterior teeth only. The characteristics of these are similar to those of class II and are, therefore, not illustrated. Class VI preparations involve small cuspal surfaces, in which the enamel rods are vertically oriented.

terial so that the restoration will rest on sound tissue. In this way not only does the restoration have a firm base, but the possibility for reinfection is lessened.

The areas most likely to become infected with caries-producing bacteria are those in which lamellae, pits, and fissures are located. Tooth surfaces that are hard to reach when brushing and cleaning are likely to suffer decay. Special attention should therefore be directed toward keeping the spaces between the teeth and around the necks of the teeth clean. These areas are very susceptible to caries attack in younger teeth. In older teeth, which are less subject to caries, such areas should be kept clean because gingival disease (periodontal infection) may result. In older persons, tooth loss is not generally due to tooth decay but to periodontal disease. The latter, which generally involve the gingiva, are primarily a result of poor oral health habits.

CHAPTER 6

Dentin

Dentin is a calcified connective tissue that encases the coronal and radicular pulp, which underlies enamel and cementum and forms the supporting base of the enamel (see Fig. 4–32). Dentin resembles bone in that the collagen component is almost exclusively of type I. Bone and dentin are also similar in germ layer origin (ectomesenchyme) and in chemical and physical characteristics. Morphologic differences occur, however, in their cell constituents. In bone, osteocytes are housed in lacunae and their processes extend into canaliculi. In dentin, only the odontoblastic processes are contained in tubules within a calcified matrix. Odontoblastic cell bodies are not included in calcified matrix; rather, they form the most peripheral layer of the dental pulp, a location characteristic of "blast" (Gr. blastos (germ)/developing) cells of other hard tissues. Dentin is slightly harder and more mineralized than bone, but is much less mineralized than enamel. Enamel is therefore more radiopaque than dentin or bone. Be-

cause dentin is much less calcified than enamel x-rays penetrate more readily, a property known as radiolucency. Polarized microscopy reveals that dentin exhibits limited birefringence.

Chemical Characteristics

Definitive dentin is made up of about 10% water, 20% organic material, and 70% mineral by weight; by volume, it is 33% water, 45% organic material, and 22% mineral. The organic portion of the matrix is predominantly type I collagen and ground substance, proteins, mostly proteoglycans and glycosaminoglycans. Collagen is in the form of fibrils. The inorganic components combine to form thousands of unit cells; their formula is that of crystal apatite, $Ca_{10}(PO_4)_6(OH)_2$. These cells average 3.5 nm in diameter, and may be as long as 100 nm. Hence, they are smaller than those of enamel and are more similar to those of cementum and bone. A unit cell is the smallest portion of apatite in a crystalline lattice.

The three components of dentin can be separated by leaching with acid (decalcification) or by incineration. In decalcification, the apatite crystals are removed so that the organic framework is retained. In incineration, the organic material is burned and water is driven off, leaving the mineral "skeleton" of matrix. Ground sections are untreated, except for cutting and polishing and, in these, the organic and inorganic materials remain more or less intact.

STRUCTURAL COMPONENTS

The basic morphologic components of dentin are the odontoblastic processes and calcified matrix. Dentin, a connective tissue, consists of few cells (processes) and an abundance of intercellular material (matrix). The latter constitutes the

PHYSICOCHEMICAL FEATURES

Physical Characteristics

Root dentin is continuous with that of the crown. The quantity and thickness of dentin in primary teeth are about half those of their permanent successors. In permanent teeth dentin is pale yellow and somewhat transparent; in primary teeth it is paler. Dentin of teeth in which the pulps have been destroyed or removed darkens, with the overlying translucent enamel reflecting the change so that the crowns appear "off color." Dentin of primary teeth is softer than that of permanent ones. In both, the tissue is elastic. This is a valuable property, because it provides stability to the overlying enamel. Be-

165

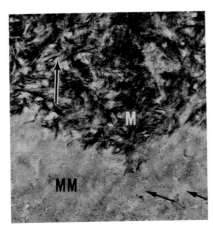

Fig. 6–1. Electron micrograph showing areas of matrix maturation (MM) and mineralization (M) of mature matrix in dentinogenesis. Note that, with matrix maturation, the structural features of collagen fibrils are obscured and with mineralization completely lose their identity. Seeding of the plate- and needle-shaped apatite crystals occurs on fibrils *(arrow).* (Selected area from Fig. 4–25; epon-embedded, uranyl acetate stain; ×20,000)

bulk of the tissue. The mineral portion of dentin represents between 22% and 25% of its volume but 70 to 80% of its total weight.

Dentin Matrix Classification

The matrix occupying the interspace between odontoblastic processes is composed of collagen fibrils (82%) embedded in a ground substance of noncollagenous protein (12%), including glycosaminoglycans and proteoglycans. The matrix synthesized initially is totally organic. It is soon mineralized by apatite crystals, which are seeded sequentially on, in, and between the fibrils (Fig. 6–1). The fibrils may have differing diameters, and may exhibit periodic bandings of less than 64 nm. The coronal dentin is classified as either mantle or circumpulpal.

Mantle Dentin. The matrix produced first, which contributes to the dentinoenamel junction, is known as mantle dentin. It occupies the space originally occupied by the basal lamina and basement membrane. Some have suggested that the thickness of the mantle dentin may approach 20 μm. Accordingly, its matrix contains aperiodic fibrils (filaments), type I collagen fibrils with periodicities of 64 nm. These are collected into bundles and are known as Korff's fibers. The aperiodic filaments course perpendicularly through the basement membrane and the basal lamina to the base of the preameloblast

(see Fig. 4–22). The Korff's fibers form fanshaped bundles at the prospective dental segment of the dentinoenamel junction (see Figs. 4–23 and 4–24). These, the predominant fibers of the mantle dentin matrix, mask the aperiodic filaments. The Korff's fiber ranges from 0.1 to 0.2 μm in diameter, with 64-nm periods.

Circumpulpal Dentin. All the dentin elaborated after the conversion of basement membrane into mantle dentin is known as circumpulpal dentin. This matrix is synthesized by the mature odontoblast and differs from the mantle dentin in that aperiodic filaments and Korff's fibers are mostly absent. When the Korff's fibers are present they are oriented parallel to the major odontoblastic process as shown in mantle dentin (Fig. 6–2; also see Fig. 4–24). The dominant fiber type of the circumpulpal matrix is the collagen fibril, produced in situ by the odontoblast. These fibrils are much smaller (50 nm diameter) and more densely packed than the Korff's fiber. They neither aggregate into bundles nor are preferentially oriented. Rather, they course throughout the matrix randomly to form a meshwork.

The ground substance is similar in composition to that of mantle dentin, but it is a product exclusively produced by odontoblasts. Matrix vesicles are not produced, and alternative methods of mineralization occur. Further, circumpulpal dentin is somewhat more calcified than mantle dentin.

Peritubular and Intertubular Dentin

Dentin matrices may be classified further, according to their relationship to the dentinal tubules, as peritubular or intertubular (Fig. 6–3). The matrix lining of the dentinal tubules is identified as peritubular dentin, while that located between the tubules is known as intertubular dentin. When ground sections of dentin are examined, a light circular band is observed that surrounds the darker solid center. The dark centers are lumina of the dentinal tubules, which in the process of grinding and polishing have become packed with debris. The light doughnut-shaped areas surrounding the lumina are the peritubular dentin, and the interspaces between the latter constitute the intertubular dentin (Fig. 6–3A).

Peritubular Dentin. The dentin matrix comprising the wall of the tubule differs from the

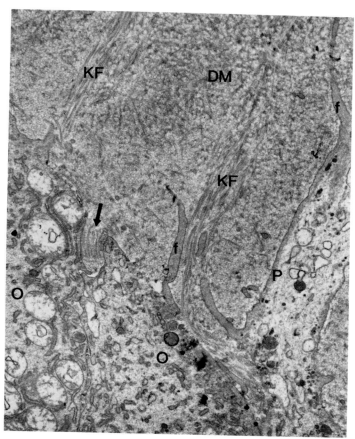

Fig. 6–2. Mantle dentinogenesis, showing segment of odontoblast (O) and mantle predentin matrix (DM). Note the interrelationship among the odontoblast process (P), filopodia (f), von Korff's fibers (KF), and the odontoblast cell body (O). Also observe the relationship of the von Korff's fibers *(arrow)* to the odontoblast. (Uranyl acetate and lead citrate stain; ×6,000) (Sisca, R.F., and Provenza, D.V.: Initial dentin formation in human deciduous teeth, an electron microscopic study. Calcif. Tiss. Res., 9:1, 1972)

intertubular matrix in thickness and mineralization. The thickness of the peritubular dentin toward the dentinoenamel junction is about 0.75 μm, while the thickness of that more toward the pulp is about 0.50 μm. The peritubular dentin is more highly calcified, about 9% greater than the intertubular matrix. Conversely, the collagen (organic) component is greatly reduced, and is comprised mostly of the terminals of the collagenous elements of the intertubular territory. The hypermineralized condition of the peritubular matrix is substantiated by its reaction to acid. The peritubular area is so highly calcified that it is severely affected by acid action, even

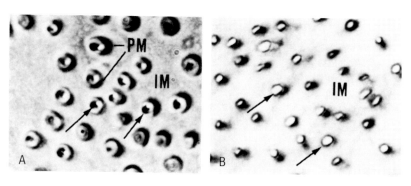

Fig. 6–3. Photomicrographs of dentin matrices. In ground section, *A*, the dentinal tubules *(black arrows)* are surrounded by doughnut-shaped darker zones, the peritubular matrix (PM), with an intervening intertubular matrix (IM). *B*, Dentinal tubules *(black arrows)* in this decalcified section are artifactually enlarged because much of the hypermineralized peritubular matrix has been dissolved by acid treatment (IM, intertubular matrix). (×800)

to the extent that most of the tissue is obliterated, as in the case of enamel. Accordingly, comparison of ground and decalcified serial sections reveals that, in the latter, lumen diameters are artifactually enlarged by the addition of the space previously occupied by the hypermineralized peritubular area (Fig. 6–3B). Electron microscopic examination of dentin reveals that the apatite crystals seeded on the fibrils are small needle-shaped or plate-shaped crystals (Fig. 6–1). The crystals render the peritubular area coarsely granular.

Electron microscopic studies reveal that the internal face of peritubular dentin is surfaced by an organic microlayer, the *lamina limitans*, rich in glycosaminoglycans and containing tubulin. Unlike predentin, its filament component is extremely fine and very sparse. Terminals of the collagen fibrils from the intertubular matrix generally do not arrive here (Fig. 4–25). This microspace separates the nonmineralized lining of the peritubular dentin and the odontoblast process, and provides for movement and the exchange of macromolecules between the tissues. Processes of odontoblasts communicate with those of their neighbors via gap junctions. Accordingly, the tubules housing the processes form an elaborate network through the dentin for the circulation of fluids. Circulation becomes obstructed by further elaboration of peritubular dentin or sclerosis of the tissue (described below).

Peritubular dentinogenesis, although occurring at greatly reduced rates, continues as long as odontoblastic processes occupy the lumina of the tubules. As the peritubular dentin matrix becomes calcified, the processes are compressed and the tubules become smaller. In time the lumina disappear, and are replaced by hypercalcified dentin continuous with that of the peritubular territory. Large segments of dentin affected in this manner are rendered transparent. When viewed by optical microscopy, the hypermineralized areas appear dark with reflected light and light with transmitted light. These dentin segments are known as sclerotic dentin (see below).

Intertubular Dentin. The greater part of dentin is occupied by intertubular matrix. This is especially the case toward the enamel, where the odontoblastic processes are of smaller diameter and the tubules housing them are cor-

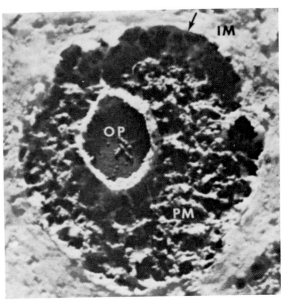

Fig. 6–4. Electron micrograph of acid-treated dentin, showing pits and craters remaining as a result of removal of some apatite crystals from the hypermineralized peritubular dentin (PM). The intertubular matrix (IM), which contains much less mineral, is less affected and does not demonstrate pits and crevices. The arrow indicates the area once thought to be the sheath of Neumann (OP, odontoblastic process in tubule). (Scott, D., and Nylen, M.: Concepts in dental histology. Ann. N.Y. Acad. Sci., 85:143, 1960.)

respondingly of decreased bore. The matrix, which comprises about half of the dentin volume, is made up of a dense meshwork of collagen fibrils embedded in the ground substance. The fibril diameter ranges from 0.05 to 0.2 μm. They exhibit periodic cross banding at 64-nm intervals. Apatite crystals about 0.1 μm long are seeded parallel to the fibril length. Collagen fibril masses course through the matrix and around the tubules, appearing to be somewhat obliquely or perpendicularly oriented to them.

Sheath of Neumann. The junctional zone between the peritubular and intertubular dentin reacts differently to stains and to treatment with acid or alkali. Based on these differences, some investigators believed that the two matrices were separated by a type of membrane, which they called the sheath of Neumann. Electron microscopic studies do not confirm the existence of a junctional sheath; rather, the two matrices have been shown to have no definable boundaries but to intermingle freely (Fig. 6–4). On the other hand, a lining membrane (discussed above) is found at the tubule wall (Fig. 6–5). It

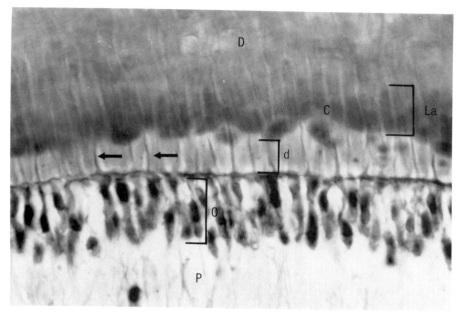

Fig. 6–5. Dentinopulpal junction showing odontoblast layer (O) at the periphery of the pulp (P). Predentin (d) shows the formation of calcospherites (C), which in mature dentin (D) are homogenized via uniform mineralization. Note the odontoblast processes *(arrows)* in the dentinal tubules extending from the odontoblast cell body through the predentin to the dentin. Observe the paucity of cellular elements in the subodontoblastic layer, or cell-poor zone of the dental pulp (La, lamella of calcifying dentin). (Hematoxylin and eosin stained; ×500)

has been suggested that this lining of nonmineralized organic matrix functions as a protective barrier for the cell processes and as a medium for metabolic exchange.

Dentinal Tubules

Dentin matrix contains numerous canals of varying sizes called dentinal tubules. Because these contain(ed) the odontoblastic processes, the tubules must reflect the size, shape, and path(s) of the odontoblastic processes. Thus, the larger tubules containing the largest processes have diameters up to 4 μm and are found toward the pulp where they originated from the odontoblastic cell body. The tubules of the more central dentin are up to 2.0 μm and the smallest ones are about 1 μm in diameter, located more toward the enamel (Fig. 6–6). Dentinal tubules may exhibit branching processes, which housed the collaterals of the principal process. These are smaller than the parent process (diameters mostly under 1 μm), and they are known as filopodia. They occur more frequently near the dentinoenamel junction (Fig. 6–6).

The tubules approaching the pulp are more numerous (40,000 to 80,000 per square milli-meter), are of greater diameter, and are closer together than those of mantle and the more peripheral circumpulpal dentin, which averages about 20,000 tubules per square millimeter. These differences are due to the long attenuated odontoblast processes, which have diameters of about 900 nm in mantle dentin and which increase in diameter to 2.5 μm with confluency with the odontoblast cell body. The latter, then, is clearly within the resolution of optical microscopy (Figs. 6–5 and 7–5). This increased concentration of the tubules per unit of area near the pulp is a product of continued dentinogenesis and crowding of the odontoblasts in a pulp chamber that becomes progressively more confining.

Enamel insulted by fractures, cracks, caries invasion, and iatrogenic dentistry provides access to the dentin. Damage to the dentin is facilitated by and rapidly spread through the unoccluded tubules. If unchecked, it may progress to involve the pulp. Drugs and chemicals, as well as microorganisms, may arrive at and affect the dental pulp through these anatomic pathways.

The paths of the large tubules suggest that the odontoblastic processes are aligned differently

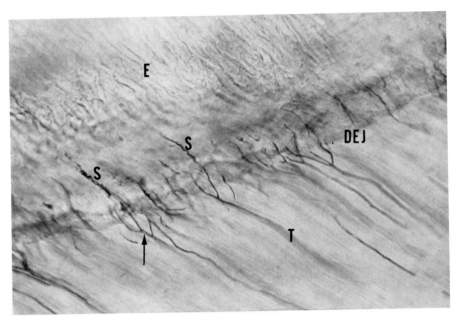

Fig. 6–6. Dentinoenamel junction (DEJ) showing dentinal tubules (T) and their branches *(arrow)* (E, enamel; S, spindles). (Ground section; ×320)

in the tooth. At low magnifications it may be noted that the tubule paths of the radicular and coronal dentin are different. Those of the apical segment of root are straight, and, as the cervix is approached, a gentle S-shaped curve is observed—the primary curvature (Fig. 6–7). Primary curvatures are comprised of two slight arcs that bend in opposite directions, forming a mirror image of the letter S. The direction of the external arc is occlusal, and that of the inner is apical. Maximum curvatures are seen in the dentin at the slope of the crown. Under the incisal edge or cusp, however, the tubule paths are straighter and the tubule segments approaching the pulp are straight, forming right angles to the pulp.

At higher magnifications, tubules viewed in longitudinal section appear to pursue gentle undulating paths, which form the secondary curvatures (Fig. 6–8). Depending on location, a tubule may display 200 or more secondary curvatures from the pulp to the dentinoenamel junction. It is believed that secondary curvatures represent the twisted path of the odontoblasts as they grow toward the pulp during dentinogenesis.

Odontoblastic Process

These are protoplasmic extensions of the odontoblastic cell body. The segment of greatest diameter is that which is confluent with the central cell mass, and the thinnest portion is the tapered terminal at the dentinoenamel junction. At intervals small branches (filopodia) may emanate from the major (parent) process. These generally end in the matrix a short distance from their site of origin. Occasionally filopodia divide, forming fine filamentous extensions. This occurs more frequently in the mantle dentin area—that is, near the dentinoenamel junction (Fig. 4–24). Electron microscopic studies reveal that the plasma membrane of the odontoblastic cell body, its major process, and the filopodia are continuous. The cytoplasm of the filopodia appears denser and organelle-free, while the cytoplasm of the parent process appears less dense, with few organelles and inclusions. The process segment joining the cell body contains more organelles, with the greatest number present during the active periods of dentinogenesis. The organelles present at this time are mainly vesicles of diverse types and mitochondria.

Predentin (Dentinoid)

A thin layer 5 to 30 μm wide separating the calcified dentin surface from the distal surface of the odontoblast is known as predentin, or dentinoid. Once produced, it persists as a definitive layer at the dentin-odontoblast junction (Figs.

Fig. 6–7. Primary curvature of dentinal tubules are shown best by empty tubules (dead tracts), into which debris has been ground during processing. Note the sigmoid path of the dead tracts beginning at the dentinoenamel junction (DEJ) and progressing toward the arrow at the pulp chamber (PC). The convex segments of the arcs are coronally directed *(white arrows)*, while the concave arc segments *(black arrows)* are apically directed. (Ground section; ×32)

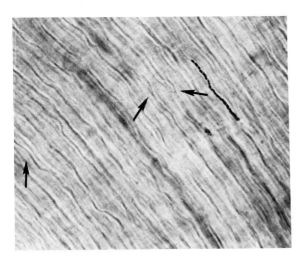

Fig. 6–8. Secondary curvatures of the dentinal tubules reflect wavy paths of the odontoblast processes. Note the filopodial branches *(arrows)*. (Ground section; ×200)

6–5, 6–9, and 6–10). Predentin is composed of collagen fibrils, odontoblastic processes (proximal ends), nerve fibers, capillary loops, and possibly lymphatic channels invested in ground substance. The latter tends to obscure the identity of its structural elements. Special techniques, including electron microscopy, allow for their visualization. Predentin is believed to function as a medium of exchange for metabolic products, as an immediate source for dentin production, and as a protective barrier against dentin resorption.

Nerves and Dentin Sensitivity

Pain is the exclusive response to dental physicomechanical and chemical stimulations. Thus, dental restorative procedures affecting the dentinal tubules, such as probling, cutting, instrumentation, and use of air and water sprays, may occasion pain. Although the precise localization of pain among teeth is not always well defined, pain in any given tooth, can be expected to be greatest at the dentinoenamel and dentinopulpal junctions.

Much speculation and controversy exists regarding the mechanisms involved in dentin sensivity. Many histologists believe that the odontoblastic processes, which in the definitive tooth do not extend beyond the pulpal half of the dentinal tubules, occupy their lumina totally except for a microspace. This provides for circulation of tissue fluids, thus acting as an exchange medium for materials between the soft and hard tissues. The microspaces are thought to be too small to house nerve fibers. Electron microscopic studies, however, have provided strong evidence for their presence in the tubules, especially in the more pulpward segment of the dentin where the tubule diameter is greatest (Fig. 2–67).

It has also been suggested that the odontoblastic processes possess a highly developed property of irritability and that, having been stimulated, they transmit the impulse to the odontoblastic cell bodies. The occurrence of gap junctions on the proximal cell membranes of the odontoblasts is especially significant, because they may provide communication and electronic coupling sites, not only with adjacent cells but with the nerve and vascular elements in the subodontoblastic zone. Impulses may therefore pass from the odontoblasts to the network of nerve

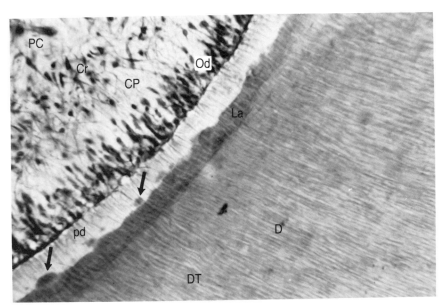

Fig. 6—9. Transverse section of the peripheral regions of the dental pulp: cell-rich (Cr), cell-poor (CP), and odontoblast (Od) zones, and pulp core (PC). Note that the calcification process is well ordered, as evidence by the amalgamation of calcospherites *(arrows),* thus forming a lamellar front (La) of mineralizing dentin. Contrast this with Figure 6—10, in which mineralization is not uniform, as evidenced by the spotted appearance of the dentin due to calcospherites (pd, predentin; D, dentin; DT, dentinal tubules). (Hematoxylin and eosin stain; ×100)

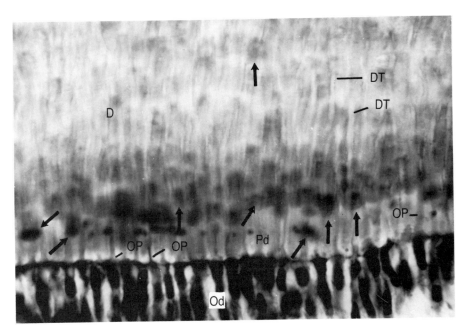

Fig. 6—10. Irregular calcification resulting in interglobular dentin. The calcospherites *(arrows),* as formed initially, are widely separated by interglobular spaces so that a lamellar front is not produced. The lag in calcification is registered in decalcified sections of mature dentin as "spotted" mature dentin (D) (Pd, predentin; Od, odontoblasts; DT, dentinal tubule; OP, odontoblastic process). (Hematoxylin and eosin stain; ×450)

endings (Raschkow plexus) that surrounds the cell bodies.

Finally, because the odontoblastic processes are known not to extend beyond the pulpal one-third or one-half of the dentin, another concept has been proposed. Data derived from in vitro and in vivo studies have indicated that tissue fluids flow to or from the pulp chamber. It has been determined that the capillary force exerted on the fluid may result in an acquired flow rate of up to 2 to 4 mm/sec. It is therefore probable that fluid flow through the tubule microspace may induce pain by stimulating the nerve endings in the tubule or in the odontoblastic layer adversely. For example, dehydrating substances or air blasts directed against the cut tubules produce an outward flow of tissue fluids, while heat produces an inward or pulpward flow. It has been suggested that even a minor outward flow of several micrometers may produce pain by disturbing the hydrodynamic balance of the system.

Thus, three hypotheses have been proposed for the perception of dental pain: direct response of sensory receptors in the dentinal tubules; reception of stimuli by odontoblastic processes, and the transmission of the impulses via their cell bodies to the subodontoblastic plexus of Raschkow; and change in hydrostatic pressure in the dentinal tubules perceived by the subodontoblastic nerve endings. However, because nociceptive nerve endings are believed to be activated by the release of chemical modulators as a consequence tissue damage, it is likely that this is the operative mechanism in teeth. Such a theory would tend to integrate the first and second proposed hypotheses.

STRUCTURAL PATTERNS

The structural patterns observed in dentin are produced by various factors:

1. Daily deposition of matrix
2. Recurrent waves of dentin matrix production
3. Involvement of several daily matrix increments in mineralization
4. Initial mineralization as spheres, which grow by peripheral accumulation of apatite crystals
5. Nonsynchronous growth and fusion of neighboring calcifying spheres

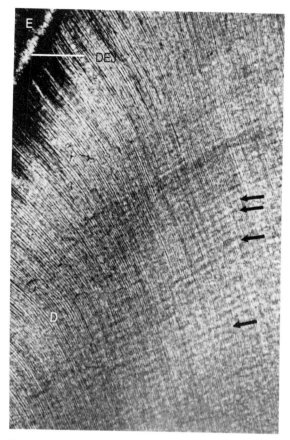

Fig. 6–11. Imbrication (incremental) lines of von Ebner in dentin (D) are indicated by arrows (DEJ, dentinoenamel junction; E, enamel). (Ground section; × 52)

6. Unequal and uneven intensities of initial mineralization throughout the dentin
7. Variation in calcium metabolism
8. Deviation in directional paths of the dentinal tubules

Incremental Lines

The daily increments of dentin deposited vary from 4 to 8 μm thick. Thickness variations occur in different areas of a given tooth, as well as in different teeth. Dentinogenesis is interrupted by rest periods that are registered as delicate markings known as incremental lines, imbrication lines, or lines of von Ebner. By etching the ground sections slightly with acid, the markings are resolved more clearly (Fig. 6–11).

Contour Lines of Owen

Dentin deposition occurs as incremental bands that begin at the crest of the dental pa-

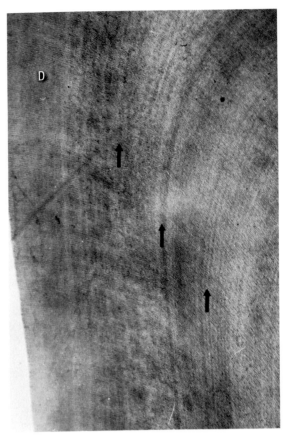

Fig. 6–12. Coronal dentin (D) sectioned longitudinally, showing contour lines of Owen *(arrows)*. (Hematoxylin and eosin stain; ×50)

pilla. Thereafter activity occurs bidirectionally, over the slope of the papilla toward the cervical loop and pulpward. In the latter, dentin deposition toward the dental papilla (prospective pulp) progressively reduces the size of the pulp chamber. Rings of matrix representing about 4 days growth (16 + μm) enter the mineralization period simultaneously. Calcification phases consequently exhibit lags of several days, reflected as sweeping bands outlining the growth pattern of the coronal or radicular dentin. These markings are known as the contour lines of Owen (Fig. 6–12).

Some investigators believe that the contour lines of Owen are caused by disturbances in the calcification process or in calcium metabolism. The most prominent contour line is produced during the period between birth and the first few days of life, identified as the neonatal line (Fig. 5–13). It ceases with the adjustment of the infant to its new environment. The enamel counterparts to the contour of Owen lines are the lines of Retzius (Fig. 5–11). The widths of the contour lines of Owen are not only governed by the size and number of daily increments affected, but also by the duration of the disturbing influence on the calcification process or on calcium metabolism.

Interglobular Dentin

Electron microscopic examination of dentin matrix mineralization reveals the deposition of needle-shaped or plate-shaped apatite crystals on the organic components (Fig. 6–1). Extension of the original calcification sites occurs throughout peripheral growth. The enlarging bodies join neighboring ones, so that more and more of the matrix is calcified. When these areas become large enough to be observed with optical microscopy they appear spherical, and are known as calcospherites (Figs. 6–5, 6–9, and 6–10). Growth and fusion of the calcospherites result mostly in the formation of a calcifying linear front, which in decalcified sections appears lamellar-like (Fig. 6–5). As orderly calcification is completed, mineral homogenization succeeds in obliterating or reducing the lamellar effects of the matrix (Fig. 6–9). When lags in the calcification rate prevent the occurrence of orderly uninterrupted sequences, there is no fusion of calcospherites to form a linear front. In decalcified sections, the dentin that represents lags in the mineralization rate appears spotted (Fig. 6–10). The darker spots (calcospherites, or calcified globules) are the more calcified areas, while the lighter intervening areas are less mineralized (hypocalcified). Dentin areas characterized by this spotted appearance are known as interglobular dentin, and the lighter irregular spaces between the calcospherites are called interglobular spaces. In some areas of dentin, a lag in fusion or disturbance in calcification may be localized. In adjacent areas, the process of calcification may occur normally. Decalcified sections of dentin possessing variations in activity may exhibit both linear bands and calcospherites.

Interglobular dentin is encountered most often in coronal circumpulpal dentin (Fig. 6–13). In the root, it is found immediately beneath the granular layer of Tomes (Fig. 6–14). Regions in which interglobular dentin occur are

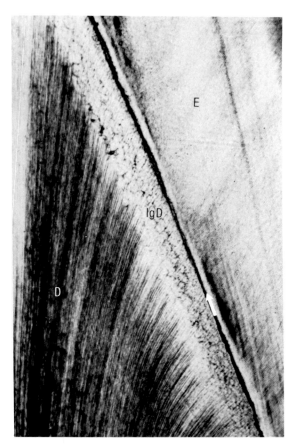

Fig. 6–13. Segment of crown showing enamel (E), dentinoenamel junction *(arrow)*, interglobular dentin (IgD), and dead tracts in dentin (D). (Ground section; ×50)

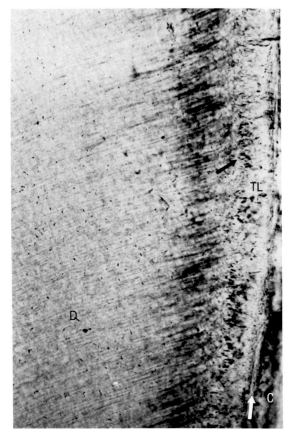

Fig. 6–14. Segment of root showing cementum (C), cementodentin junction *(white arrow)*, and granular layer of Tomes (TL). The latter *(black arrow)* often bears a resemblance to interglobular dentin. (D, dentin). (Ground section; ×50)

expectedly associated with the contour lines of Owen. Peritubular dentin cannot occur with interglobular dentin, because this represents a faulty process resulting in hypocalcified matrix.

As indicated, the spotted effect of interglobular dentin is easily resolved in decalcified, hematoxylin-and-eosin stained sections, in which the calcospherites and lamellae are blue and the intervening areas are pink. In ground sections, however, the appearance of interglobular dentin is quite different. In the preparation of ground sections, much of the tooth's moisture is lost during cutting, grinding, and polishing procedures. Areas of interglobular dentin are severely affected by tissue shrinkage, producing cracks and pits. With optical microscopy these appear dark because of the debris and air trapped in them (Fig. 6–13).

Granular Layer of Tomes

The initial few microns of root dentin deposited, particularly in the cervical half, are coarsely granular and are therefore, known as the granular layer of Tomes (Figs. 6–14 and 6–15). This layer has also been reported to be present in hypomineralized cervical dentin. The location and morphology of the spaces are similar to those of interglobular dentin, leading to the concept that the Tomes' granular layer represents an interruption in matrix mineralization. It has also been suggested that the granular texture is due to the fibers in the area, which are coarser and more poorly calcified. It has also been suggested that its resolution in ground sections is a result of light refraction. More recently and perhaps more accurately, this layer has been determined to consist of spaces reflecting looped branched

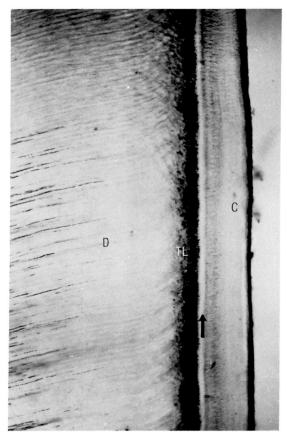

Fig. 6–15. Longitudinal ground section of root, showing dentin (D), granular layer of Tomes (TL), and cementum (C). The hyaline layer of Hopewell-Smith located between the cementum and granular layer of Tomes is indicated by a black arrow. (×80)

terminals of the odontoblast processes. Looping produces a more porous matrix, which in ground section appears to be granular. Strangely enough, this granular layer is not seen in electron micrographs or in decalcified stained sections.

Hyaline Layer of Hopewell-Smith

This glassy (hyaline) layer, about 10 μm thick and hypermineralized, intervenes between the cementum and the granular layer of Tomes (Fig. 6–15). It too is mainly restricted to the cervical half of the root, and is more distinct in poorly calcified teeth. The exact origin of this layer has not been determined. Because it is the first tissue to appear at the prospective cementodentinal junction, some oral histologists believed that it is probably a product of the odontoblasts.

As such, it is possibly a special type of dentin, devoid of tubules. It has also been suggested that the hyaline layer is produced by Hertwig's root sheath, or even precementoblasts. Thus, it can be speculated that it is either a kind of intermediate cementum or "enameloid" substance, the latter being produced by elements of the epithelial root sheath. The function of the hyaline layer is thought to be one of uniting the dentin to cementum. The function of the hyaline layer of Hopewell-Smith is thought to be one of uniting the dentin to cementum.

PRIMARY AND SECONDARY DENTIN

Dentin develops continuously from the period of odontoblastic differentiation in the enamel organ until the pulp dies or is removed. For convenience, dentin can be classified as having three stages: primary; secondary; and tertiary, also known as reparative or irregular dentin.

Primary Dentin

Prenatal Primary Dentin. Mantle dentin and circumpulpal dentin produced prenatally may be classified as developmental dentin. Unlike postnatal enamel, which differs in the degree of mineralization, developmental primary dentin does not appear to be qualitatively different from dentin produced after birth (i.e., produced beyond the neonatal dentin) except that the fibrils are greatly reduced in diameter (0.05 μm). Notable differences in the directional paths of the tubules are not observed; rather, the tubules continue imperceptibly into the postnatal primary dentin.

Postnatal Primary Dentin. The dentin produced from the neonatal line to the time that the tooth assumes its occlusal position, when it is completely formed, is known as primary dentin. Dentinogenic activity continues after the tooth meets its antagonists, although at a greatly reduced pace and quantity. For the most part, dentinogenesis is not localized; rather, it occurs generally over the entire pulp chamber. The stimuli for primary and possibly even for secondary dentin formation are mild and are not necessarily associated with normal physiologic activity, because dentinogenesis has been shown to occur even in unerupted teeth. This dentin is not unlike that produced previously.

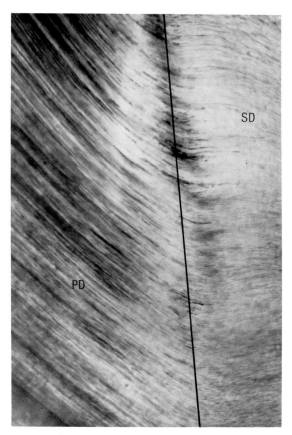

Fig. 6–16. Secondary dentin. The black line indicates the demarcation zone (line) or locus at which the paths of the dentinal tubules of the primary dentin (PD) are changed by the production of secondary dentin (SD). (Ground section; ×80)

Secondary Dentin

With wear of the biting and chewing surfaces, and especially with changes in diet and harsher stimuli odontoblasts continue to produce more rapid and greater deposits of dentin. The pulp chamber is therefore reduced in size (pulp recession) and the odontoblasts become crowded. This condition is believed to cause the odontoblasts to shift slightly in position, so that the dentinal tubules of the matrix deposited subsequently reflect the alteration in the tubule path. This is recorded permanently as a demarcation line, formed by the slight inclination of the dentinal tubules (Fig. 6–16). Because secondary dentin formation occurs more often on the ceiling and floor of the pulp chamber, the morphology of the tissue no longer reflects the profile of the tooth. This is particularly the case

with molars. Except for the slight shift in tubule path, there are no other substantial differences. Pronounced differences are not observed in secondary dentin.

Secondary dentin may be summarized as follows: it is usually deposited over the internal surface of the pulp chamber; its elaboration is induced by normal physiologic stimuli; its physicochemical properties are similar to those of primary dentin; and, except for a slight initial inclination of the tubule path, its course is regular.

Tertiary Dentin

This type may be described as reparative or tertiary dentin because of substantial differences in deposition site(s), as well as other aspects of its morphology. Acutely stimulated, odontoblasts as in severe caries attack or operative procedures (burr and chisel actions), respond by depositing irregular secondary or reparative dentin (Fig. 6–17). Involvement of dentin segments more closely located to the odontoblasts, especially in older pulps, exhibits greater sensitivity to stimuli, tending to react more violently. It has been estimated that as much as 3.5 μm of matrix may be deposited daily. Reparative dentin is limited to the insulted tissue and is characterized by fewer tubules, which pursue irregular and tortuous paths. In instances in which the stimulation is lethally severe, cells, probably fibroblasts, that underlie the destroyed odontoblasts are recruited for matrix production. Under such conditions tubules are wanting because these cells do not possess long protoplasmic processes, as do the odontoblasts. Additionally, the cells may be trapped in their matrix and die, leaving a more porous dentin, osteodentin. Disintegration and autolysis of the distal processes result in empty dentin tubules known as dead tracts (Fig. 6–18; also see Fig. 5–20). These are discussed below.

ALTERATIONS

Dentin changes produced by aging or external stimuli of varying intensities include the formation of tertiary dentin, tubule closure, dead tracts, and sclerotic (transparent) dentin. Tertiary dentin formation has been discussed above.

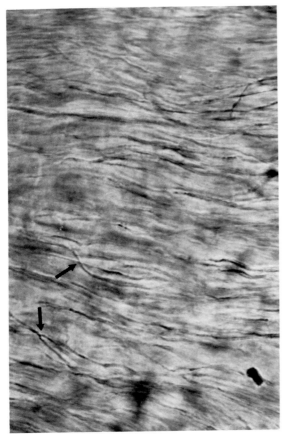

Fig. 6—17. Reparative ("irregular," or tertiary) dentin. Note the extremely irregular path of the dentinal tubules *(arrows),* reflecting the gyrations of the odontoblasts under excessive stress. (Ground section; ×320)

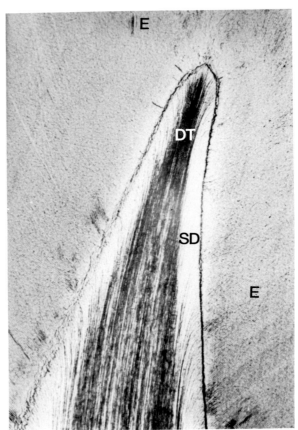

Fig. 6—18. Incisor crown segment showing enamel (E), dead tracts (DT), and sclerotic dentin (SD). (Ground section; ×50)

Tubule Closure

Dentinogenesis progresses pulpward for the life of the tooth, provided that the pulp is present (i.e., the tooth is "vital"). Similarly, dentin will continue to be deposited in the dentin tubules as long as processes are present. Accordingly, tubule closure by the production of peritubular dentin may occur naturally with age. Some investigators believe that peritubular dentin in young teeth forms a narrow band around the lumen of the tubule. With age the width of the peritubular collar increases, thereby progressively narrowing the bore of the tubule until it is closed. It is obvious that the odontoblastic processes are eventually strangulated and die, and are removed. The smaller tubules at the dentinoenamel and cementodentinal junctions are the first to become occluded. In older teeth this phenomenon may extend to the larger tubules located more pulpward, near the odontoblast. Tubule occlusion is a protective sealing mechanism, isolating the odontoblastic cell bodies and the pulp from injurious external agents and bacteria.

Dead Tracts

It has been noted that lethal stimulation of odontoblasts results in the formation of dead tracts. Not all dead tracts, however, result from destroyed odontoblasts. Some stimuli may not be lethal, but may be strong enough to cause retraction of the odontoblastic process. In such a case the tubule segment lacking the process is also called a dead tract, and is obviously shorter. It has been suggested that, in aging dentin, only those tubules occupying the pulpal third of this tissue house processes.

Death of the odontoblasts is not always due to external causes. It is entirely possible that,

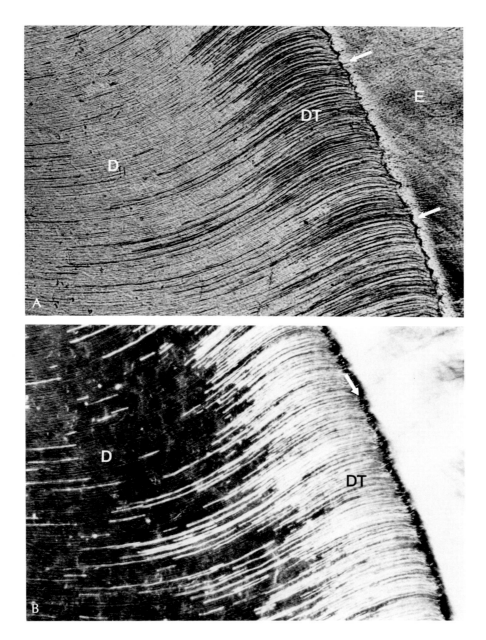

Fig. 6–19. Dead tracts. *A,* Dead tracts (DT) of dentin (D) as observed with transmitted light appear dark against a light background. (×80) *B,* With reflected light the tracts (DT) appear light against a dark background (E, enamel; dentinoenamel junction *(white arrow);* D, dentin). (Ground sections; ×80)

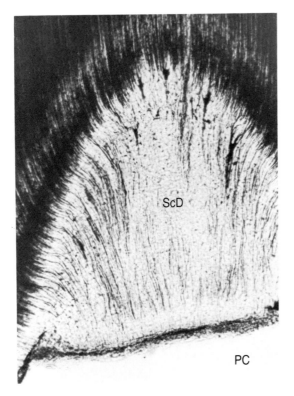

ScD

PC

Fig. 6–20. Sclerotic dentin (ScD) in ground section (PC, pulp chamber). (×50)

with deposition of primary and secondary dentin, and with the attendant diminution in pulp chamber volume, the odontoblasts become crowded and the older and less vital cells do not survive because of being stifled in such a restricted environment. In this regard, it has been estimated that the odontoblastic population is gradually reduced by 50% between the ages of 20 and 70 years.

In ground section, dead tracts appear dark with transmitted light because of the air and debris packed into the tubules. With refected light they appear light (Fig. 6–19).

Sclerotic (Transparent) Dentin

Ground sections of teeth may exhibit bands of dentin that appear glassy under transmitted light. These areas of sclerotic, or transparent dentin represent regions in which the dead tracts have been transformed into protective barriers of hypermineralized dentin (Figs. 6–18 and 6–20; also see Fig. 5–20). Some investigators believe that organic materials of the dead odontoblast processes act as sites for apatite crystal deposition. Because groups or tubules are filled with highly calcified material, this sclerotic dentin is similar to that produced in peritubular dentin formations. Sclerotic dentin is most often produced under very thin enamel, such as that in pits and fissures. Sclerotic dentin is more resistant to caries attack but, because of its increased mineralization, it is very brittle and less permeable. Sclerosis tends to increase pulp longevity.

CHAPTER 7

Dental Pulp (Endodontium)

The dental pulp is also known as the endodontium. Endodontics is an important branch of dentistry that is concerned with the causes, prevention, diagnosis, and treatment of diseases of the pulp. The lay person might call it the "nerve," which is an obvious misnomer because dental pulp in most respects is a primitive connective tissue resembling mucoid connective tissue. The dental pulp forms the substance of the crown's core as the coronal pulp and of the root as the radicular pulp. The pulp is surrounded completely by a layer of odontoblasts, the innermost boundary of the dentin.

FUNCTIONS

The four primary functions attributed to the dental pulp include formative, nutritive, sensory, and protective. The first is performed by the developing tooth organ, and the others by the definitive organ.

Formation. The forms and shape of the crown and root are established by the initial layer of dentin deposited. In the crown the first layer is mantle dentin and, in the root, it is the granular layer of Tomes. Dentin synthesis ceases only in the absence of the dental pulp. As long as odontoblasts or other matrix-producing cells are present, the tooth may engage in dentinogenesis.

Nutrition. Dentin is avascular, except for capillaries looping around the odontoblastic cell bodies, where they may contact the predentin. The dentin thus relies on the blood vessels of the pulp for its nutritive and metabolic needs. Consequently, there is a rich blood supply to the pulp.

Sensation. Myelinated and nonmyelinated nerve fibers are found in the dental pulp. Some are associated with the smooth muscle fibers of the blood vessels, while others travel independently and end as a dense network (plexus) around the odontoblasts. All stimuli (e.g., heat, cold) received by the nerve endings are interpreted as pain. No distinction can be made between hot and cold.

Protection. The protective cells of the pulp include odontoblasts, which form reparative dentin, and macrophages and lymphocytes, which combat inflammation, remove debris, and possibly engage in cell immune reactions.

Tertiary (reparative) and to a lesser extent secondary dentinogenesis are defensive reactions separating the pulp from offensive external agents by interpositioning a protective layer of dentin. These external factors may range from those of natural wear to caries attack. The extent to which the pulp will react to external stimuli depends on the type, severity, and duration of the insult. Similarly, severe dental restorative procedures are in themselves insulting, and dental pulps will react to some operative procedures and some restorative materials more than others.

MORPHOLOGY

The pulp is a specialized type of mucoid connective tissue. Its gelatinous nature allows the pulp to be extirpated without appreciable loss of form. The larger portion is contained in the crown. Pulp volume varies with tooth type and with the arch on which it is located. For example, for the maxillary permanent teeth, the volume ranges from 0.012 cm^3 for an incisor to 0.068 cm^3 for a first molar. In comparison, the volume of a mandibular incisor is 0.006 cm^3 and that of a mandibular first molar is 0.053 cm^3. Except for the third permanent molars, the pulps of all other maxillary teeth are larger.

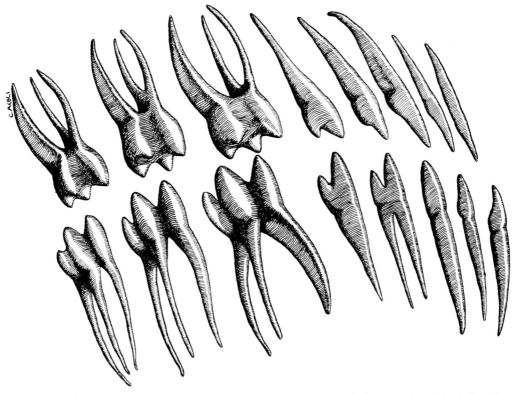

Fig. 7—1. Dental pulp diagrams of maxillary and mandibular dentition central incisor to molars. The delineation of the coronal and radicular pulps is marked by a constriction occurring at the level of the cervix (cementoenamel junction). The gelatinous (mucoid) nature of the pulpal connective tissue provides for the more intact extirpation of these organs. (Adapted from Bhaskar, S.N.: Orban's Oral Histology and Embryology. 10th Ed. St. Louis, C.V. Mosby, 1986.)

Coronal Pulp

The profile of pulps of young teeth reflects the shape of the crown more accurately. In crowns, the contours of the cusps and incisal edges are followed faithfully (Fig. 7—1). Extensions of the central pulp into the cusps and edges are known as pulp horns (Fig. 7—2). The number of pulp horns in a tooth usually corresponds to the number of cusps. The coronal pulp achieves maximum volume and most accurately reproduces the form of the crown in the period during which the tooth entered the oral cavity. Thereafter, secondary and especially tertiary dentin production reduces the chamber volume, thus altering its morphology. Tertiary dentin and calcified bodies, such as denticles, drastically change the form of the pulp chamber (pulp recession). Dentin formation in molars occurs rapidly on the floor of the pulp, less on the ceiling, and least on the lateral walls. The shape of the pulp is therefore altered more rapidly in its vertical axis.

Root Pulp

The coronoradicular junction is markedly narrowed, and is known as the cervix, or neck, of the tooth (Fig. 7—1). The cervix and crown are in the oral cavity, but the conical roots are submerged in bony sockets anchored by the principal fibers of the periodontal ligament. The pulps of newly developed roots similarly reflect their external morphology. Although generally long and attenuated, roots vary in length and size (Fig. 7—1). Although some roots are straight, they are more often gently curved, especially near the tip or apex. The inner wall of the root is composed of dentin and the surface is made up of cementum. Dentin and cementum are uninterrupted from the cervix to the apex, except for occasional small channels that course from the periodontal tissue to the root pulp. These

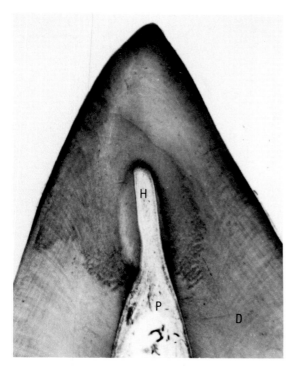

Fig. 7–2. The pulp horn (H) is a connective tissue extension of the central pulp (P) tissue (D, dentin). (Hematoxylin and eosin stain; ×8)

small channels are variously referred to as lateral canals, accessory canals, secondary canals, or apical ramifications (Fig. 7–3). Structurally, the soft tissue contained in these is similar to that of the radicular pulp. It is not, however, derived from the dental papilla, as is the dental pulp. Rather, it originates from the connective, vascular, and nervous tissues of the dental follicle, which invade the developing root through premature breaks in Hertwig's epithelial root sheath.

The volume of the radicular pulp is also greatest immediately after eruption. Although the radicular tissue is gelatinous, as is the coronal pulp, it differs from the coronal pulp in that its connective tissue cell population is minimal. It is composed mainly of arteries, veins, lymphatics, and nerves. These and the meager population of connective tissue cells are invested in a delicate fibrillar ground substance, containing glycosaminoglycans and proteoglycans. The odontoblastic layer is present and active, especially during its developmental period when the root canal is widest. Thereafter dentinogenic activity is greatly retarded or abated, reducing the

possibility of complete root canal occlusion, which would sever the lifeline of the coronal pulp.

Apical Foramen

The entrance to the root canal is known as the apical foramen (pl., foramina) (Fig. 7–4A). Through this opening the nerves and vascular and lymphatic channels enter and exit the pulp (see Fig. 8–1). The size and location of the opening(s) are not always the same. Foramina are largest immediately on completion of their development. The average diameter of a foramen is about 0.4 mm for a permanent mandibular tooth, and about 0.3 mm for a maxillary tooth.

A minor constriction occurs at the cementodentinal junction about 0.5 mm from the apical foramen. It is to this point that root canal therapy is directed. Its diminished diameter is consistent with smaller pulp volumes. Because roots may lengthen throughout the life of the tooth their apical foramina may become smaller, and may shift position. These changes occur in accord with the new growth stimulated by tooth movements during and after eruption (vertical, mesial, tipping). Thus, in some teeth, the apical foramina are located at the root tip (Fig. 7–4A), but more often they shift laterally (Fig. 7–4B).

HISTOLOGY

The microstructure of the dental pulp changes continuously from its developmental stage throughout adult life. The pulp originates from ectomesenchyme of the dental papilla. In pulps of young teeth the changes are minimal, except for the establishment of blood, lymph, and nerve supplies. As discussed below, changes induced by external stimuli and aging involve tissue volume, morphology, and structural composition.

Dental Papilla

The dental papilla, or developing pulp, consists of a peripheral layer of (pre)odontoblasts and a core of mesenchymal cells and fibroblasts, with appropriate vascular and nerve elements supported by a meshwork of type III collagen also known as reticular (argyrophilic) fibrils. In the early bell stage, blood vessels develop in situ in the dental papilla near the prospective dentinogenic layer. These are joined by others arriving from the dental follicle. The number of

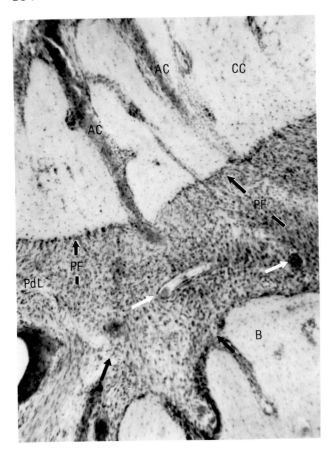

Fig. 7–3. Apical segment of root showing cellular cementum (CC) and accessory canals (AC). The periodontal ligament (PdL) demonstrates the orientation of the apical principal fibers (PF) and blood vessels *(white arrows)*. Bone (B) constitutes the floor (fundus) of the socket. Note the emerging blood vessels *(black arrows)*. (Hematoxylin and eosin stain; × 100)

blood vessels contained in the dental papilla increases rapidly with the initiation of dentinogenesis. The precise period for the arrival of the nerves is undetermined. Similarly, little is known about the development of lymphatic drainage of the pulp. The cellular composition of the papilla is mostly (pre)odontoblasts or their cell precursors, mesenchymal and fibroblast-like cells (discussed in Chapter 4).

Young Mature Pulps

Young coronal pulps not engaged in dentinogenesis are comprised of four regions, of which the largest is the central core. The other layers are located at the periphery of the pulp, and occupy about a 100-μm space. These include the odontoblastic layer, the outermost limit of the pulp, its subjacent cell-poor zone of Weil, and its underlying cell-rich layer (Fig. 7–5). Neither the cell-rich nor cell-poor zones are morphologically constant features. Odontoblasts, however, persist for the life of the pulp, but do not necessarily engage continuously in dentinogenesis.

Odontoblastic Layer. Anatomically, odontoblasts are an integral part of the dentin, as osteoblasts and chondroblasts are to bone and cartilage. Their cytologic features as preodontoblasts and functional odontoblasts have been discussed elsewhere. It will be recalled that these cells range in shape from low to tall columnar, with the latter most often associated with the active period of dentin formation. Low columar and especially cuboidal odontoblasts possess diminished organelle populations. Dentin synthesis is either abated or halted.

Zone of Weil. This region, up to 40 μm in width, is located pulpward to the odontoblasts and is known as the zone of Weil or the cell-poor zone (Fig. 7–5). It has also been referred to as the cell-free zone, which is inaccurate because nerve cells and those forming the walls of blood and lymph channels are always present. Additionally, mesenchymal cells, fibroblasts, and macrophages may occasionally be present,

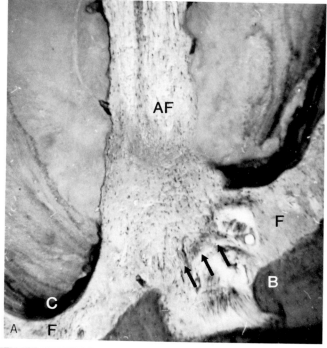

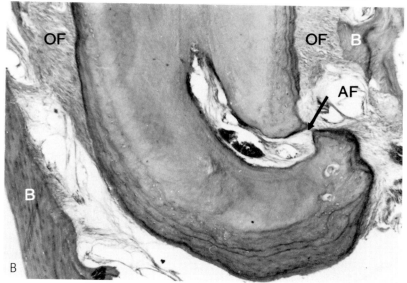

Fig. 7–4. Root apices. *A,* Root apex showing apical foramen (AF), apical fibers (F) attached to cellular cementum (C), and bone (B) forming the floor of the socket. Note that the blood vessels and nerves emerge from the floor of the socket to gain access to the periodontal ligament and root canal *(arrows). B,* Root tip showing lateral location of the apical foramen (AF), oblique fibers (OF), and bone (B). (Hematoxylin and eosin stain; ×50)

particularly as part of the capillary perithelium. These elements must pass through the zone of Weil to arrive in the odontoblastic and (pre)dentin layers. Nerve fibers are naked, having shed their sheaths in the cell-rich zone. Fibroblasts produce and maintain the fibrils of the

area. Mesenchymal cells, mostly in the form of pericytes, are pericapillary in location, and are reserve stem cells. The macrophages have a protective function.

Cell-Rich Zone. The region pulpward to the zone of Weil is richly populated with cells, and

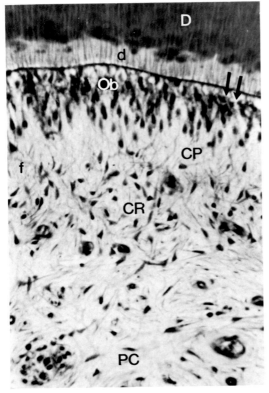

Fig. 7–5. Transverse section of pulp showing pulp core (PC), cell-rich zone (CR), cell-poor zone (CP), odontoblasts (Ob) and calcospherites in dentinoid (d) and dentin (D). The vacuolated area of the odontoblastic zone *(arrows)* may be artifactually produced (f, Raschkow's plexus). (Hematoxylin and eosin stain; × 320)

is designated as the cell-rich zone (Fig. 7–5). This layer may also be present in the root pulp, although here it is less prominent. The cell-rich zone is not always clearly demonstrated, even in the coronal pulp. In older teeth, which have fewer cells in the core, the cell-rich zone may be more distinct.

The prominence of this layer is not uniform throughout the pulp. Under special conditions, as in inflammation, the cell-rich zone and surrounding areas may become dominated by large numbers of defender cells that have infiltrated the area. The cell-rich zone accordingly becomes less defined. In the case of matrix synthesis, however, components of the cell-rich zone migrate toward the odontoblastic layer, and hence do not comprise a distinct layer. The components of the cell-rich zone are similar to those of adjacent regions.

Pulp Core. The central connective tissue mass

of the tooth, the pulp core (or pulp proper), contains most of the cellular elements, blood and lymphatic vessels, and nerves, which are invested in a mucoid ground substance containing a meshwork of collagen types I and III (reticular fibers) (Fig. 7–5).

Cells in the pulp are mostly fibroblasts. Mesenchymal cells, when present, are located in the perivascular tissue. The cytomorphologic features of these cell types are similar to those found in other loose connective tissues of the body. Normally there are few defense cells in the pulp; these include histiocytes, plasma cells, lymphocytes, eosinophils, and macrophages. When greater protection is required, defense cells increase in number. This occurs either by their migration from other areas of the body or by the differentiation of local stem cells. The cytomorphologic features of these cells are the same as those of other tissues.

Ground substances invests the components of pulpal connective tissue throughout the crown and root. The unformed amorphous components, in water suspension, contain chondroitin sulfate, glycoproteins, glycosaminoglycans, and hyaluronic acid. These provide for the diffusion of metabolites throughout the tissue, as well as to and from the blood and lymphatic channels. Pulps of the crown are more gelatinous than those of the roots.

Fibers, the formed intercellular elements, are arranged as a meshwork invested in the ground substance. The fibrils of collagen types I and III are found in respective proportions of 55:45. The latter fibrils are not clearly defined with optical microscopy when routine laboratory stains are employed, but they are readily seen with silver staining techniques. Thus, these are termed "argyrophilic" (reticular) fibrils. With electron microscopy they demonstrate the characteristic 64-nm cross banding; they are therefore collagenous, and are also referred to as type III collagen. Elastic and oxytalan fibers are also present. Oxytalan fibers, which may be visualized only with special staining procedures, are detected only in the dental papilla and only during the formative period of the pulp. If they are preelastic fibers, as suggested by some investigators, their disappearance is to be expected, because elastic fibers are not found in the pulp as separate entities. Rather, they are associated only with the

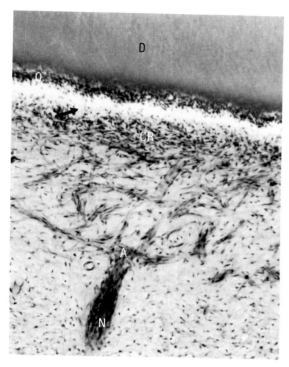

Fig. 7–6. Periphery of pulp sectioned obliquely and showing artery (A) with perivascular nerve bundle (N). Note the branching of the artery and the separation of nerves from the perivascular bundle. These participate in the formation of vascular and nerve plexuses in the subodontoblastic area (CR, cell-rich zone; cell-poor zone *(arrow)*; O, odontoblasts; D, dentin). Because of the angle of sectioning, the widths of these zones are distorted. (Hematoxylin and eosin stain; ×80)

walls of arterioles. The fiber content of the radicular pulp is greater than that of the coronal pulp. This is especially significant in the removal of pulps in which the more gelatinous fragile coronal pulps are removed by spoons or burs (pulpotomy), and the more fibrous apical pulps are removed by broaches (pulpectomy). Qualitative and quantitative changes in fiber features and population occur with age (see below, Age Changes).

Blood vessels, arterioles and venules, enter and leave the root canal system by the apical foramina and by accessory canals. It is interesting to note that, in the development of the root, both dentin and cementum accommodate the vascular channels, as in the formation of the root and accessory canals.

The arteriole(s), measuring up to 160 μm, enter the root canal. In the root they neither branch frequently nor form elaborate capillary

labyrinths, because the composition of radicular tissue is minimally cellular and mostly fibrillar. On entering the pulp chamber, however, their lumina increase in diameter and they branch repeatedly. Most of the arteriole branches travel toward the pulp margin, where they form a dense capillary network beneath the odontoblastic layer (Fig. 7–6). In the latter, the diameter of the capillaries ranges from 4 to 8 μm. Some of these pass through the cell-rich and cell-poor zones to loop around the cell bodies of the odontoblasts and return to join the venules of the subodontoblastic area.

Capillary beds in the pulp core, which accommodate its cells and nerves, are not as dense as those in the periphery. As a mucoid connective tissue, however, the pulp possesses an unusually rich blood supply and varied vessel types, including those of microcirculation (precapillaries and metarterioles). This is probably due to the high metabolic requirements of the odontoblastic layer, especially during dentin matrix synthesis. To this end, vessels looping around the odontoblastic cell bodies may possess fenestrated walls, thereby increasing the efficiency of these vascular channels. It is interesting to note that, with dentin deposition, the capillaries located in the predentin-odontoblast interface are shut down, and other capillary elements are established de novo in the area.

Venules drain the subodontoblastic and pulp core capillary plexuses. Venules merge with others while descending to the root canal, where they exit by the foramina as one or more larger venules. The largest venules are found in the basal segment (floor) of the pulp chamber. Arteriovenous anastomoses are also present as pulp components.

Lymph vessels are structurally indistinguishable from blood vessels microscopically because the capillaries and venules of the pulp are morphologically atypical. Some oral histologists believe that lymphatic channels are not a component of the pulp. Research studies employing perfusion, topical application, and injection techniques using radioisotopes and other recoverable tracer substances suggest, however, that lymph vessels are present. The vessels are believed to originate as cul-de-sacs, loops, or ballooned terminals in the odontoblastic area, and accompany the capillaries. Their exit paths from the coronal pulp correspond with those of the venules. According to some histologists,

lymphatic channels differ structurally from capillaries in that their lumina are wider, occasionally exhibiting serially arranged dilations and
constrictions. The endothelial cells are larger
and sometimes discontinuous. Similarly, the
basal lamina may be discontinuous and frequently absent. At the base of the socket, near
the apical foramen, lymph channels draining the
tooth meet and fuse with those draining the periodontal ligament.

The paths pursued by nerves and arterioles
are similar, and tend to branch synchronously.
Both arterioles and nerves may branch several
times in anticipation of entering the apical foramen. One of their branches veers laterally to
innervate and vascularize the fundus of the
socket. Others travel to and enter the apical foramen. Branching of the nerves and arterioles in
the root is infrequent, or rare.

Two organizational nerve units are found in
the pulp. The first is the typical nerve bundle,
or fascicle, which is composed of nerve fibers,
connectve tissue fibrils, Schwann cells, and mi-
crocirculatory vessels (Fig. 7–7). The second is
peculiar to the dental pulp in that the nerves
surround the arteriole in varying degrees, forming a type of perivascular neuroadventitia.

Myelinated and nonmyelinated nerves are
found in the pulp. These elements are structurally the same as those in other tissues. The nonmyelinated fibers stimulate the smooth muscles
of the blood vessels to contract, and thus control
the size of the vascular channels. Contracted
vessels, with their smaller lumina, reduce the
flow of blood. Nonmyelinated fibers may become separated from the nerve bundle associated with a given blood vessel to travel to the
muscle coat of another blood vessel, which they
will innervate. These nerve fibers end as very
small bead-shaped or thorn-shaped processes on
the surface of the smooth muscle cell.

Myelinated fibers are more numerous in the
pulp. To arrive at their final destination, the
marginal area of the pulp, the fibers fan out from
the parent bundles located in the pulp core. As
they approach the cell-poor zone, the myelin
sheath is shed. Each fiber then engages in a
series of branchings to produce a dense network
known as the plexus of Raschkow (Fig. 7–5).
This has been shown to develop posteruptively.
Some of the branches pass between the odontoblasts to enter the predentin; others extend a

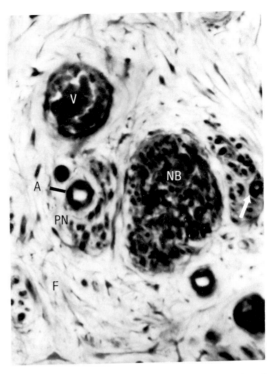

Fig. 7–7. Segment of pulp core showing nerve bundle
(NB), arteriole (A), perivascular neuroadventitia (PN), ven-
ule (V), and fibers (F). Vasa nervorum (blood vessels of
nerves) are indicated by the white arrow. (Hematoxylin and
eosin stained; ×320)

short distance into the dentin tubules with the
odontoblastic process. The two are in very close
proximity (about 20 nm), a factor possibly implicated in dentin sensitivity. Although nerve
fibers may generally be distinguished from the
odontoblastic process by the presence of abundant neurofilaments and neurotubules, a more
specific feature is the presence of vesicles and
mitochondria that are not normal constituents
of the odontoblastic process.

AGE CHANGES

Dimensional, as well as structural changes occur in the pulp throughout its life. The volume
of the pulpal tissue is greatest, and thus the
coronal chamber is largest, when the tooth assumes its occlusal position in the oral cavity.
Dentin formation is a continuous process, and
is accelerated during periods of increased stimulation. Dentin becomes progressively thicker
and the pulp chamber and connective tissue diminish, until the chamber is almost obliterated.

Although diminution of the volume and diameter of the root canals occurs, the proportions of reduction are rarely comparable to those of the coronal chamber. The progressive thickening of the dentin produces profound changes in the peripheral plexuses of blood vessels and nerves.

With the encroachment of the pulp by dentin, the odontoblasts are progressively reduced in number and the vascular supply in the area is commensurately decreased. That is, the capillaries "close down" or collapse, and new capillary plexuses are established beneath the relocated odontoblastic zone. The peripheral nerves experience a similar fate. The diminished general blood supply to the pulp influences metabolic exchanges, affecting the overall cell population. It has been suggested that the cell number is halved from the second to the seventh decades of life. The decrease in the number of cells results in increased fiber density, which may become excessive (Fig. 7–8).

CLINICAL CONSIDERATIONS

Although normal environmental stimuli produce changes in the coronal pulp, severe or acute stimulation of the pulp—for example, that produced by caries, cavity preparations, or in- jury—may cause atropic changes in the pulpal connective tissue.

A thorough knowledge of the morphology of the human dental pulp and root complex is required to perform both diagnostic and therapeutic procedures. Dental pulp morphology changes as a function of time, restorative procedures, tooth trauma, and other physicochemical tissue insults. Considerations of various entities affecting the coronal and radicular dental pulp will therefore be presented.

Secondary and Tertiary Dentin

As teeth are subjected to various restorative procedures, trauma, attrition, and aging, the pulp chamber and root canal systems become smaller. Figure 7–9 demonstrates a mandibular first molar of a young patient whose pulp chamber and root canal systems were quite large. In contrast, Figure 7–10 shows a mandibular first molar of an elderly patient, who had undergone extensive restorative procedures that resulted in a small pulp chamber and calcified root canals.

Deposition of secondary and tertiary dentin by odontoblasts presents specific clinical problems. Because the pulp chamber is considerably smaller, special care must be exercised in the initial step of root canal therapy to prevent compromising the pulp chamber floor. Because pul-

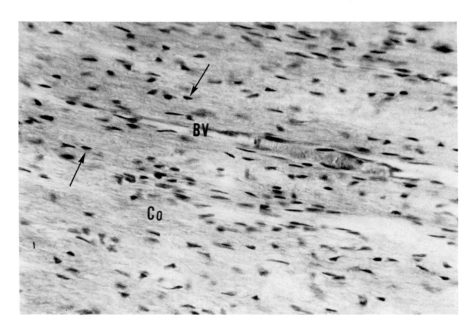

Fig. 7–8. Pulp fibrosis. Dominating the field are collagen fibers (Co). Also shown are fibroblast nuclei *(arrows)* and a blood vessel (BV). (Hematoxylin and eosin stain; ×320)

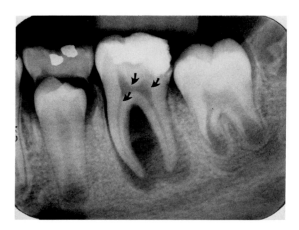

Fig. 7–9. Radiograph of a mandibular first molar in a young patient whose pulp chamber and root canal systems are quite large *(arrows)*.

pal tissue is minimal, the clinician must proceed with extreme caution to find the coronal soft tissue.

Furthermore, in endodontic therapy, highly calcified root canals are much more difficult to negotiate with files or cleaning instruments. If canals are completely calcified, instrumentation to the apices is impossible.

Pulp Stones

Pulp stones, calcific deposits of varying size, are located in the pulp chamber. They are potential problem sources for the clinician. Pulp stones can be seen radiographically as small multiple radiopaque areas within the pulp chamber (Fig. 7–11). Some are quite large, obliterating the pulp entirely (Fig. 7–12). For proper execution of root canal treatment, calcific deposits must be removed. Pulp stones may be so small that they obstruct the orifice of the root canal, thereby almost preventing access to and cleaning of the canal system.

Lateral Canals

These are 90° branches of the root canal, and are more prevalent in anterior teeth. Occasionally they can be demonstrated radiographically. If the dental pulp becomes necrotic lateral canals may be a source of infection, possibly leading to inflammation of the soft and hard tissues adjacent to the site at which they exit the tooth. Extensive involvement may spread to and result in bone destruction, which can sometimes be detected radiographically (Fig. 7–13). When root canal therapy has been completed, the canals are often filled with gutta percha, cement, or both, thereby resolving the pathosis associ-

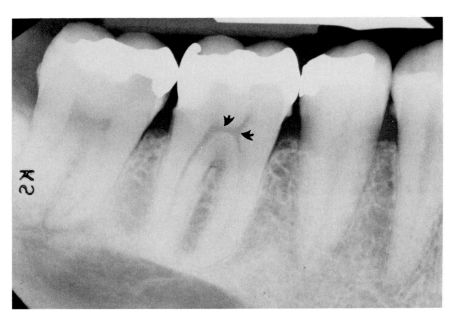

Fig. 7–10. Radiograph of a mandibular first molar from an elderly patient. Note the small pulp chamber and calcified root canals *(arrows)*.

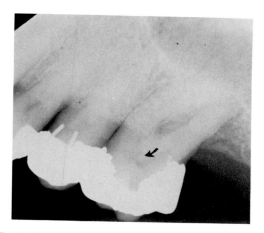

Fig. 7–11. Radiograph of a molar with pulp stones in the chamber *(arrow)*.

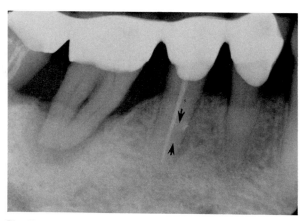

Fig. 7–13. Radiograph of a bicuspid tooth with lateral canals which have been filled with cement during the obturation of the root canal *(arrow)*. Note the radiolucency lateral to the canals which denotes bone destruction.

ated with the lateral canal. Because these canals are present in both a mesiodistal and bucco(facio)lingual plane, the canals may sometimes be detected radiographically (Figs. 7–14 and 7–15). Furthermore, periodontal procedures, including scaling and root planing, can affect direct communication between the oral environment and the pulpal tissue via the lateral canals. This may result in inflammation and eventual pulp necrosis, thereby necessitating root canal treatment or extraction of the affected tooth.

Furcation Canals

These lateral channels exit the pulp chamber by way of the tooth furcation (Fig. 7–16). They can vary in number from tooth to tooth, and are

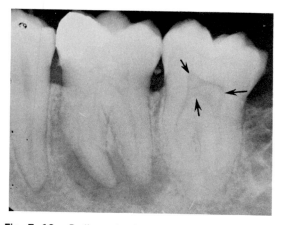

Fig. 7–12. Radiograph of a molar with multiple pulp stones. Note that the mineral deposits have obliterated most of the pulp chamber *(arrows)*.

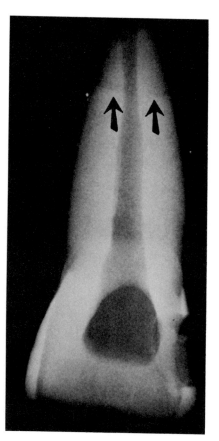

Fig. 7–14. Radiograph of a central incisor with multiple lateral canals *(arrows)*.

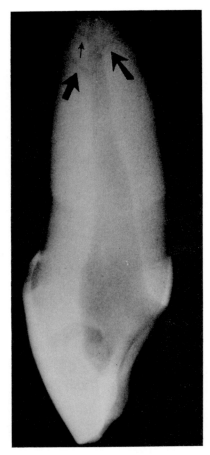

Fig. 7–15. Mesial view of tooth shown in Figure 7–14. Note that the lateral canals *(arrows)* are more distinct at this angle.

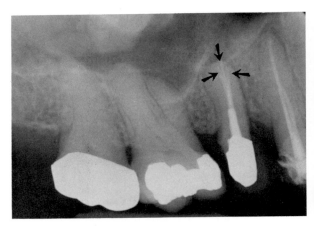

Fig. 7–17. Radiograph of an apical delta in a bicuspid. Note the three canals, which have been obturated with cement and/or gutta percha *(arrows)*.

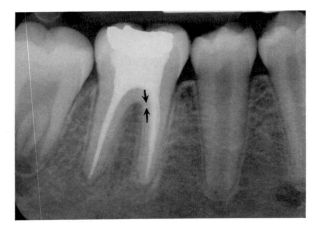

Fig. 7–16. Radiograph of a lower molar with a furcation canal that has been filled with cement *(arrows)*.

Fig. 7–18. Radiograph of a maxillary central incisor with a severe amount of internal resorption at the root apex *(arrows)*.

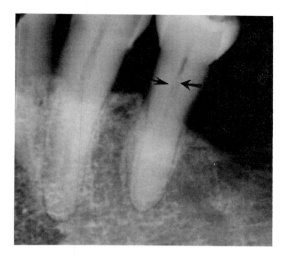

Fig. 7–19. Radiograph of a mandibular bicuspid, demonstrating a fast break. Note that the radiolucent canal becomes more radiopaque at the site at which the fast break occurs *(arrows)*.

most prevalent in primary molars and less frequently found in permanent molars. Most often the canals cannot be detected radiographically, unless root canal therapy has been completed and the furcation canal has been filled with a radiopaque material.

Apical Delta

In the apical segment of the root, the main root canal system sometimes branches into several channels. The branching is usually not at a right angle to the main canal system; rather, the angle is less acute (Fig. 7–17). This phenomenon occurs mostly in the mandibular premolars, and is referred to as an apical delta. The configuration presented in an apical segment of the root can present problems for the clinician because of the difficulty in cleaning and obturating these accessory canals adequately. Necrotic tissue remaining in the small channels represents a potential source of inflammation and abscess, and may result in continued discomfort and pain even after completion of root canal therapy.

Internal Resorption

If a pulp becomes irritated and inflamed secondary to restorative procedures or trauma to the tooth, internal resorption may ensue. Internal resorption is defined as a physiopathologic process that results in the loss of the internal dentinal wall. Because the resorptive process can proceed rapidly, treatment must be initiated

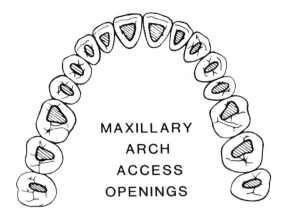

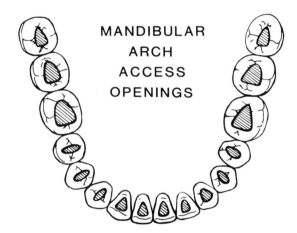

Fig. 7–20. Diagrammatic representation of the shape of the access openings in both the maxillary and mandibular arches.

immediately. Internal resorption may occur in the walls of both the pulp chamber and root canal (Fig. 7–18).

The process of internal resorption occurs only in vital teeth. If left untreated, it may eventually progress through the dentin and cementum, communicating with the periodontium. Root canal therapy is the preferred treatment for internal resorption, because removal of the inflamed pulpal tissue abates the resorptive process.

Fast Break

In some root canal systems the main canal may divide into two or more canals, sometimes referred to as a fast break (Fig. 7–19). These are generally found in the coronal two-thirds of the root canal system, although they may be more apically located. Such a division of the root canal is not to be confused with an apical delta. Rather, the fast break is a major branching of the main root canal into two (sometimes more) canals, in contradistinction to apical deltas, which are multiple minute apical channels of the main canal. Radiographically the main canal becomes more radiopaque at the site of division into the two canals, as indicated by the arrow in Figure 7–19. The clinician must be aware of the potential for splitting of the main canal, because successful endodontic treatment depends on the thorough cleaning and filling of all canals.

Pulp Chamber Morphology

The morphology of the pulp chamber varies with the specific tooth group. Both the size and morphology of the pulp chamber are important to the clinician, because these factors determine how and where access to the coronal soft tissue may be achieved for root canal therapy. Access to the pulp is usually accomplished by removal of either the lingual or occlusal tooth structure; thus, an understanding of the basic shape of the various pulp chambers is essential (Fig. 7–20).

Anterior teeth have pulp chambers of triangular shape that vary in size, a function of the physical dimension of the clinical crown. Bicuspid pulp chambers are typically ovoid. The pulp chamber of maxillary molars is triangular to ovoid in shape, while the mandibular molars have a triangular to square shape. Because many factors influence the size and shape of the pulp chamber, the clinician must be cognizant of variations and, therefore, determine access to the pulp chamber relative to both the clinical and radiographic presentations of the tooth.

CHAPTER 8

Attachment Apparatus

Four structures comprise the tooth's attachment apparatus: the cementum; ligaments (e.g., gingival, periodontal, and transseptal, or interdental); alveolar process; and attachment epithelium and its associated structures. These are collectively known as the periodontium. The tooth is secured in the alveolus (socket) by the periodontal ligament. Its collagen fiber bundles are inserted into the cementum of the tooth on the one side and into the bony lining (cribriform plate) of the alveolar process on the other. The fibers originating in the periodontal ligament and inserted in cementum and/or bone as Sharpey's fibers are *extrinsically* produced, as opposed to those synthesized by cementoblasts or osteoblasts in matrix formation and which are *intrinsically* produced. Additionally, the tooth is in intimate relationship with the gingiva and with adjacent teeth via separate ligaments—the gingival and transseptal, respectively. These structures are protectively sealed from the external environment (oral cavity) by the attachment epithelium (Fig. 8–1). This chapter will present a discussion of these major four parts.

CEMENTUM

The root is covered by cementum, a mineralized tissue similar to bone in chemical composition, organic framework, ground substance, crystal composition, development, and reorganizational potential.

Location

Cementum forms the surface of roots, extending from the cervix (cementoenamel junction) to their tip(s). Internally, cementum joins root dentin at the cementodentinal junction and, externally, it interfaces with the periodontal ligament. Cementum of the cervical third of the root is acellular (cell-free), while that of the apical two-thirds is mostly cellular.

Physicochemical Features

Cementum is more bone-like in that it is less mineralized and is softer and more radiolucent than other dental hard tissues. Cementum is more yellow than enamel, but paler and more transparent than dentin. Cellular cementum is more permeable than acellular cementum, probably because of its greater organic content. Calcification of cementum progresses with age and results in decreased tissue permeability.

Organic and inorganic components constitute cementum. Collagen fibers, the principal organic product, are cemented in the matrix by proteoglycans. The ratio of organic to inorganic material and water varies due to differences in age and in the proportion of acellular to cellular cementum, and to contamination of cementum samples by soft tissues, especially of the periodontium. Thus, values of the organic content (by weight) have been given as ranging from 23% to 26%. Mineral crystal apatite constitutes about 65% and water about 12% of the matrix (by weight). Crystals are seeded and grow on and between the collagen fibrils. Calcium and phosphorus are the dominant minerals, although trace amounts of copper, fluorine, iron, lead, magnesium, potassium, silicon, sodium, and zinc may also be present.

Functions

The most important function of cementum is that of securing the root in its socket, which is accomplished by providing anchor sites for the terminals of the collagen fiber bundles in the periodontal ligament. The terminals of the collagen fibers are inserted into the cribriform plate of the alveolar process and into cementum as

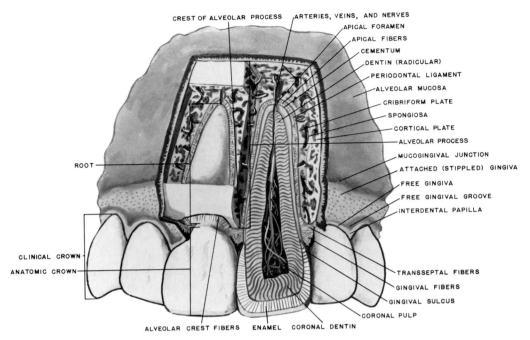

CREST OF ALVEOLAR PROCESS — ARTERIES, VEINS, AND NERVES
APICAL FORAMEN
APICAL FIBERS
CEMENTUM
DENTIN (RADICULAR)
PERIODONTAL LIGAMENT
ALVEOLAR MUCOSA
CRIBRIFORM PLATE
SPONGIOSA
CORTICAL PLATE
ALVEOLAR PROCESS
MUCOGINGIVAL JUNCTION
ATTACHED (STIPPLED) GINGIVA
FREE GINGIVA
FREE GINGIVAL GROOVE
INTERDENTAL PAPILLA

ROOT

CLINICAL CROWN
ANATOMIC CROWN

TRANSSEPTAL FIBERS
GINGIVAL FIBERS
GINGIVAL SULCUS
CORONAL PULP

ALVEOLAR CREST FIBERS   ENAMEL   CORONAL DENTIN

Fig. 8–1. Diagram of anterior teeth and supporting structures, showing the interrelationships of the dental hard and soft tissues associated with the attachment apparatus: alveolar process, cementum, periodontal ligament, gingival, and attachment epithelium at different sectional levels.

cemental Sharpey's fibers. Another function involves cementogenesis, which assists in the maintenance of occlusal functional relationships by apical deposition (see Figs. 4–35 and 7–4). Cementogenesis provides for fiber reattachment and relocation in mesial drifting and other tooth movements by surface deposition (See Fig. 8–30). This also serves to maintain the width of the periodontal ligament.

Other functions attributed to cementogenic activity include root repair of horizontal fractures, walling in of filled canals, isolating necrotic pulps by occluding the apical foramen, protecting underlying dentin, mediating the effect of osteoclasia, and stimulating osteogenesis.

Cementoenamel Junction Relationships

In most cases (60%), cementum overlaps enamel at the neck of the tooth. Simple butt association may occur (30% of cases) and, in some teeth (10%), the two tissues do not meet, leaving the dentin exposed.

More than one relationship may be present on a given tooth. Cementum overlapping enamel is a condition caused by a break in the continuity of the reduced enamel epithelium that overlies the cervicalmost enamel. The path

provided by the break allows for the passage of dental follicle components to reach the enamel, where cementoblasts are formed and deposit acellular cementum. This matrix, although devoid of cementocytes, is composed of collagen that exhibits periodicities. The second condition is one in which the tissue-producing sequence occurs uninterrupted and in order—that is, cessation of amelogenesis followed immediately by cementogenesis. The third and least frequently seen relationship, in which enamel and cementum are separated, is caused by a delay in the tissue-forming sequence. In this situation the root sheath maintains continuity with the cervicalmost segment of the reduced enamel epithelium and, although amelogenesis has ceased, cementogenesis is prevented because the root sheath remains adherent to the underlying dentin. Cementogenesis is initiated ultimately, somewhat removed from the enamel, and only after breaks occur in the root sheath. With tooth eruption and disintegration and removal of the successors of the reduced enamel epithelium, the cervicalmost dentin is left exposed and unprotected.

Another condition also caused by local degeneration of the reduced enamel epithelium and

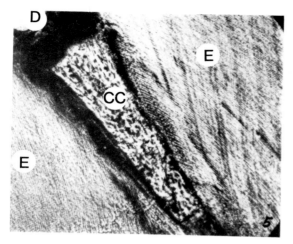

Fig. 8–2. Occlusal fissure containing coronal cementum (CC) and bordered by enamel (E). The coronal cementum is covered by debris (D). (Ground section) (Boyle, P.E.: Kronfeld's Histopathology of the Teeth and Their Surrounding Structures. 4th Ed. Philadelphia, Lea & Febiger, 1955.)

not limited to the cementoenamel junction is the formation of afibrillar cementum. This process also involves the migration of follicular connective tissue into the spaces created by the degenerated epithelium, or by discontinuities in the epithelium. Electron microscopic studies have shown that fibrillogenic cells pass through the gaps to enamel, where they synthesize an electron-dense fibrillar reticulum. The components of the latter lack 64-nm periodicity, and the tissue is therefore designated as afibrillar cementum. This tissue induces the differentiating cementoblasts to deposit a fibrillar matrix, covering the afibrillar cementum. The fibrils produced by the cementoblasts, however, are characteristically type I collagen. Afibrillar cementum may be found in occlusal fissures and in other areas of the crown in which discontinuities have occurred in the reduced enamel epithelium. At these sites, however, the term "coronal cementum" may be applied (Fig. 8–2).

Thickness

From the cementoenamel junction apicalward, the thickness of cementum increases. Because cementogenesis does not cease when the tooth arrives at its occlusal position, the greatest thickness is achieved in the apical third of the root. Thus, cementoblasts are active for the life of the tooth, especially those located more apicalward. It has been estimated that the width of

cementum from the first to the seventh decades of life increases threefold. The thickness of cementum at the tip of the root may be in excess of 700 μm (see Fig. 7–4). At root bifurcations cementum has been observed to be even thicker. Cementum is progressively thinner toward the cervix of the tooth, with thicknesses of 10 μm or less at the cementoenamel junction. Roots of impacted teeth are those that do not erupt to assume functional positions in the oral cavity. Because these teeth are in reduced function and anchoring tissues are not needed, they are covered only by an extremely thin layer of cementum.

Stages of Cementogenesis

The events in cementogenesis occur sequentially, and include the formation of discontinuities in Hertwig's root sheath, migration of dental follicle components to the adjacent radicular dentin, differentiation of cementoblasts, and cementum matrix production (involving fibrillogenesis, matrix maturation, and mineralization). As a result of these activities, three layers persist in the definitive tissue: the fibrillogenic layer of cementoblasts, the cementoid layer (mature matrix), and the cementum (calcified matrix). The dental follicle is converted into the periodontal ligament, and the components of Hertwig's epithelial root sheath disintegrate or reorganize into cell rests of Malassez.

Hertwig's Epithelial Root Sheath and the Dental Follicle

These structures bear special relationships in cementogenesis. For example, in anticipation of cementum formation, the connective tissue of the dental follicle increases in quantity at the level of the prospective cervix. Elongation and organizational changes of the cervical loop epithelium result in the conversion of the tissue into a bilaminar layer, the epithelial root sheath (Figs. 4–33 and 4–42). With dentin matrix mineralization and its shrinkage the cells of the root sheath pull away from the dentin, causing breaks and disrupting the continuity of the epithelial sheath. This discontinuity produces gaps and provides sites for the invasion of connective tissue elements from the dental follicle toward the dentin. These include vascular buds, their perivascular mesenchymal cells, and extrinsic collagenous elements. The mesenchymal cells differentiate to form a cementoblastic layer juxtaposed to the newly formed radicular dentin.

Cementoblasts

Cementoblasts are derived from mesenchymal cells, although it may be that, in the definitive tissue, especially during periods of stress, fibroblasts are mobilized and pressed into fibrillogenic activity. The cementoblastic layer may consist of one or many cell layers. In the former cells are usually rounded or cuboidal, while in the latter they are more squamoid (Figs. 8–3 and 8–4, also see Fig. 4–29). Numerous processes project from the cell body. In cells actively producing cementum matrix, the processes are fewer and longer. It has been suggested that the organelle-poor processes communicate with the processes of neighboring cementoblasts and young cementocytes, possibly via nexus.

Because the cell processes and their environment possess similar optical properties, they are not readily observed with ordinary light microscopic techniques. The cuboidal cells range in diameter from 8 to 12 μm, the nuclei are centrally located, and the cytoplasm is basophilic, indicating an abundance of rough endoplasmic reticulum (rER) (Fig. 8–4). Electron microscopy reveals an extensive network of rER, free ribosomes, Golgi apparatus, and numerous vesicular bodies (Fig. 8–5). Thus, ultrastructural features of the cementoblasts are similar to those of other fibrillogenic cells.

Cementoid (Precementum)

Cementoblasts produce intrinsic fibrils during the fibrillogenic stage and with the extrinsic Sharpey's fibers arriving from the presumptive periodontal ligament form the mixed fiber population of the cementum matrix. The fibrillar components synthesized by the cementoblasts exhibit the periodicity and dimensional characteristics of type I collagen. Cementoblasts additionally synthesize ground substance, which causes matrix maturation; the result is designated as cementoid, or precementum. Ground substance tends to mask the fibrils.

Thus, the fibrils in the first-formed cementum matrix are of two sources: those produced by fibroblast or mesenchymal cells in the dental follicle, and those produced in situ by cementoblasts. Development and maturation of the cementum matrix succeeds in pushing the epithelial cells of the root sheath further from the root. In the connective tissue of the future periodontal ligament, they are destroyed and removed or form epithelial cell rests (Fig. 8–6); also see Fig. 4–29). Additionally, the ends of the cemental Sharpey's fibers (bundles) from the maturing

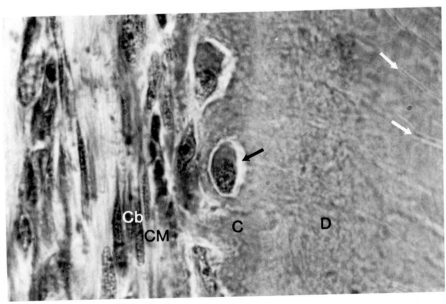

Fig. 8–3. Cementogenesis. The root segment shows dentin (D) and the dentinal tubules *(white arrows).* The external surface of the root shows a multicellular cementoblastic layer (Cb), developing cementum matrix (CM), and several trapped cementocytes (C). (Hematoxylin and eosin stain; ×800)

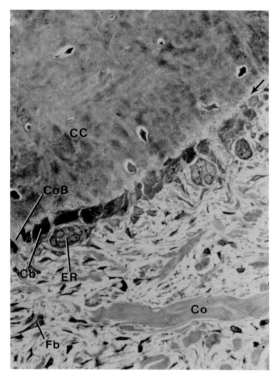

Fig. 8–4. Photomicrograph showing cellular cementum (CC), layer of precementum *(arrow)*, cuboidal cementoblast (Cb), epithelial rest of Malassez (ER), and collagen fiber bundle (CoB) inserted into the cementum matrix. The periodontal tissue shows secondary collagen fibers, as well as bundles (Co), fibroblasts (Fb), and other cells. (Toluidine blue stain; ×580) (Courtesy of L. McCleary.)

periodontal connective tissue are introduced in the developing matrix. With matrix calcification the collagen bundles form guy lines, thus stabilizing and anchoring the emergeing tooth in its developing alveolus (Figs. 8–4 and 8–5).

Precementum (cementoid) persists for the life of the tooth. As the most recently produced matrix, precementum separates the calcified matix from the periodontal tissue. With hematoxylin and eosin stain, cementoid is seen to be eosinophilic and nonmineralized, composed mostly of collagen fibrils masked by ground substance. The average width of the precementum layer is about 8 μm. Precementum is only a few microns thick toward the cervix, but it achieves maximum width in the apices. As a persisting layer, precementum also separates the cementoblasts from the calcified matrix (Fig. 8–7). Some histologists believe that the precemental layer tends to resist cementoclasia.

Mineralization of the matrix follows matrix maturation. The calcifying front may be several microns removed from the maturing matrix, particularly during accelerated growth periods. The needle-shaped and plate-shaped apatite crystals are smaller than those of other calcified dental tissues. The crystals are seeded on and grow parallel to the long axis of the fibrils. Thus, their orientational relationships to the fiber are similar to those found in bone and dentin. With matrix mineralization, the intrinsic fibrils are completely mineralized, while only the margins of the neighboring Sharpey's fibers are calcified; their centers remain noncalcified.

Types

There are two forms of cementum, cellular and acellular.

Acellular Cementum

The matrix produced for the upper third or half of the root does not contain cementocytes, and is known as acellular cementum (Fig. 8–8). Some oral histologists are of the opinion that the extrinsic fibers form the principal fiber population of acellular cementum and that here they are impregnated with apatite crystals throughout. The remaining one-half to two-thirds of the root, except for the initial layer forming the cementodentin junction, is cellular (Fig. 8–9). Acellular lamellae of cementum may be found intervening between cellular ones. These represent periods during which cementogenesis oc-

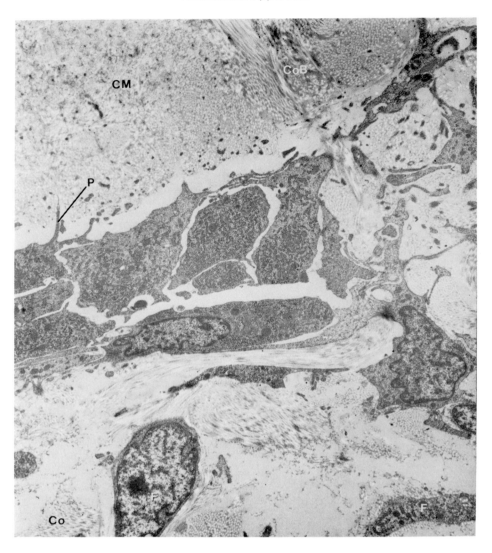

Fig. 8–5. Electron micrograph of multicellular cementoblastic layer. The cytoplasm is densely packed with organelles, including a labyrinth of rough-surfaced endoplasmic reticulum. Cementoblastic processes (P) extend into the matrix (CM), as do cemental Sharpey's fiber bundles (CoB). The latter are mostly oriented perpendicular to the cementum matrix fibrils. Within the periodontal space are fibroblasts (F) and collagen (Co). (Uranyl acetate and lead citrate stain; ×7,500) (Courtesy of L. McCleary.)

curred less rapidly, thus providing time for cementoblast retreat.

Cementum or cementum-like tissue may occur at sites other than the root surface. Such aberrant deposits are discussed elsewhere as afibrillar cementum.

Cellular Cementum

This tissue represents periods in which matrix maturation and mineralization occurred in rapid succession, or almost simultaneously. With the possible exception of the root tips of aged teeth,

the cementum adjacent to dentin, is acellular for the length of the root. Toward the apices, where the cementum is thickest, lamellae are produced at varying rates. Thus, the number of cementocytes per lamella may range from none (acellular) to many (cellular) (Fig. 8–9).

Cellular cementum differs from acellular cementum in that it is found mostly on the apical two-thirds of the root, its thickness is greater, its distribution is more variable, and its lamellae are wider. Cementocytes in some lamellae are closely packed, while in others they are few and

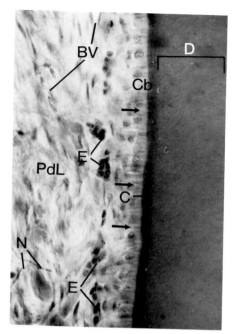

Fig. 8-6. Section of attachment apparatus showing dentin (D), acellular cementum (C), a monolayer of cuboidal cementoblasts (Cb), and periodontal ligament (PdL), containing blood vessels (BV), nerves (N), and an assortment of cells, including fibroblasts and epithelial rests (E). Note the passage of collagen fibers *(arrows)* from the periodontal ligament through the cementoblastic layer for insertion into the cementum. (Hematoxylin and eosin stain; ×320)

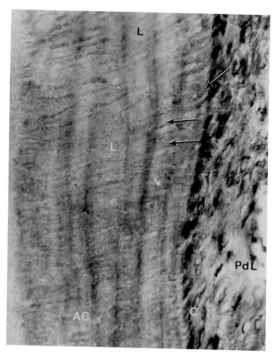

Fig. 8-8. Collagen fibers *(arrows)* from the periodontal ligament (PdL) pass between cementoblasts (C) and become perpendicularly inserted into the lamellae (L) of acellular cementum (AC). (Hematoxylin and eosin stain; ×320)

widely separated. Cementum that has been produced rapidly exhibits wider lamellae and greater cementocyte populations (Fig. 8-8).

Cementocytes. The size and shape of cementocytes vary considerably (squamoid, oval, and round). The central cell masses are contained in lacunae that range in diameter from 8 to 15 μm. Optical microscopy reveals the cytoplasm to be basophilic. The nucleus is large and eccentric, occupying most of the cell.

Electron micrographs of young cementocytes reveal their ultrastructural features to be similar to those of osteocytes. Cementocytes distantly removed from the vascular supply, however, exhibit decreased organelle populations. These observations correlate well with data derived from enzyme studies (see below). Cytoplasmic processes extend from the cell body into canaliculi. The processes are about 1 μm in diameter and may be as long as 15 μm. In a single plane, 30 or more processes may be observed. Although cementocyte cell bodies occupy most of the lacunar spaces, it is doubtful whether their processes remain in the canaliculi. Rather, they are probably withdrawn or contracted into the la-

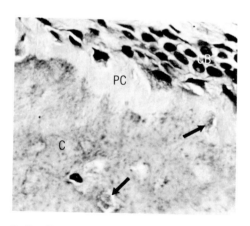

Fig. 8-7. Root segment showing multicellular cementoblastic layer (CB), cementoid/precementum (PC), and cellular cementum (C). Note cementocytes in the lacunae *(arrows)*. (×500) (Mjör, I.A., and Fejerskov, O.: Histology of the Human Tooth. 2nd Ed. Copenhagen, Scandinavian University Books, 1979.)

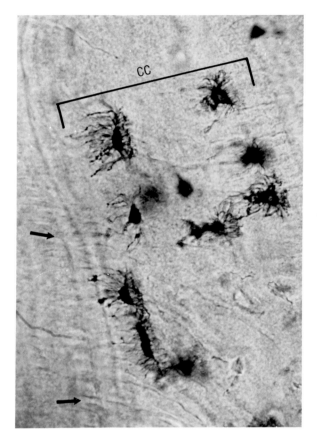

Fig. 8—9. Ground section of root segment taken near the apex, showing lamella of cellular cementum (CC) between lamellae of acellular cementum. Insertion of collagen fiber bundles is indicated by arrows. Note the differences in the shape of the lacunae and the direction of the canaliculi. The latter are directed toward the periodontal ligament (not shown), which is the nutritional source. (×320)

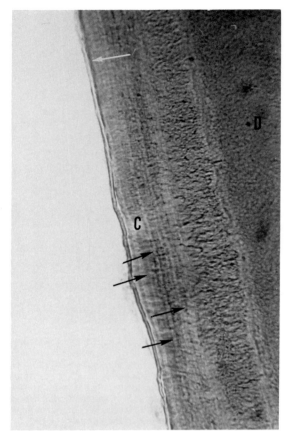

Fig. 8—10. Ground section of acellular cementum (C) and dentin (D). Incremental lines of the outer lamella are indicated by black arrows. A resting line is shown near the root surface *(white arrow).* (×430)

cunae (Fig. 8–3). Canaliculi, either empty or occupied, function as conduits through which materials may circulate. Some of the canaliculi are directed toward the dentin, but most are oriented toward the periodontal tissue. The latter is the vascular source that accommodates the metabolic needs of cementum.

Studies of enzyme activity, considered to be an index of metabolic activity, reveal that the cementoblasts exhibit the most vigorous action, becoming less active as young cementocytes (nearer the precementum) and least active in the older cementocytes (nearer the dentin). The latter are often moribund, and eventually die and disintegrate to leave empty lacunae. It has been estimated that only those cementocytes less than 100 μm from the vascular supply in the perio-

dontal ligament remain viable. Those further removed exhibit distinct signs of degeneration.

Growth Patterns

Cementum matrix, especially that produced during eruption, is deposited in two planes—apically from the cementoenamel junction toward the prospective fundus of the socket, and laterally from the dentin toward the periodontal tissue.

Incremental Lines

Cyclic activity of cementogenesis is revealed as incremental or imbrication lines which conform to the contour of the root (Fig. 8–10). These appear as very fine dark lines bordering the wider light bands. Microradiographs of the incremental lines appear as fine, light, x-ray-dense lines between the darker bands. The former are thought to be areas comprised mostly of ground substance (thus, they are fibril-poor), while the matrix of the darker bands is fibrillar.

Lamellae

Groups of cyclic increments constitute lamellae (Fig. 8–10). Their widths are not uniform in a given root; rather, they are influenced by the intensity, focus, and duration of the cementogenic stimuli. Acute stimuli induce wide, cementocyte-rich lamellae, while weak stimuli promote the production of narrow, cementocyte-poor or cementocyte-free lamellae. In the former, more than one cementocyte may occupy a lacuna.

Cementum production is characterized by intervals of inactivity of indefinite duration ("rest periods"). These periods of synthesis abatement are recorded in the cementum as somewhat straight lines known as resting lines (Fig. 8–10).

As the lamellae increase in number, cementum appropriates more and more of the territory of the periodontal ligament. This occasions the insertion of a greater number of fibers into the more recently deposited cementum. The older, more deeply attached fibers tend to become mineralized, and are incorporated into the matrix. Thus, the uncalcified fibers of the surface are the more functionally competent ones.

Intermediate Cementum

Between the granular layer of Tomes and the acellular cementum in the apical one-half to two-thirds of premolar and molar roots may be found a tissue that, although calcified, does not bear the morphologic characteristics of either dentin or cementum. Because of its location it is called intermediate cementum (Fig. 8–11). Some believe that this tissue represents premature breaks in the root sheath and the passage of dental follicle connective tissue toward the dentin. This connective tissue eventually becomes isolated, trapped, and calcified in the area between dentin and acellular cementum. Others believe that this tissue is composed of root sheath epithelium that has not been pushed into and relocated in the periodontal tissue. The epithelial root sheath components are isolated and become trapped during synthesis and calcification of the cementum matrix. Intermediate cementum occurs only in patches, and is never lamellar.

Disorders

There are a number of conditions that may affect the cementum adversely, such as cementoclasia, hypercementosis, and aberrant cementum.

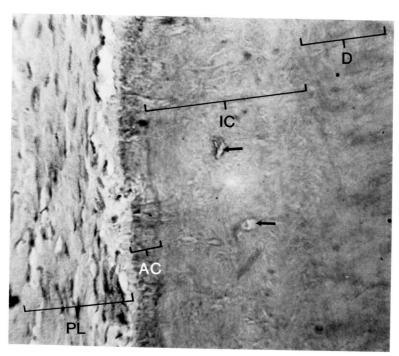

Fig. 8–11. Intermediate cementum (IC) bordered externally by acellular cementum (AC) and internally by dentin (D). Arrows indicate the cellular elements in the intermediate cementum (PL, periodontal ligament). (Hematoxylin and eosin stained; ×430)

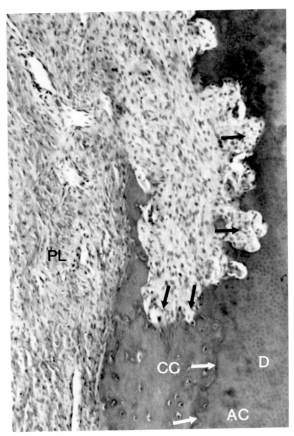

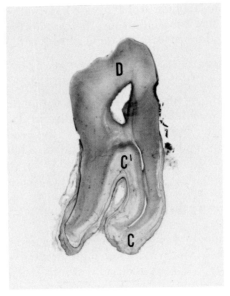

Fig. 8–13. Section of molar showing excessive deposit of cementum on the roots, especially on the apices (C) and bifurcation (C'); (D, dentin).

Fig. 8–12. Odontoclasia (cementoclasia, dentinoclasia), with Howship's lacunae *(arrows)* in dentin and cementum. Note that some of the lacunae contain "clast" cells, (CC, cellular cementum; D, dentin; PL, periodontal ligament; (AC, acellular cementum; resorption line, *white arrows*). (Hematoxylin and eosin stain; ×50)

Cementoclasia

Except for root resorption activities in the exfoliation of primary teeth, cementoclasia is not considered to be of normal occurrence. Rather, it results from continuously insulting stimuli such as those produced by certain pathologic conditions (e.g., cysts, infections, tumors), extraordinary masticatory forces, or inordinate pressures of orthodontic appliances. When the stimuli are acutely persistent, even the dentin may be resorbed (Fig. 8–12). The mechanism involved in cementum resorption is similar to that of resorption of other hard tissues. The resorbed surface of the root exhibits excavated bays, called Howship's lacunae, that are occasionally occupied by cementoclasts (Fig. 8–12). The cytomorphologic features of cementoclasts

are also similar to those described for osteoclasts.

Removal of the stimuli results in the cessation of resorption, disappearance of cementoclasts, reappearance of cementoblasts, and deposition of cementum, which is usually of the cellular variety. In hematoxylin and eosin stained sections the resorption boundary is marked by an intense blue line of irregular contour, known as the resorption line, which indicates the path of resorption (Fig. 8–12).

Hypercementosis

The condition characterized by extraordinary thickness of cementum is known as hypercementosis, or cementum hyperplasia (Fig. 8–13). It may be restricted to a small area, or it may involve the entire root(s). Hypercementosis is often observed on chronically inflamed root apices to form a collar-like band. Apices of teeth lacking antagonists may also exhibit hypercementosis. Hypercementosis may take the form of spikes or spurs, spheres, collars, or ledges. Roots subjected to extraordinary activity may develop cervically oriented spurs, which are believed to provide more surface area for fiber attachment.

Aberrant Cementum

Unusual structures or tissues occurring in unusual locations and form are said to be aberrant. Cementogenesis may occur in unusual places, producing aberrant cementum. Two atypical locations for cementum are the crown, where it is known as afibrillar or coronal cementum, and the periodontal ligament, where it is known as cementicle.

Coronal (afibrillar) cementum may be located aberrantly in the neck of the tooth or in occlusal fissures (Fig. 8–2). At these sites, breaks in the reduced enamel epithelium induce cementogenic activity. Cells and fibrils from the adjacent connective tissue pass through the discontinuities to arrive at the enamel, where mesenchymal or similar fibrillogenic cells synthesize and secrete material, which in the ground substance forms a fibril matrix. These fibrils do not bear the 64-nm markings characteristic of collagen; thus, this coronal cementum is also designated as afibrillar. Although it may occur in occlusal fissures, coronal (afibrillar) cementum occurs mostly in the cervical region of the tooth.

Cementicles are calcific bodies located in the periodontal ligament. The materials acting as the primary centers of calcification include degenerating cell remnants of Hertwig's epithelial rooth sheath and phleboliths (vein stones). These are treated as foreign bodies, and are isolated by calcification. Cementicles may be diversely located in the periodontal ligament, and are rarely larger than 0.2 mm. As free bodies, cementicles are located totally in the periodontal tissue (8–14A). As cementum is added to the surface of the root, and encroachment on the periodontal tissue occurs, cementum grows closer and eventually fuses with the free cementicles to form attached cementicles (Fig. 8–14B). With continued cementum apposition the attached cementicles are incorporated totally into the cementum, and are known as interstitial cementicles.

ALVEOLAR PROCESS

The bony extensions of the body of the maxilla or mandible, which form crypts in which roots are anchored, are known as alveolar processes (Fig 8–1). Roots, the periodontal ligament, and alveoli form a synarthrosis, or immovable joint.

Functions

The primary function of the alveolar process is the formation of bony sockets for root anchorage. Of equal importance is its protective function for soft tissues (e.g., nerves, arteries, veins, lymphatic vessels) that travel through and exit the bone. In their crestward path to and from the subgingival connective tissue, branches of the nerves, arterioles, venules, and lymphatics, which accommodate various levels of the periodontal ligament, exit and enter the cribriform plate of the process. The soft tissues in the marrow spaces of the alveolar process serve a hematopoietic function. The bony processes also function as repositories for calcium salts. Additionally, haversian systems "opened" by osteoclasia release the soft tissue of their canals into the periodontal ligament. This tissue acts as reserve tissue for the spent indefinite (loose connective) spaces of the periodontal ligament. Finally, the alveolar processes contribute to the esthetic features of the face.

Physicochemical Properties

Living bone is pink in color; it is composed of 21% organic matter, 71% inorganic material, and 8% water. The organic material is responsible for its tough, elastic, and resilient properties. The organic portion, as in cementum, is comprised mostly of collagen, embedded in a ground substance of glycoproteins and proteoglycans. When the mineral salts are leached from bone by acid treatment it loses its hardness, but the morphologic and anatomic details are retained by the organic components. The organic bone residue, however, is leather-like and flexible. The inorganic portion, which provides bone with its rigidity, is present as crystalline apatite, which is made up of 85% calcium phosphate, 10% calcium carbonate, and 5% other mineral salts.

Development

Development of the skeletal structures (maxilla, premaxilla, and mandible) and their associated facial primordia and dental and vestibular laminae is initiated about the same time, when the embryo is 7 weeks of age and its crown-to-rump length is between 16 and 20 mm. Development is completed during adolescence, when the maxilla and mandible have attained their

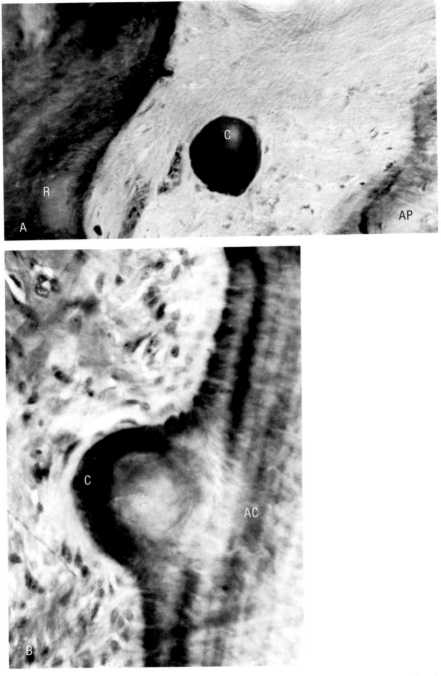

Fig. 8–14. Cementicles. *A,* Free cementicle (C) in periodontal ligament. Cell rests of Malassez are located between the cementicle and the root (R), (AP, bone of alveolar process). (Hematoxylin and eosin stain; ×120) *B,* Attached cementicle (C). Note that collagen fibers of the periodontal ligament are inserted into the lamellae of acellular cementum (AC). (Hematoxylin and eosin stain; ×320)

definitive dimensions. Growth of the alveolar processes and roots occurs simultaneously—that is, with the completion of anatomic crowns. Synchronous growth of the roots, their bony sockets, and connecting periodontal ligament continues until the tooth acquires its functional position, meeting its antogonist. The alveolar processes are of intramembranous bone development that require a mesenchymal environment. Spicules of bone develop, grow, and fuse with others to form an elaborate network (see Figs. 2–45, 4–8, and 4–10). These are not permanent bone structures, because alterations in functional stresses induce adaptive reorientation, or remodeling, through the interaction of osteoclasia and osteogenesis. Remodeling is a natural process, which occurs throughout life in most bones. This activity is especially important in the alveolar process during periods when primary teeth are shed and replaced by permanent ones. The alveolar processes, with their associated roots, are resorbed, and new bony processes are produced to form crypts accommodating the larger, longer roots of their permanent successors.

Structure

During the development of the alveolar process, two tables of compact bone with an intervening diploe of spongy bone are produced (Fig. 8–1). The outer tables are found on the vestibular and lingual aspects, and the inner table forms the wall of the socket. The outer tables are designated as cortical plates and the inner one as the cribriform plate. The body beams (trabeculae, or trajectories) make up the spongiosa, which constitute the diploe or core of membrane bone. The trabeculae are the first to be produced, and the compact tables are formed last.

Roots of teeth are separated from those of adjacent teeth by spongy bone partitions and by the appropriate bony plates. The partitioning bones are called interdental septae (Figs. 8–1 and 8–15A). Roots of multirooted teeth are also separated by a bony partition, the interradicular septum, and they are comprised only of the cribriform plate, which encases trabeculae of spongy bone. Thus, each root possesses a bony socket and anchoring tissue (periodontal ligament) (Fig. 8–15).

Cortical Plate

The cortical plate is composed of compact bone. Its lamellar systems include outer or periosteal, inner or endosteal, haversian, and interstitial lamellae. The lamellae of the first two are parallel to the long axis of the alveolar process (see Fig. 2–52). The haversian systems have no definite orientation and the interstitial lamellae, which are remnants of all lamellar systems, show definite arrangement only if they are of periosteal or endosteal lamellar origins.

The thickness of the cortical plate varies with the arch involved, the position on the arch, and the cortical plate involved (i.e., vestibular or oral) (Fig. 8–16). For example, the cortical plates of the mandibular arch are thicker than those of the maxillary arch. Cortical plates of the molars are thicker than those of the bicuspids, and those of the latter are thicker than those of incisors. The processes of incisors may be so thin that they consist only of a narrow layer of compact bone (see Fig. 4–32). The oral cortical plates are thicker than the vestibular ones.

The vestibular cortical plate of the maxillary arch contains many perforations representing the external openings of the Volkmann's canals, which permit nerves and blood and lymphatic vessels to enter and leave the alveolar process. The perforations of the mandibular cortical plate, although fewer, are larger.

Cribriform Plate

This bone forms the socket wall, and it is sometimes known as the alveolar bone proper. The name "cribriform" (sieve-like) is used because the bone is perforated by numerous Volkmann's canals, through which nerves and blood and lymph vessels travel to gain access to the various levels of the periodontal ligament (Fig. 8–17).

Endosteal lamellae of the cribriform plate are arranged in layers that conform to the shape of the adjacent marrow spaces. Other lamellae belong to haversian systems, or to their remnants. The outer or periosteal lamellae, which face the periodontal ligament, are those into which the principal fiber bundles (Sharpey's fibers) are inserted (Fig. 8–18). Because of the great number of Sharpey's fibers and because the lamellae are less conspicuous, this bone is also referred to as bundle bone.

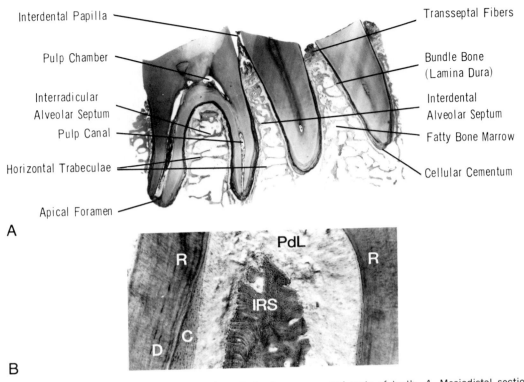

Interdental Papilla

Pulp Chamber

Interradicular
Alveolar Septum

Pulp Canal

Horizontal Trabeculae

Apical Foramen

Transseptal Fibers

Bundle Bone
(Lamina Dura)

Interdental
Alveolar Septum

Fatty Bone Marrow

Cellular Cementum

A

PdL

R

R

IRS

C

D

B

Fig. 8–15. Micrographs showing interrelationships of alveolar process and roots of teeth. *A,* Mesiodistal section of mandibular segment of the alveolar process from the premolar and first molar areas showing trabecular patterns of interdental and interradicular septa. (Decalcified, hematoxylin and eosin stain; enlarged from ×3) *B,* Interradicular septum (IRS) separating the two roots (R), (PdL, periodontal ligament; C, cellular cementum; D, dentin). (Hematoxylin and eosin stain; ×30)

Bundle bone exhibits additional matrix features, including fibril orientation, fibril quantity, mineral content, patency of lamellae, and roentgenographic features. The matrix fibrils are fewer than in other bone, and they are oriented at right angles to the Sharpey's fibers. These two features cause the lamellae to be less distinct. Bundle bone lacks trabeculae and their intervening marrow (spongiosa); hence it is composed only of compact lamellae, which render this tissue more radiopaque. In the presence of large amounts of this tissue, a distinct layer referred to as the lamina dura (dense or hard layer) may be seen roentgenographically (Fig. 8–19).

In many areas of the cribriform plate bundle bone may be the only bone present. On the other hand, bone is constantly undergoing change, as evidenced by the presence of resting and resorption lines (Fig. 8–20). (The factors inducing changes in the cribriform plate will be presented later in this chapter.) Occasionally, resorption of the cribriform plate may advance

to involve the haversian systems (Figs. 8–20 and 8–21). Some investigators believe that "opening" haversian systems provide the periodontal ligament with the loose connective tissue for the interstitial spaces (see below).

Spongiosa

Except in extremely thin alveolar processes, such as those of incisors, spongy bone is always present to some degree (Fig. 8–22). In some posterior teeth trabeculae may be found around the entire process. In others they may be found only on the lingual or only on the vestibular side. If the tooth is inclined lingually, the spongiosa may be reduced or absent. If it is tilted toward the cheek or lip, that aspect of the process will exhibit few or no trabeculae. The spongy appearance is due to the numerous marrow spaces formed by the trabecular network.

Depending on age, the marrow spaces may be filled with red or yellow marrow. In the very young the marrow is red because it is a hema-

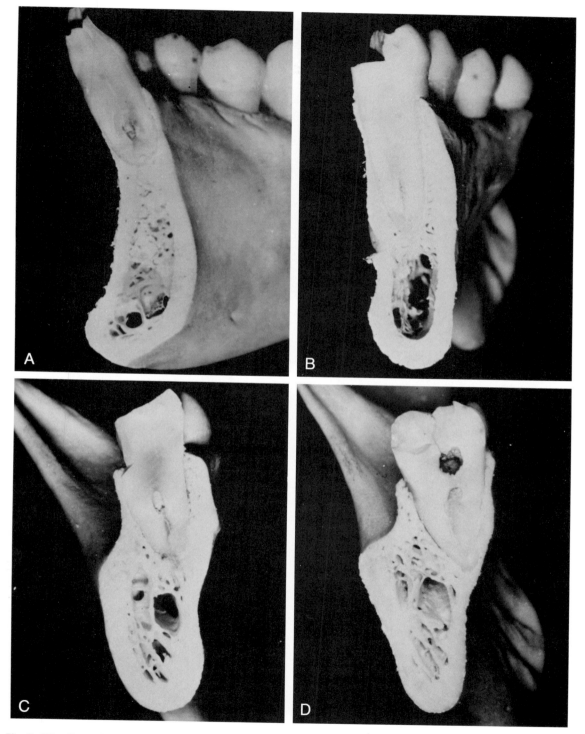

Fig. 8–16. Buccolingual sections of the mandible and maxilla, showing trabeculation. Comparison of the thicknesses of the alveolar bone can be made. Mandibular arch: lateral incisor *(A)*; second bicuspid *(B)*; second molar *(C)*; third molar *(D)*. Maxillary arch: lateral incisor *(E)*; second molar *(F)*. (Courtesy of C.T. Middleton, D.M.D.)

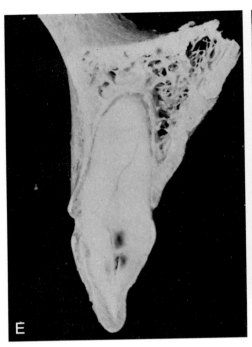

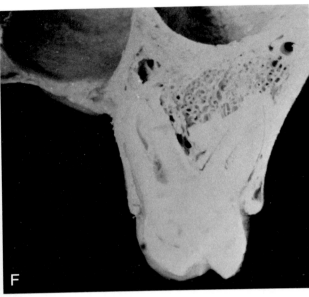

Fig. 8–16 (continued).

topoietic tissue producing erythrocytes, platelets, and granular leukocytes. In older persons
the tissue is converted to fat, and the marrow is
thus yellow (Fig. 8–20).

Spicules of embryonic and early development
are mostly thin, spike-shaped bone segments.
Prenatally these are nonlamellar but, with age,
the young spicules are removed and replaced by
mature ones, which exhibit lamellar systems. A
paucity of spicules is compensated for by either
an increase in spicule thickness or by an increase
in the thickness of the adjacent cortical plates.
With additional stresses the size of these bone
structures increases, and their orientation is
changed. Spicules might then be designated as
trabeculae. The latter, formed prenatally, are
also nonlamellar.

Trabeculae are larger beams of bone that form
supporting frameworks in alveolar processes
called trajectories. The size, shape, and orientation of the trajectories vary, depending on the
functional forces they accommodate. The common forms of trajectories include flat plates,
straight and curved bars, and tubules.

Two systems of trajectories are found in the
bones of the jaws, those of the body of the mandible and maxilla and those of their processes.

Trajectory arrangements of the bodies of these
bones are governed by masticatory muscles. Trabecular orientation for the alveolar processes is
influenced by the functional activity of the teeth.
Stimuli are transmitted from the cementum
across the principal fiber groups of the periodontal ligament to the cribriform plate.

Trabeculae of the alveolar process extend from
the cribriform to the cortical plate (Fig. 8–22).
They span the space between cribriform plates
in the interdental septae. The horizontally arranged trabeculae often possess supporting
branches, which are disposed obliquely or vertically. These branches provide additional
strength and support for the horizontal trajectories. The distribution and alignment of the trabeculae in the interradicular septae are essentially the same as for those on other parts of the
process.

The cribriform plate experiences remodeling
changes consistent with the varied changing
forces of the stomatognathic apparatus. These
involve osteoclasia and osteogenesis. To mediate
altered forces, osteoclasts effect erosion (resorption) of lamellae, thus producing concavities
known as Howship's lacunae (absorption/resorp-
tion concavities or bays). Having removed the

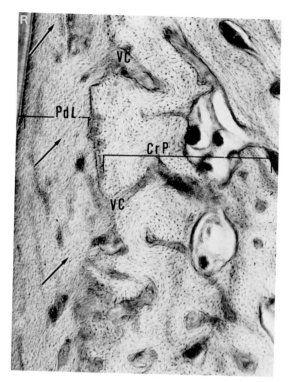

Fig. 8–17. Cribriform plate (CrP) of the alveolar process. The many openings provide for the entrance and exit of blood, lymph, and nerve elements to and from the periodontal ligament. The openings lead into Volkmann canals (VC), which communicate with haversian canals. Shown are principal fibers, arrows; PdL, periodontal ligament; R, root. (Hematoxylin and eosin stain; ×50)

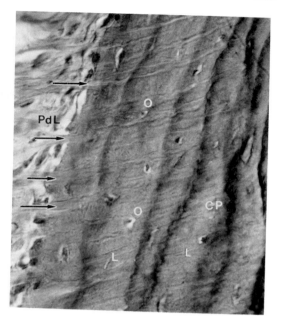

Fig. 8–18. Cribriform plate (CP) showing lamellae (L) with osteocytes (O) and insertion of Sharpey's fibers *(arrows)* from the principal fiber bundles of the periodontal ligament (PdL). (Hematoxylin and eosin stain; ×320)

bone to accommodate the new forces, osteoclasts are replaced by osteoblasts, which line the resorption concavities to deposit osteoid, thus reversing the process of bone removal for bone deposition. The site at which the reversal process occurs is known as the reversal zone and is delineated by a pronounced cementing line (Fig. 8–20). The interface between calcified bone and osteoid is called the frontier line or mineralizing front.

LIGAMENTS OF THE PERIODONTIUM

Fibers of the periodontium are contained in three ligaments the gingival, interdental (transseptal), and periodontal. Some histologists classify the transseptal fibers with the gingival group. Delineation between ligaments is not determined clearly, and the course of the fiber path must be examined to ascertain their classification.

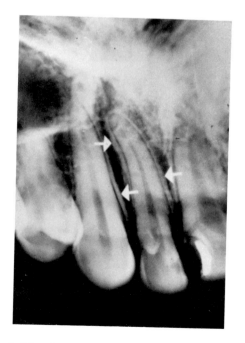

Fig. 8–19. Roentgenograph of lamina dura *(white arrows)*.

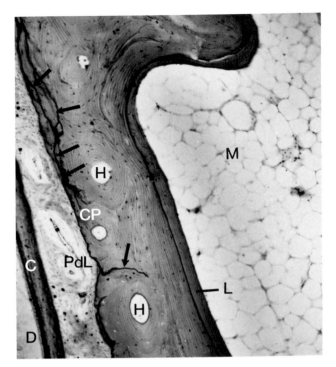

Fig. 8–20. Segment of attachment apparatus, showing radicular dentin (D), cementum (C), and periodontal ligament (PdL), with interstitial space and cribriform plate (CP). The latter exhibits osteones (H), interstitial lamellae, and remodeling reversal sites marked by cementing lines *(black arrows).* Also shown are endosteal lamellae (L) and yellow marrow (M), consisting mostly of fat cells. (Hematoxylin and eosin stain; ×80)

Development

The periodontal and other ligaments, in fact all the structures associated with the attachment apparatus, exclusive of the attachment epithelium, are derived from ectomesenchyme of the dental follicle (see Fig. 4–10 and Table 2–1). As the soft tissue is condensed by the growth of the alveolar process on the one side of the socket and by cementum on the other, the character of the periodontal tissue changes from loose (areolar; see Fig. 4–27) to dense, irregularly arranged fibrous connective tissue (see Fig. 4–27). Finally, when the tooth arrives at its occlusal position and becomes fully functional, the tissue is organized into three ligaments (regularly arranged dense fibrous tissue), the gingival, transseptal, and periodontal. The gingival ligament

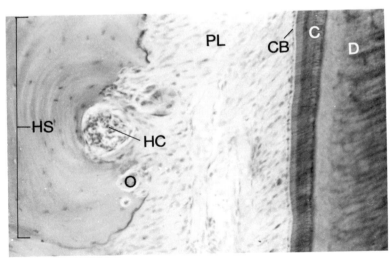

Fig. 8–21. Opening of a haversian system, (HS), releasing the loose connective of its haversian canal (HC) into the periodontal ligament (PL) to form an interstitial space (CB, cementoblast; O, osteoclast in Howship's lacuna; C, cementum; D, dentin). (Hematoxylin and eosin stain; enlarged from ×100) (Courtesy of Zander.)

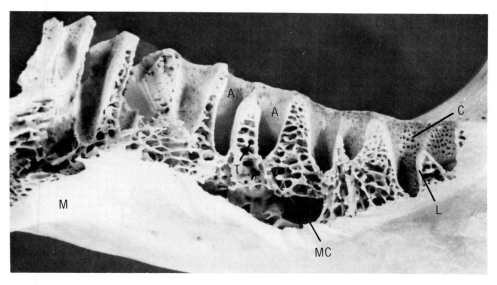

Fig. 8—22. Sagittal section from cuspid through third molar region showing the interdental and interradicular septa. The porous cribriform plate (C) is clearly shown. Note the trabecular pattern in the various regions of this section, and also the decreased size of absence of the spongiosa in the incisor-bearing segments of the mandible (A, aveolar sockets; T, trabeculae of interradicular process; M, medial mandibular surface; L, bundle bone; MC, mandibular canal). (Enlarged from ×3)

is comprised of four fiber bundle groups, the transseptal (interdental) fiber ligament consists of only one fiber group, and the periodontal ligament consist of five fiber groups (see below, Principal Fiber Groups). Complete organization of the principal fiber bundles, as functional groups, progresses from the cervix of the tooth to its apex. Thus, the first fiber bundles to complete development are those of the gingival ligament, and the last are those of the apical group.

Anticipating eruption, the cytologic features of the fibroblasts in the periodontal tissue increase in number, especially those associated with fibril synthesis. As a consequence of their activity the fibril population is greatly increased. Just before the crown emerges into the oral cavity, fibroblasts of the cervical third of the root become oriented obliquely toward the cementum. This is followed by organization of the fibers into bundles, which are the presumptive alveolar crest fibers (Fig. 8–23A–E). When the crown penetrates the gingival epithelium to gain entrance to the oral cavity, only the alveolar crest fibers are present. More apically, light microscopic examination of the periodontal tissue reveals fine collagen fibers extending from the surface of the cementum into the soft tissue. Later, their counterparts are observed extending from the bone surface. The collagen fibers

from these two sources (bone and cementum) grow toward one another, meet, coil, and fuse.

It was once believed that the central segment of the periodontal ligament, where the dental and bone fibers meet, was the site of fiber bundle engagement and disengagement, facilitating the varied tooth movements. This area was also considered to be one of extraordinary metabolic activity, and was designated as the intermediate plexus. Radioactive labeling studies have indicated that high metabolic activity is not restricted to the more central or intermediate area but exists generally throughout the ligament. Furthermore, with tooth movement, the more metabolically active sites are those of fiber bundle attachment in the cementum or alveolar process. Additionally, studies involving tracing the path of individual collagen fibers from the tooth surface to the cribriform plate have not indicated the presence of a plexus.

With occlusal contact, organization of the fiber bundles of the cervical third of the tooth (alveolar crest and horizontal) is completed, while those of the middle third (oblique) are in the formative stage of development. As root growth continues, maturation of the principal fibers continues apicalward. The last of the fiber groups to organize and differentiate are the apical fibers, which occurs with the acquisition of functional competency by the tooth.

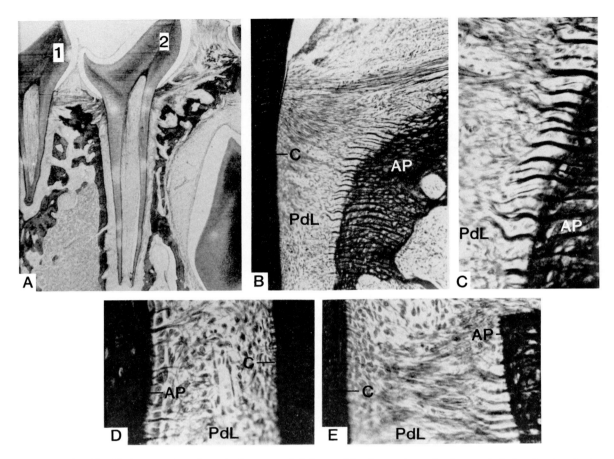

Fig. 8–23. Mandibular premolars shown in five sequential stages of development, ranging from period of occlusal contact to full articulation. *A,* At the cervix of tooth 1, the dentogingival and transseptal fibers of the interdental group of the gingival ligament are well organized. In tooth 2, the fibers located more apical to the oblique fibers are less organized. The collagenous elements of the periodontal tissue around the developing root tip anticipate differentiation to form the apical fiber bundles. These will be organized with complete development of the root tip. (Mallory's connective tissue stain; ×30) *B,* Section of tooth 2. The fibers are well visualized using silver impregnation techniques. Segments of the transseptal fibers are visualized coursing from the cementum over the crest of the alveolar process. More apically, the cemental fibers veer toward the alveolar bone, where they join the seemingly well-spaced osseous fibers. Note that the latter are inserted deep into the bone. (Original magnification, ×150) *C,* Section of tooth 2 stained by silver impregnation showing intercalation onset of cemental and osseous fibers. The latter appear to disentangle as they disperse, joining the more delicate and closely packed fibers from the cementum. (Original magnification, ×250) *D,* Section of periodontal space at midroot level, stained by silver impregnation. Note that the more apically located connective tissue is even less organized than that shown above. The osseous fibers are dense at sites of bone insertion, but they tend to splay rootward as they meet the short cemental fiber tufts. (Original magnification, ×125) *E,* In contrasting this tissue segment with that in *D* (same stain and magnification), note that the functional collagenous elements of the midroot level (tooth 1) pursue an uninterrupted path from the cementum to the cribriform plate, where they join and entwine with the perforating fibers of the bone AP, alveolar process; C, cementum; PdL, periodontal ligament. (Grant, D., and Bernick, S.J.: Formation of the periodontal ligament. J. Periodontal., *43*:17, 1972.)

Principal Fiber Groups

The ligaments of the periodontium are similar to others of the body in fiber composition and arrangement. They differ, however, in that the fibers are collected into many different functional groups (Table 8–1). Furthermore, as a ligament, the periodontal tissue is most unusual in its abundant blood, lymph, and nerve supplies,

which travel throughout the ligament via the loose connective tissue spaces known as interstitial spaces. Additionally, cells other than fibroblasts make up the cell population.

Gingival Ligament

The gingival ligament is comprised of four principal fiber groups: dentogingival; alveolo-

Table 8–1. Principal Fiber Groups

Name	Path	Function	Plane of Cut
Gingival Ligament Dentogingival Alveologingival Circular Dentoperiosteal	All pass from cementum to gingiva or periosteum	Support gingiva Attach gingiva Secure functional competency of teeth and their attachment complex	Vestibulo-oral Mesiodistal Transverse
Interdental Ligament Transseptal	Cementum to cementum of adjacent teeth	Maintain relationships of adjoining teeth	Mesiodistal Vestibulo-oral
Periodontal Ligament Alveolar crest	Cementum to alveolus Cementum to crest	Retain teeth in socket Oppose lateral forces	Mesiodistal Vestibulo-oral
Horizontal	Cementum to top third	Restrain lateral tooth movement	Mesiodistal Transverse Vestibulo-oral
Oblique	Cementum to mid two-thirds	Resist axial forces	Mesiodistal Transverse Vestibulo-oral
Apical	Apex to crypt fundus	Prevent tooth tipping Resist luxation Protect blood, lymph, and nerve supplies	Mesiodistal Transverse Vestibulo-oral
Interradicular	Cementum to crest of interradicular septum	Aid in resisting tipping and torquing	Mesiodistal

gingival, circular, and dentoperiosteal (Figs. 8–24B,C; 8–25; and 8–26).

Dentogingival principal fibers are inserted in the cementum at the cervix and fan out, with their distal terminals ending in the connective tissue of the attached and free gingivae (Figs. 8–24, 8–25, and 8–26). These are the most numerous of the gingival groups. The plane of sectioning best suited for their resolution is given in Table 8–1.

Alveologingival principal fibers are inserted in the crest of the alveolar process. From this site, the fibers splay out through the connective tissue to end in the free gingiva (Figs. 8–24B,C, and 8–25).

Circular principal fibers course around the neck of the tooth as a circumferential band. They function as drawstrings promoting the maintenance and intimate association between the free gingiva and the tooth (Fig. 8–24B,C). Of the gingival groups, the circular principal fibers are least in both caliber and number. It is not uncommon for elements of a given bundle to become disassociated and join with those of another group, thus providing additional strength and support to the gingiva.

Dentoperiosteal principal fibers have one terminal embedded in the cervical cementum. The other terminal, after passing through connective

tissue and periosteum of the alveolar crest, is anchored to the cortical plate of the alveolar crest (Figs. 8–24B,C, and 8–26). Some fibers may extend into the neighboring tissue to become intercalated with the collagen and muscle fibers of the oral vestibule. Some may even extend to the sublingual sulcus.

Among the *functions* of the gingival ligament are those of providing support for gingivae, including the interproximal gingiva, providing gingival attachment to teeth, and securing the functional competency of teeth and the components of their attachment mechanism (Table 8–1).

Interdental (Transseptal) Ligament

Transseptal principal fibers, a synonym for the interdental ligament, pursue a path from the cementum of one tooth over the interdental bony septum to become anchored to the cementum of adjacent teeth (Figs. 8–23A,B, 8–24A and 8–27). They are observed clearly in mesiodistal and transverse sections. Functionally, they provide support for the interproximal gingiva and hold adjacent teeth together. They also assist in preserving the integrity of teeth in the arch (Table 8–1).

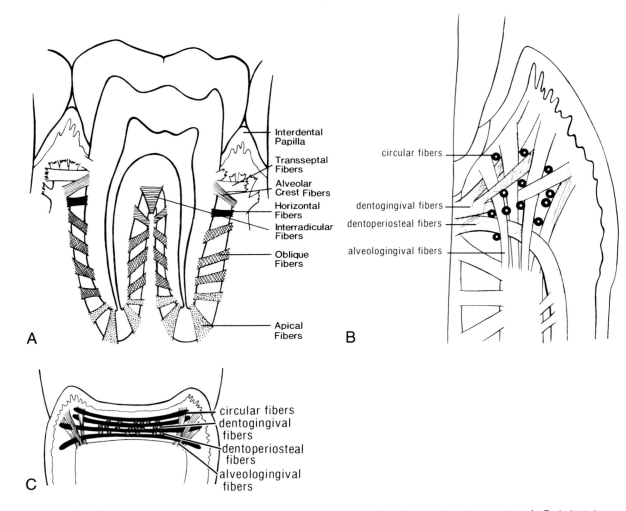

Fig. 8–24. Diagrammatic representation of the ligaments associated with the attachment apparatus. *A,* Periodontal ligament fiber groups: alveolar crest, horizontal, oblique, apical, and interradicular. Note the relationship of the transseptal fiber bundle of the interdental ligament to the alveolar crest fibers bundles. *B,* Principal fiber groups of the gingival ligament in labial section; circular, dentogingival, dentoperiosteal, and alveologingival. *C,* Gingival principal fiber groups visualized interproximally (vestibulo-oral aspect). (Redrawn from Freeman, E.: Periodontium. *In* Oral Histology: Development, Structure and Function. Edited by A.R. Ten Cate. St. Louis, C.V. Mosby, 1980.)

Periodontal Ligament

The periodontal ligament, a regularly arranged dense fibrous connective tissue, classified as a ligament because the collagen fiber bundles are arranged into functional groups. The periodontal ligament and its adjacent tissue (cementum and alveolar process) are all of ectomesenchymal origin. Their common origins and their structural and functional interrelationships, including limited movement, support their collective classification as the attachment apparatus. The functionl relationships of these tissues result in a specific type of joint, known as a synarthrosis.

Location. The periodontal ligament is that soft tissue contained in the territory between the alveolar process and the cementum of the root (Figs. 8–25 and 8–26). It extends longitudinally from the crest of the alveolar process to the fundus of the socket (Fig. 8–1).

Width. The periodontal ligament ranges in width from 0.10 to 0.4 mm. It is widest at its cervical and apical extremities and narrowest at the midroot region. The latter acts as a fulcrum for functional tooth movements. The width of

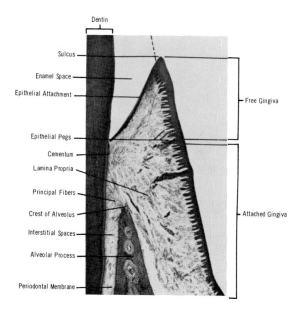

Dentin

Sulcus

Enamel Space

Epithelial Attachment

Free Gingiva

Epithelial Pegs

Cementum

Lamina Propria

Principal Fibers

Crest of Alveolus

Attached Gingiva

Interstitial Spaces

Alveolar Process

Periodontal Membrane

Fig. 8—25. Photomicrograph showing the relationships of the gingiva to the attachment apparatus of the tooth at the level of the alveolar crest and cervix. (Hematoxylin and eosin stain; ×34)

the ligament increases with age. During the midteens it averages 0.2 mm, and decreases in the decades that follow to arrive at an average width of 0.18 mm in the fifth decade of life and beyond. Although the width of the periodontal ligament varies with age and other factors, it is generally wider when healthy and fully functioning. The periodontal ligament is widest, however, in hyperfunction. Pathologic width changes of up to 0.68 + mm may occur.

Functions. The periodontal ligament contributes to the maintenance of a healthy functional attachment system for teeth. In this regard specific activities are involved including anchorage of teeth in their sockets, development and reorganization of the hard tissues of the attachment apparatus, development and maintenance of the soft tissues of the attachment apparatus, protection of vascular, lymphatic, and nerve elements that emerge from the central canal to accomodate the tissue at the base of the alveolus, provision of a nerve supply to structures of the ligaments, and accommodation of the metabolic needs of the tissue via vascular channels. Tissue regeneration in the periodontal ligament is believed by some investigators to be effected by fibroblasts and/or mesenchymal cells.

The cells participating in the development of

the alveolar process and cementum are derived from the parent tissues of the periodontal ligament (dental follicle and periodontal membrane). Changes in the definitive alveolar process and cementum are adaptational, and the stimuli for these changes are transmitted through the periodontal ligament.

The periodontal tissue serves as both an anchoring ligament and a separating membrane. As a ligament the periodontium tightly secures the tooth in the socket and, as a separating tissue, it prevents the fusion of the cementum to bone. If fusion should occur, an abnormal condition known as ankylosis results. Thus, shock resulting from the forces of mastication is lessened by the separating tissue. Normal masticatory forces are conveyed to the alveolar process as pulling forces; these are essential in maintaining the alveolar process in a healthy condition.

The collagen fibers responsible for securing teeth in the alveoli (sockets) are aggregated into numerous bundles (prinicpal fiber) passing from the cementum to the alveolar process. Thus, there are greater numbers of principal fibers in the periodontal ligament than in either or both of the other ligaments (i.e., gingival and interdental or transseptal). For some time it was thought that the Sharpey's fibers from the cementum ended in the cribriform plate as alveolar (perforating) Sharpey's fibers. Evidence based on animal studies has shown that the fibers continue through the spongiosa to join the principal fibers of the cribriform plate of adjacent teeth, so continuity of the fiber path exists from tooth to tooth. In humans, tooth to tooth continuity is generally not accepted. The anchoring fibers of teeth form coiled springs and, although they pass through a bony (mineralized) environment, their cores remain mostly uncalcified and functional.

The periodontal ligament houses and protects blood and lymph channels, thereby providing for its needs as well as for those of associated tissues (cementum, cribriform plate, and gingiva).

The soft tissue and bone at the base of the socket are prevented from being crushed because the tooth is "suspended" in the socket. The central canal, carrying the parent vessels and nerves, is similarly protected by the suspensory action of the periodontal ligament and

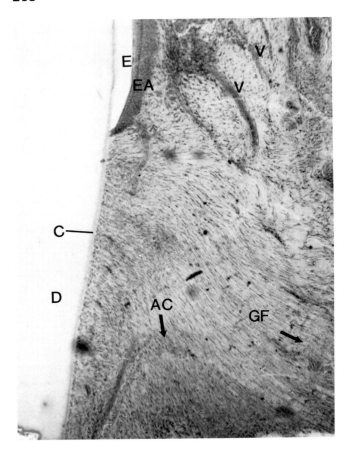

Fig. 8–26. Section of tooth and attachment apparatus at the level of the cervix showing dentin (D), enamel (E), attachment epithelium (EA), and underlying connective tissue, with its vascular supply (V). Shown is the relationship of the alveolar crest fibers (AC), which are inserted in the cementum (C) and veer toward the crest of the alveolar process, where they are anchored as Sharpey's fibers. The dentoperiosteal fibers (GF) of the gingival ligament, are inserted in the more cervically acellular cementum. (Hematoxylin and eosin stained; ×100)

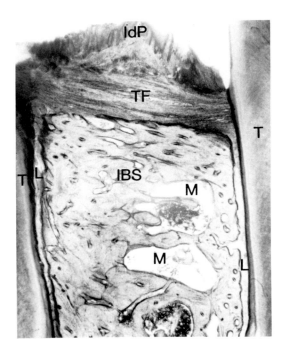

Fig. 8–27. Interdental bony septum (IBS) separating adjacent teeth (T), with transseptal fibers (TF) located below the interdental papilla (IdP) and overlying the septum (M, marrow spaces; L, periodontal ligament). (Hematoxylin and eosin stain; ×80)

by the limited movement of the tooth (see Fig. 4–32).

Nerve endings (proprioceptors) of the periodontal ligament receive stimuli that are translated into information related to such factors as the forces of mastication, chewing movements, and food texture.

Components. The collagen fibers of the periodontal ligament contribute to the definite, or principal, groups. All are attached to the cementum on the one side and to the alveolar process on the other. The fiber groups do not assume straight paths from one hard tissue to the other, nor are they taut. Rather, they course in wavy paths, thereby providing for limited tooth movement. From the cervical cementum apicalward to the fundus of the socket, four principal fiber groups constitute the periodontal ligament: alveolar crest, horizontal, oblique, apical and, in multirooted teeth a fifth group is present, interradicular fibers (Figs. 8–15, 8–24A and 8–28; Table 8–1).

Alveolar crest principal fibers are attached to the cervical cementum, and are directed apicalward to join and become inserted in the alveolar crest with the periosteal fibers (Figs. 8–23B, 8–24A, 8–25, 8–26 and 8–29). It is not uncommon for these fibers to become separated from parent groups to intermingle with those of the gingival ligament. They are observed in vestibulo-oral and mesiodistal sections. The alveolar crest fibers serve to oppose lateral forces, and secure the tooth in its alveolus (Table 8–1).

Horizontal principal fibers are apical to the alveolar crest group. They occupy the coronal third of the root, and span the periodontal space from the cementum to the cribriform plate (Figs. 8–24A and 8–29). They may be seen in any sectional plane, forming thick bundles that are oriented parallel to the occlusal surface. Their principal function is that of counteracting lateral tooth movements.

Oblique principal fibers occupy the middle and apical thirds of the root, and are the most numerous of the fibers of the periodontal group. They course diagonally at an angle up to 45° from the cementum occlusalward to the alveolar process (Figs. 8–24A, 8–30, and 8–31). They may be observed in any sectional plane, and their orientation in the periodontal space anchors and suspends the tooth in its socket. Additionally, these fibers tend to resist the pressures of chewing and biting. The pulling (tension) forces on the alveolar process also help to maintain the functional competency of the bone (Table 8–1).

Apical principal fibers fan out from the apex of the root to the bone-forming fundus of the socket (Figs. 8–24A and 8–28). They are visible irrespective of the plane of sectioning. These fibers stabilize the tooth by preventing tipping. Additionally, they resist forces of luxation to protect the vascular, lymphatic, and nerve elements emerging from or entering the central canal en route to the apical foramen.

Interradicular principal fibers are present in multirooted teeth only. From their attachment in the cementum at furcations, these fibers extend through the periodontal tissue to become anchored in the crest of the interradicular septum (Figs. 8–15, 8–24A, and 8–32). They are best observed in mesiodistal sections. They are credited with resisting tooth tipping, torquing, and luxation. Gingival recession associated with age may result in the loss of fibers of this group if the condition progresses to expose furcations.

Secondary Fiber Groups

Collagenous elements are dispersed throughout the periodontal tissue. These are independent of the principal fiber groups, and are known as secondary fibers. It is likely that they are formed in situ by fibroblasts. As reserve fibers they probably await engagement with neighboring principal groups or with other structures that require a fibrous substructure. Secondary fibers are mostly oriented coronoapically, and are associated with the vascular, lymphatic, and nerve elements accommodated in the interstitial spaces.

Reticular fibers of the periodontal tissue are associated mostly with the vessels of microcirculation, where they form a delicate perithelium.

Elastic fibers of the periodontal ligament are associated with the walls of the larger afferent vessels. Here they provide elasticity for the arteriole wall.

Oxytalan fibers are believed by some histologists to be precursors of elastic fibers. In the periodontium they appear with root development, and are eventually observed along the full length of the periodontium. They are inserted into both cementum and the alveolar process. Although they probably do not course the full

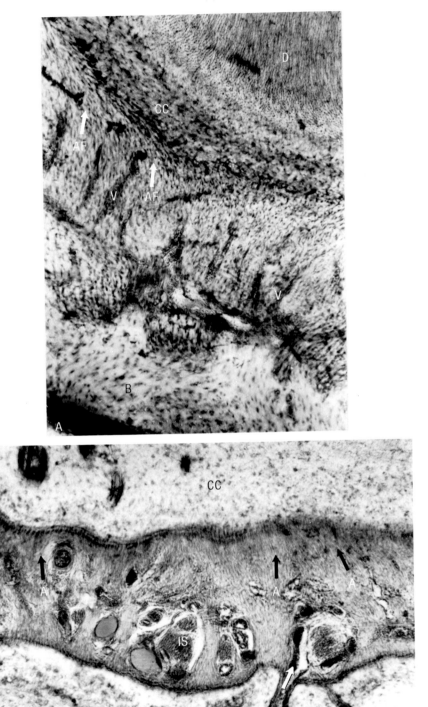

Fig. 8–28. Apical fibers. *A,* Section of root tip showing dentin (D), cellular cementum (CC), apical fibers (AF, *white arrows*), and bone (B) at the fundus of the socket. As the blood vessels exit the floor of the socket, one branch travels to the apical foramen and the others form an anastomotic network (V) around the root tip. (Hematoxylin and eosin stain; ×80) *B,* Fundus of the alveolus showing blood vessel *(white arrow)* emerging from the bony floor (B) of the socket. Within the periodontal ligament are shown the apical fibers (AF, *black arrows*), and the interstitial space (IS) (CC, cellular cementum). (Hematoxylin and eosin stain; ×50)

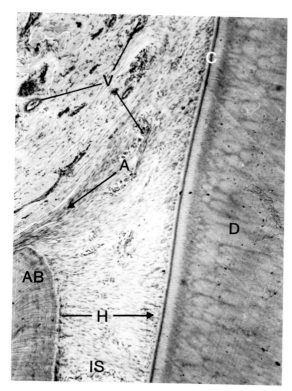

Fig. 8–29. Section of periodontal ligament showing two principal fiber bundles, alveolar crest (A) and horizontal (H), which course from the acellular cementum (C) to the lateral aspect of the alveolar process (AB). The blood vessels (V) that emerged from the crest of the alveolar bone (AB) branch to form a rich supply for the connective tissue associated with the gingival ligament, (D, dentin; IS, interstitial space). (Hematoxylin and eosin stain; ×100)

width of the periodontal tissue their free ends appear to be splayed, intermingling with the collagenous elements. They are not observed in bundles and do not possess the ordered orientation seen in the principal fibers. Because of their location, some investigators believe that they possess a suspensory function. Areas of the periodontal ligament subjected to greater stresses are reported to contain more and larger oxytalan fibers.

Structural Considerations

Between and among the principal fiber groups are the interstitial spaces, in which are found cells of the periodontal ligament, secondary fibers (see above), and vascular, lymph, and nerve elements. These will be discussed in this section.

Ground Substance

The cellular and fibrillar components of the periodontal ligament exist in ground substance produced mostly by fibroblasts. As in the case of other connective tissues, it is composed mostly of water (70%) bound to proteoglycans and glycosaminoglycans. Fibronectin and laminin are also present in the ligament. The former is identified by immunocytological methods employing fluorescent antibodies. It is a fibrous protein associated with the basal lamina, collagen, and cells, especially fibroblasts. Principal among its functions is its binding or adhesion potential—cell to cell, fibril to fibril, and possibly cell to fibril.

Of the main glycosaminoglycans (chondroitin sulfate, heparan sulfate, dermatin sulfate, hyaluronic acid, and keratin sulfate), only hyaluronic acid is found independently in the ground substance. It is of special significance in the periodontal ligament because of its propensity to combine with water, thus facilitating the transport of pathogens and metabolites through the ground substance. The sulcular epithelium, which secures the periodontal tissue from the external environment, is especially vulnerable. Certain microorganisms may produce the enzyme hyaluronidase which under certain conditions hydrolyzes hyaluronic acid which promotes the invasion and spreading of microorganisms and their products, thus resulting in pathogenesis.

Laminin is principally found in the basal lamina and is concerned with the attachment of epithelium to type IV collagen located in the substrata.

Cellular Components

The cell types found in the periodontal tissue include osteoblasts and cementoblasts, and their associated "-clast" cells. Mesenchymal cells, fibroblasts (fibrocytes), mast cells, macrophages, and other leukocytic phagocytes and epithelial rests are also located in the periodontium. Although these cells have been discussed in Chapter 2, only those with activities specific for this tissue will be reviewed here.

Cementoblasts and osteoblasts and their counterparts (e.g., cementoclasts and osteoclasts) have been discussed earlier in this chapter. In hard tissue synthesis and resorption they

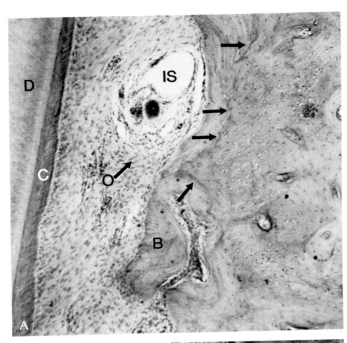

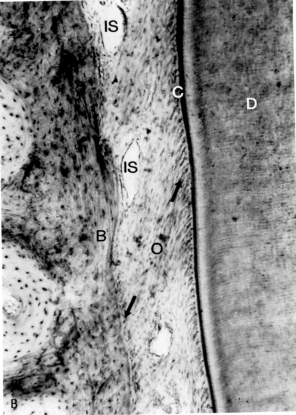

Fig. 8–30. Physiologic mesial drift. *A,* Periodontal ligament at the level of the oblique fibers showing the activities associated with the mesial aspect of the tooth, such as the following: the opening of an osteone to liberate its soft tissue into the periodontal tissue as a round interstitial space (IS) in the periodontal tissue; osteoclasia, as indicated by cementing lines *(black arrows)*; and slack oblique fibers (O), (D, dentin; C, cementum; B, alveolar bone). *B,* Periodontal ligament at the distal aspect of tooth, showing ovoid or flattened interstitial spaces (IS), taut oblique fiber bundles (O), and osteogenic activity on bundle bone (B), (C, cementum; D, dentin). (Hematoxylin and eosin stain; ×80)

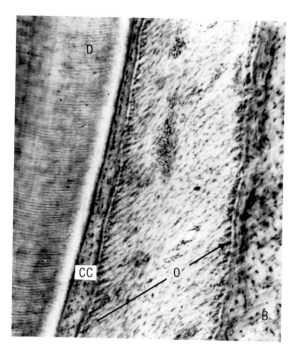

Fig. 8–31. Section of periodontal ligament, showing oblique fiber bundles (O). They are inserted apicalward in cellular cementum (CC) and extend obliquely to the alveolar bone (B), where they are inserted as Shapey's fibers, (D, dentin). (Hematoxylin and eosin stain; ×80)

also serve to maintain the dimensional aspects of the periodontal space. These important functions are considered below. Their actions are probably induced intrinsically in the periodontal ligament. The precise nature of the stimuli (i.e., chemical, physical, or both) is a matter of speculation. Some investigators believe that the stimuli are primarily physical, as evidenced by bone resorption and apposition in the orthodontic movement of teeth, and by physiologic mesial drifting of teeth. These are naturally occurring phenomena in the healthy attachment apparatus, just as osteogenesis or cementogenesis and odontoclasia (cementum and dentin) occur naturally during tooth eruption. It is entirely possible that chemical stimuli may also be involved in these activities. On the other hand, these cells' actions may be caused by cemental fracture or other insults, including the inflammation that accompanies periodontal disease.

Whether or not mesenchymal cells are discrete components of the periodontal ligament has been questioned. It has been suggested that they are either pericytes or, more likely, inactive (resting—G_0 phase) fibroblasts, cemento-

blasts, or osteoblasts awaiting activities of a successive stage in their life history.

Fibroblasts are the most numerous cells of the periodontal ligament. They are reported to be among the most valuable cells of the periodontium, and the functions attributed to them are numerous and varied. These include contraction, synthesis of collagen and elastin proteins, synthesis of glycoproteins and proteoglycans, synthesis of collagenase and probably elastase, and phagocytosis of foreign material and collagen. The last function further enhances the cell's potential for maintaining tissue homeostasis.

The histocytologic features of fibroblasts have been discussed elsewhere. Noteworthy is the fact that the sizes of fibroblasts obtained from chronic inflammation of the periodontium are dramatically larger. Furthermore, abnormalities such as discontinuities in the plasma membrane, formation of surface blebs, dilation of the mitochondria, and dispersed chromatin have been observed in fibroblasts from inflamed periodontal tissues.

Of particular interest is the recent finding that phenotypically distinct fibroblasts with functionally different subpopulations may occur in the periodontium. These are cytomorphologically indistinguishable, yet their functions may be varied. For example, one subpopulation may engage in collagen synthesis while another may produce collagenase; it is even possible that different subpopulations produce different types of collagen. Phenotypically stable and functionally varying fibroblast subpopulations have been reported in three tissues: skin, gingiva, and periodontal ligament.

Epithelial rests of Malassez are cell groups on the cemental aspect of the periodontal ligament. They may extend from the cervix to the apex as cords or other configurations, and they are generally more numerous toward the apex. These cell clusters are known as epithelial or cell rests of Malassez, and they are remnants of Hertwig's epithelial root sheath (Fig. 8–33; also see Fig. 4–29).

The morphologic aspects of the epithelial rests differ with the plane of sectioning. If longitudinal and transverse cuts are made, the cell rests are revealed as long cords or cell clusters (Fig. 8–33). If they are cut tangentially, they appear as a network. Electron micrographs show that they contain many organelles (Fig. 8–34). This

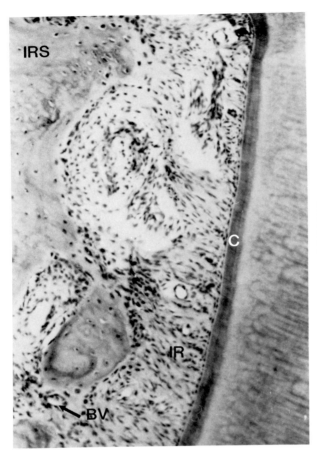

Fig. 8–32. Section showing interradicular fiber bundles (IR) traveling from the cementum (C) to the crest of the interdental septum (IRS). Blood vessels (BV, *black arrow*) emerge from the bone to supply the periodontal tissue (Hematoxylin and eosin stain; ×80)

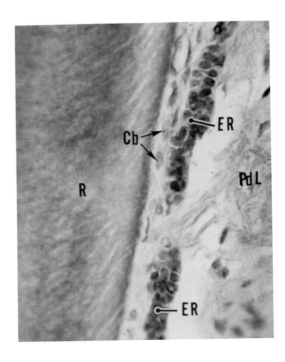

Fig. 8–33. Root (R) aspect of the periodontal ligament (PdL), showing cementoblasts (Cb) and epithelial rests of Malassez (ER). (Hematoxylin and eosin stain; ×80)

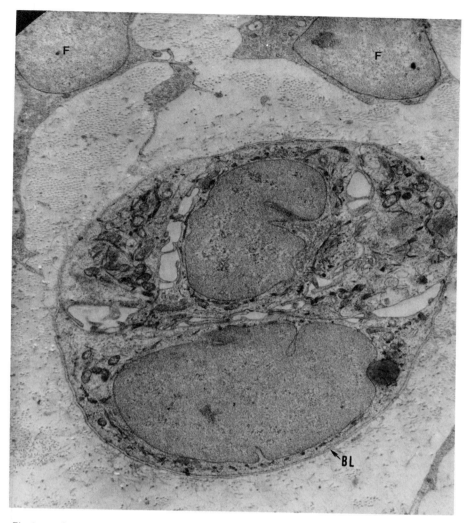

Fig. 8–34. Electron micrograph of an epithelial rest of Malassez, with neighboring fibroblasts (F) and their processes in periodontal tissue. The epithelial rest is surrounded by a basal lamina (BL). (Uranyl acetate stain; ×8,000) (Valderhaug, J.P., and Nylen, M.U.: Function of epithelial rests as suggested by their ultrastructure. J. Periodont. Res., *1*:69. Copyright 1966, Munksgaard International Publishers, Copenhagen.)

cytologic feature supports the interpretation that the rests of Malassez may become active and participate in the formation of root cysts, tumors, and calcific bodies.

Interstitial Spaces

Dispersed among the principal fiber groups of the periodontal ligament are areas known as interstitial spaces, which consist of loose connective tissue (Figs. 8–28B and 8–30). Secondary fibers, blood vessels, lymphatics, and nerves are contained in the interstitial spaces. Other than fibroblasts, most of the cellular elements are pleuripotential cells and macrophages. The

rich vascular supply suggests that the metabolic needs of the tissues are met, especially those involved in fibrillogenesis and ground substance production.

The location, size, and shape of the interstitial spaces are variable. They may be found either centrally or laterally in the periodontal ligament. The interstitial spaces nearer the bone are generally the largest, and they tend to become smaller as they move across the periodontal space toward the cementum. Location, size, and contour of the interstitial tissues are also influenced by the organization and movement of the collagen bundles. Their contour changes with

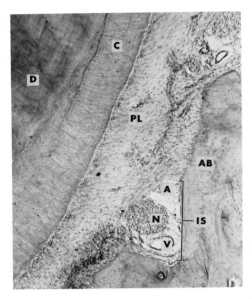

Fig. 8–35. Photomicrograph of periodontal ligament (PL) containing a newly formed interstitial space (IS), in which is seen an artery (A), vein (V), and nerve bundle (N) in loose connective tissue derived from an "opened" haversian canal of the alveolar process (AB). Insertion of collagen fiber bundles into cementum (C) is clearly visible, D, dentin). (Hematoxylin and eosin stain; × 100)

functional pressures exerted by the principal fiber groups.

During development of the periodontal ligament, the remaining loose connective tissue of the periodontal membrane, which is not organized into a ligament, constitutes the interstitial spaces. Later, as a result of the various activities of the periodontal ligament, the tissue of the interstitial spaces may become spent, and may need to be replaced. This is done by marrow, particularly that of the haversian canals.

The bone of the alveolar process is extremely plastic, and is continuously engaged in reorganization. With osteoclasia, resorption may continue into the haversian system, exposing and releasing the marrow tissue of the canal (Figs. 8–21 and 8–35). This accounts for the larger interstitial spaces located toward the bone—that is, toward their source of origin.

Vascular, Lymph, and Nerve Supply

The periodontium is unusual in that its vascular, lymph, and nerve supply is extraordinarily high for ligamentous tissue, a result of the high metabolic requirements of developmental or germinal tissues, such as osteoblasts, cemento-

blasts, and basal epithelial layers peripheral to the periodontium.

Arteries. Blood is supplied and drained from the periodontal ligament by branches of the dental, interdental, and interradicular arteries, all of which orginate from the alveolar artery. Innervation of the periodontal ligament is accommodated by the dental branches of the *alveolar nerves*.

Dental arteries emerge from the floor of the crypt and travel toward the apical foramen (see Figs. 7–4A and 8–28B). En route to the apical foramen of the tooth, one branch of the emerging arteriole(s) divides to form an anastomotic periapical network (Fig. 8–28A). The parent arteriole continues rootward, entering the apical foramen(a) and traveling through the root to arrive at the pulp chamber.

Interdental (interalveolar) arteries pass through the spongiosa, distributing numerous branches that veer toward the cribriform plate and that emerge into the periodontal ligament as perforating arteries (Figs. 8–17 and 8–36). These supply the periodontal tissue from the fundus of the socket to the level of the alveolar crest (Figs. 8–1, 8–17, 8–28A, 8–29 and 8–36). Most of the periodontal ligament is supplied by these arterioles. The perforating arteries of the maxillary periodontal ligaments are more numerous than those of the mandible (as noted previously). When the interdental arterioles arrive at the crest of the alveolar process, they exit the bone as gingival branches and form a capillary network in the connective tissue of the free gingiva (Figs. 8–1, 8–29, and 8–36).

Interradicular arteries are found only in multirooted teeth. They course through the spongiosa of the interradicular septum and emerge from the cribriform plate to supply blood to all levels of the ligament. The area of the furcation is provided with vessels from the arterioles, whih emerge from the crest of the septum.

Regardless of their origin, arterioles associated with the attachment mechanism of the teeth eventually communicate in the periodontal space, where they form an anastomotic network. Here the arterioles are between 15 and 50 μm in diameter. The larger afferent vessels travel vertically among the principal fiber groups, through the interstitial spaces, and arrive at sites where capillary networks are formed (Fig. 8–30A). In this regard the afferent blood

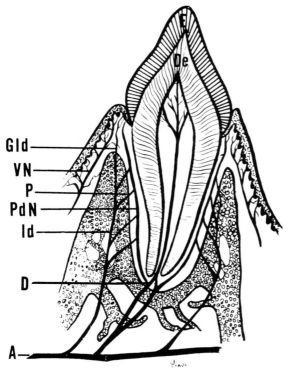

Fig. 8–36. Diagram of tooth and its supporting structures. Common paths of the vascular and nerve supplies are assumed, except where otherwise specified (A, alveolar; D, dental; GId, gingival branch of interdental; Id, interdental or interalveolar; P, perforating; VN, vestibular nerve (buccal or labial); PdN, periodontal nerve; E, enamel; DE, dentin).

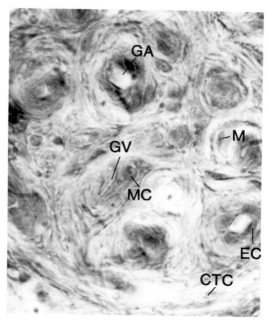

Fig. 8–37. Photomicrograph of glomus showing its arterial (GA) and venous (GV) components. Note the capsule-like (CTC) tissue of the glomus, (M, muscle; EC, endothelial cell; MC, mesenchymal cell). (Hematoxylin and eosin stain; ×980) (Provenza, D.V., Biddix, J., and Cheng, T.: Studies on the etiology of periodontosis, II. Glomera as vascular components in the periodontal membrane. Oral Surgery, *13*:157, 1960.)

supplies of the cervical and apical thirds of the periodontal ligament are greater than that of the middle third. Furthermore, the arteriole supply of the ligaments of the posterior teeth are greater than those of the anterior teeth.

Venous Drainage. Arterioles form capillary networks for the periodontal ligament and for the free and interproximal gingivae. Blood from these areas is drained by the dental, interdental, and interradicular veins. After retracing paths of the companion arterioles, they ultimately drain into the alveolar veins.

Glomera are specialized vessels that proceed along wavy or convoluted paths from companion artery to vein (Fig. 8–37). Accordingly, glomera bypass capillary networks. On the arteriolar aspect the vessels are structurally the same as the muscular arteries, while on the other side they are vein-like.

Lymphatic Drainage. These vessels are located throughout the gingivae and the perio-

dontal tissue. The pathways of lymph drainage follow those of the veins (Fig. 8–36). Thus, lymph from the gingivae empties into interdental vessels, where it is joined by that from the periodontal ligament. These flow into the alveolar lymph channels, which are additionally joined by the dental and interradicular lymph vessels.

Periodontal Nerves. Nerve paths in the periodontium are similar to those of their accompanying arteries (Fig. 8–36). The nerves serving the periodontal tissues are the dental branches of the alveolar nerves, which originate from two trigeminal nerve divisions. Branches from the alveolar nerve pass through the floor of the alveolus and become separated from the parent group to innervate the tissue at the fundus of the socket. Other branches (e.g., ascending periodontal or dental nerve) travel in the periodontal tissue toward the cervix, distributing fibers along the path. These fibers join those of the interalveolar nerve, arriving by way of the cribriform plate. Together they innervate most of the periodontal ligament. Fibers from these two sources (dental and interalveolar) are known collectively as the periodontal nerves (Fig. 8–36). Similarly, the paths of the interradicular nerve fibers follow those of their companion arterioles.

Innervation of the gingival connective tissue is derived from a number of sources. For example, the free and attached gingivae are innervated by branches of the labial, buccal, or palatal nerves. The gingivae of the sulcular (crevicular) area and of the attachment epithelium are supplied by branches of the buccal, labial, and palatal nerves, as well as by those from the interalveolar and ascending periodontal (dental) nerves (Fig. 8–36). The periodontal nerves also supply the interproximal gingivae.

Pain, pressure, proprioception, and touch of the periodontal tissue are accommodated by sensory innervation. Autonomic function involving innervation of the vascular channels is served by delicate unmyelinated nerve fibers distributed throughout the periodontal tissue. Nerve endings occur among the fiber bundles, interstitial connective tissue components, including vascular channels, and osteogenic and cementogenic layers. Depending on the site the nerve endings form tendrils, bulbs, spindles, or barbs.

Age Changes

Aging is believed by some to affect the location of the alveolar crest and horizontal and transseptal fiber groups. It is thought that, with apical migration of the dentogingival epithelial junction, the fibers of the area atrophy. Apical migration of the epithelium, if it occurs, is not a normal process. Rather, it is a result of such factors as gingival recession, inflammation, or root surface exposure. These are discussed later.

DENTOGINGIVAL EPITHELIAL JUNCTION

Because of the functional relationship of the dentogingival epithelial junction to the periodontium, particularly the gingival and interdental ligaments, the dentogingival epithelial junction is discussed as a component of the attachment apparatus, and not the oral mucosa with which it is more closely related anatomically (Fig. 8–26). The dentogingival epithelial junction consists of crevicular (sulcular) epithelium and junctional (attachment) epithelium (Fig. 8–38). The crevicular epithelium is continuous with that of the free gingiva and is sep-

arated from the tooth by a shallow depression, the gingival sulcus (Figs. 8–25 and 8–26). The attachment epithelium is located at the base of the sulcus; its dental aspect adheres by hemidesmosomes to the tooth surface, and its gingival aspect is joined by desmosomes to the cervicular segment of the crevicular (sulcular) epithelium (Fig. 8–38). The tissue source of the dentogingival junction changes developmentally as the tooth grows into occlusion (see below).

Function

The dentogingival epithelial junction, via its desmosomal (biologic) or organic attachment to the tooth, or via both, succeeds in isolating, sealing off, and protecting the periodontal tissue from foreign materials arriving in the oral cavity. Initially this occurs by way of the primary junctional (attachment) epithelium, and later by way of the secondary junctional (attachment) epithelium.

Development

The dentogingival epithelium is initially found over the entire anatomic crown. With root growth and the emergence of the crown into the oral cavity, the epithelium shifts progressively from the tips of the cusps over the slope, and then onto the cervix of the tooth, both enamel and cementum (see Figs. 4–35 and 4–36). If the enamel is deprived of its epithelial covering, cellular and fibrillar elements of the adjacent connective tissue will move into the denuded area and produce matrix to cover the unprotected enamel. These deposits have been previously described as coronal cementum or afibrillar cementum, because the matrix fibrils lack the 64-nm banding.

Later in life, the epithelium may shift apically, and may be located totally on the root (see Fig. 4–35). Concomitant with migration of the epithelium from the crown to the root, the dentogingival epithelial components change in composition from primary to secondary, as described below.

In discussing tooth eruption, the development of the dentogingival epithelial junction was described (see Chapter 4, Tooth Eruption). That is, as the crown emerges, its overlying reduced enamel epithelium advances onto and unites with the basal layer of the gingival epithelium through desmosomal connections (Figs. 8–39

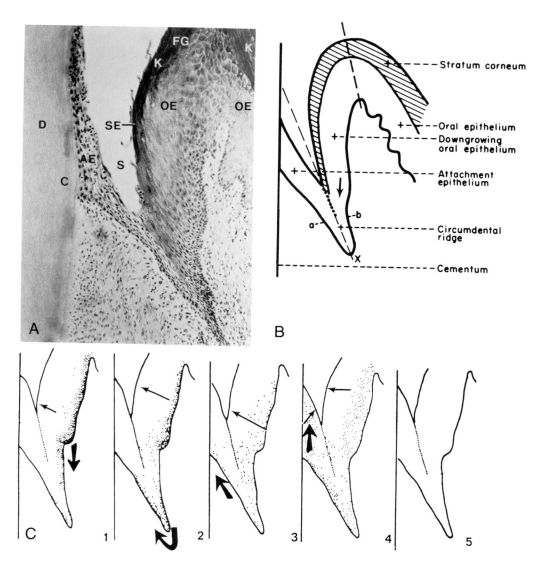

Fig. 8–38. Relationships of the attachment apparatus to dental and paradental tissues. *A,* Photomicrograph showing dentin (D), cementum (C), and attachment (junctional) epithelium (AE). The gingival (crevice) sulcus (S) is bordered by the attachment epithelium on its dental aspect and its sulcular (crevicular) epithelium (SE) on its gingival aspect. The surface layer of the oral epithelium (OE) of both the sulcular epithelium (SE) and free gingiva (FG) is cornified (K), whereas that of the attachment epithelium is nonkeratinous. (Hematoxylin and eosin stain; × 140) *B,* Diagrammatic representation of *A* showing the downward and cementward path *(arrows)* of the oral epithelium (OE) as the replacement source for the primary (attachment) junctional epithelium to form the secondary (attachment) junctional epithelium. *C,* Diagrammatic representation of the replacement path of primary attachment epithelium by the secondary attachment epithelium *(large black arrows)* surfaceward, and of the path of epithelial desquamation toward the sulcus *(small arrows).* Paths are based on tritiated thymidine tracer studies on rat incisor. The labeled tissues were processed for autoradiographic study at periods of 30 minutes (1), 6 hours (2), 24 hours (3), 72 hours (4), and 144 hours (5). (Anderson, G.C., and Stern, I.B.: The proliferation and migration of the attachment epithelium, *Periodontics, 4:*115, 1966.)

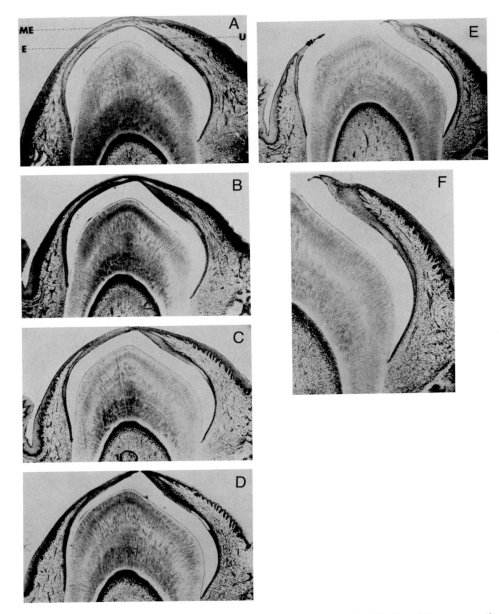

Fig. 8–39. Tooth eurption stages. *A,* The crown is covered by the reduced enamel epithelium (U), or prospective primary attachment (junctional) epithelium. Note that the tooth has grown through most of the subepithelial connective tissue and is converging onto the basal layer of the gingival epithelium (ME). The compressed tissue in the path of eruption is indicated by arrow, (E, enamel space). *B,* Immediately anticipating fusion, the connective tissue is condensed *(arrow),* and the epithelia have increased in thickness due to mitotic activity of the basal cells of gingival epithelium. *C,* Fusion of the prospective attachment and gingival epithelium has occurred, thus decreasing the thickness of the intervening connective tissue. *D,* Fused epithelia at the necrotic site has sloughed off. *E,* Later stage showing repositioning of the primary (junctional) epithelium on the crown more cervicalward. *F,* Lingual aspect of *E* at increased magnification. Note the fused epithelial layers have increased in thickness (arrow) by mitotic activity. (Carmine stain; *A–E,* enlarged from ×30, *F,* enlarged from ×80) (Churchill, H.: Meyer's Normal Histology and Histogenesis of Human Teeth and Associated Parts. Philadelphia, J.B. Lippincott, 1935.)

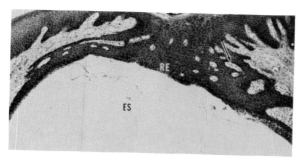

Fig. 8–40. Fusion of prospective attachment and gingival epithelia. Note the cords of gingival epithelia *(arrows)* growing into the lamina propria to arrive at and fuse with the reduced enamel epithelium/prospective primary attachment (junctional) epithelium (RE), (ES, enamel space). (Hematoxylin and eosin stain; ×50) (McHugh, W.D.: Dent. Practitioner, *11*:314, 1961.)

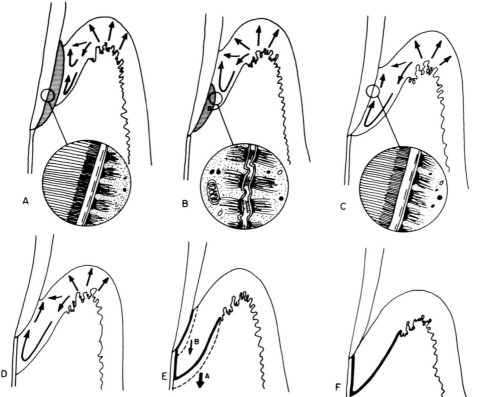

Fig. 8–41. Diagrammatic representation of migration of attached junctional epithelium. In *A* and *C*, the cohesive bonds are illustrated between the tooth and epithelium by the basal lamina and hemidesmosomes—hence, the epithelial attachment. Junctional complexes, both desmosomal and tight junctions, effect the union between the primary and secondary attachment (junctional) epithelia, as shown in *B. A,* Primary attachment (junctional) epithelium is secured to the tooth surface via the basal lamina and hemidesmosomes, both products of the postameloblasts of the reduced enamel epithelium. Fusion of the primary *(shaded)* and secondary junctional epithelium is shown. The arrows indicate the migratory path of the cells of the basal layer of gingival epithelium. Note that these cells supply both the spent gingival epithelium and the degenerating primary attachment epithelial components, thereby initiating the formation of the secondary junctional epithelium. *B,* The migratory path is shown, with progressive replacement of the primary junctional epithelium by the secondary components. The fusion zone marked X is of high mitotic activity. *C,* With degeneration and desquamation of the primary junctional epthelial components, the secondary epithelium establishes connections with the dental surface by a basal lamina hemidesmosomal complex. The migratory path of epithelium continues to be along the basal lamina (epithelial attachment). Note that the location of attachment epithelium is exclusively on the enamel. In time, it may migrate, so that the junctional epithelium is located on both enamel and cementum, as shown in *D. D,* The path of growth and proliferation is indicated by arrows. Note that the secondary attachment epithelium is located on both enamel and cementum. *E,* Relocation of the attachment epithelium to a cementum site is shown by the apical shift of the attachment epithelium from locus A to B. This is indicated by the shift of the bold black line to that of the broken line. In its new location the basal lamina is secreted de novo, thereby establishing a new epithelial attachment. *F,* Cemental location of the epithelial attachment. A more apical shift of the attachment epithelium is indicative of gingival recession, a pathologic condition. (Reproduced by permission from Grant, D.A., Stern, I.B., and Everett, F.G.: Periodontics in the Tradition of Orban and Gottlieb. 5th Ed. St. Louis, 1979, The C.V. Mosby Co.)

and 8–40). The union of the two epithelia, gingival and reduced enamel, persists as an anatomic entity, the dentogingival epithelial junction. It may be recalled that the noncellular organic layer, comprised of the basal lamina hemidesmosome-complex, is designated as the epithelial attachment. On the other hand, the overlying cellular layer, which is formed first and is composed exclusively of reduced enamel epithelium as the advancing front of the emerging tooth, is appropriately called the primary junctional, or attachment epithelium. With the emergence of the crown tips into the oral cavity, the gingival epithelium begins to replace the reduced enamel epithelium component of the primary junctional (attachment) epithelium. Thus, at this stage, the attachment epithelium is of mixed origin, reduced enamel and gingival epithelia (Fig. 8–41). In time the reduced enamel epithelium is desquamated and replaced entirely by cells originating from the gingiva. Thereafter the entire dentogingival epithelial junction, including the junctional attachment epithelium, is composed exclusively of cells originating from the basal layer of gingival epithelium. The dentogingival epithelial junction is now known as the secondary attachment epithelium. (The migratory replacement path of the basal or germinating layer of the gingival epithelium is indicated in Figures 8–38 and 8–41.)

Structure

With the acquisition of its occlusal position, the junctional epithelium spans the area from the fundus of the gingival sulcus to the neck of the tooth, at about the level of the cementoenamel junction (Figs. 8–42 and 8–43). Its greatest width (20 to 30 cell layers) is at the base of the sulcus and is attenuated toward the cervix, becoming a single cell layer thick at the cementoenamel junction (Fig. 8–43).

Intercommunicating processes of adjacent junctional epithelial cells do not occur as often as those in either the oral or crevicular epithelium. The intercellular spaces are larger and the desmosomal connections fewer, accounting for the ease with which leukocytes transmigrate and fluids are exchanged in the junctional epithelium en route to or from the gingival sulcus. In the oral cavity leukocytes are known as salivary corpuscles. Their occurrence in small numbers is

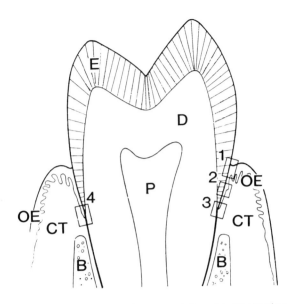

Fig. 8–42. Diagrammatic representation of tooth showing enamel (E), dentin (D), pulp chamber (P), oral epithelium (OE), connective tissue (CT), and bone (B). The four boxed areas (1–4) are enlarged in Figure 8–43 to show structural differences. (Schroeder, H., and Listgarten, M.: Fine structure of the developing epithelial attachment of human teeth. Monogr. Dev. Biol. 2:1977.)

natural, an increase would indicate a pathologic condition.

The junctional epithelium is bordered by two straight basement laminae, so pegs and ridges are wanting. Both laminae are believed to be of similar origin and composition. Toward the tooth the junctional epithelium abuts the internal basal (basement) lamina and at the connective tissue aspect the epithelium interfaces with the external basal (basement) lamina. These laminae are of characteristic structure; that is, they are comprised of a lamina lucida and lamina densa, and their communication to the tooth surface or basal epithelial cell layer is by way of hemidesmosomes. The basal lamina is arranged circumferentially about the tip of the apicalmost oriented junctional epithelium (Fig. 8–43C and D). Cells resting on the basal laminae are low columnar to cuboidal in shape. These cells, particularly those of the more apical terminal of the junctional epithelium, bear cytomorphologic features and mitotic potential expected of the basal cells in stratified squamous epithelium. The organelle populations, particularly the rER and Golgi components, are significantly higher, and the tonofilaments are reduced in number.

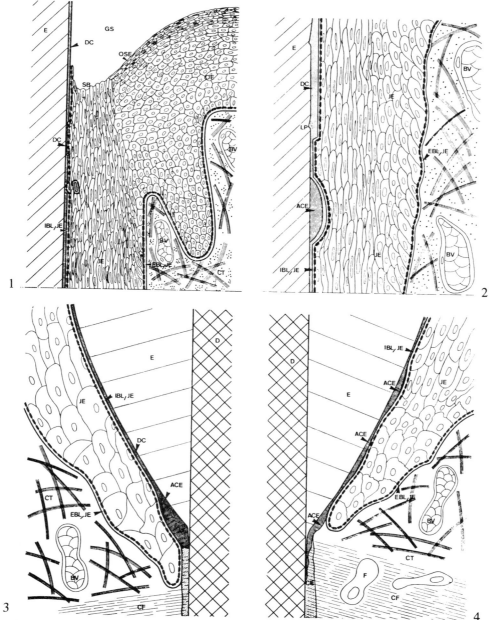

Fig. 8—43. Diagrammatic representation of the four boxed areas in Fig. 8—42, enlarged to show detail. 1, Normal healthy gingival sulcular (GS) relationships. The sulcus base (SB) is composed of the sloughing surface of the junctional epithelium (JE), which is demarcated from the connective tissue (CT) compartment by an external basement lamina (EBL, JE) and from hard substance of the tooth by an internal basement lamina (IBL, JE). The demarcation between the cells of the JE and those of the oral sulcular epithelium (OSE) is prominent. A displaced segment of dental cuticle (DC) is shown (BV, blood vessels; OE, oral epithelium; E, enamel). 2, Junctional epithelium (JE) and its relationship to the tooth surface covered by a dental cuticle (DC) and afibrillar cementum "islands" (ACE). Note the basal cells of the JE (more cuboidal) adjacent to the external basement lamina, and the more flattened suprabasal cells of the JE (LP, line of Pickerill). 3, Relationships of the most apically situated junctional epithelial cells to the hard substance of the tooth and to the connective tissue compartment. Note that the external basement lamina is continuous with the internal basement lamina around the apicalmost cell of the junctional epithelium. Attachment of the junctional epithelium to the dental cuticle is depicted. Also shown is its attachment to afibrillar cementum (ACE), and to cementum on the root surface (CF, connective tissue fiber bundles). 4, Condition similar to that in 3, but with the attachment of the junctional epithelium to the hard substance of the tooth showing islands of afibrillar cementum that have formed on the enamel surface (CE). (Schroeder, H.E., and Listgarten, M.A.: Fine structure of the developing epithelial attachment of human teeth. Monogr. Dev. Biol., 2:1977.)

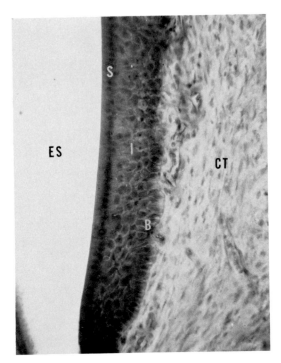

Fig. 8–44. Secondary attachment epithelium showing enamel space (ES), superficial layer (S), intermediate layer (I), basal cell layer (B), and connective tissue (CT). (Hematoxylin and eosin stained; ×200)

The turnover rate of the basal cells, as determined by autoradiographic techniques in animals, is very high.

The cells located between the two basal layers are rounded or mildly squamoid, and are oriented somewhat parallel to the tooth surface. The features of these suprabasal cells are similar to those of the intermediate layer of the nonkeratinous oral epithelium (Fig. 8–44). As they are pushed toward the gingival sulcus, where they are desquamated, they become moribund but not keratinous.

Gingival Sulcus (Crevice)

The gingival sulcus is a shallow trough surrounding the crown. It contains desquamated epithelium, leukocytes, bacteria, and various substances in solution, including antigens. The sulcular limits are defined basally by the desquamating surface of the junctional epithelium, dentolaterally by enamel and/or cementum, and epitheliolaterally by oral crevicular (sulcular) epithelium. The average depth is about 1.8 mm (the range is 0.5 to 3.0 mm). Healthy and hygienically maintained gingivae possess shallower crevices; those deeper than 3.0 mm may indicate disease, (i.e., "pocket" formation).

Crevicular (sulcular) epithelium is nonkeratinous stratified squamous and, because the basal lamina on which it rests is straight, rete ridges and pegs are absent. Junctional and crevicular epithelium are similar in certain respects; for example, both lack ridges and pegs, and both are nonkeratinous stratified squamous epithelium with basal, intermediate, and desquamating surface layers. On the other hand, there are substantial cytomorphologic differences. That is, crevicular epithelial cells are densely packed, so the intercellular spaces are smaller. The cytoplasm of crevicular cells is reduced and, except for tonofilaments, the organelles are few. Additionally, crevicular epithelium is not usually able to differentiate into junctional epithelium. This is an important clinical consideration for the surgical treatment of periodontal disease.

Crevicular epithelium also differs from its parent tissue, gingival epithelium, in being much thinner and nonkeratinous with no pegs and ridges in its healthy state. These differences are probably consequences of its sheltered position as the sulcus lining.

Primary and Secondary Cuticles

Primary (first) cuticle is the ultramicroscopic basal (basement) lamina, consisting of the lamina lucida and lamina densa, products of the post-ameloblasts. The primary cuticle is also known as the epithelial attachment, and it causes the union of the reduced enamel epithelium (primary junctional epithelium) and crown via hemidesmosomes. After replacement of the primary junctional epithelium by the secondary junctional (attachment) epithelium another cuticle is produced; because it is located on the enamel it is known as the secondary enamel cuticle. After the primary junctional epithelium migrates cervicalward onto the cementum, the primary cemental cuticle and the secondary enamel cuticle are collectively known as the dental cuticle. The cuticles are recognized as secretory products of the juxtaposed epithelium.

CLINICAL CONSIDERATIONS

There are various conditions related to the attachment apparatus that may be appropriately considered at this point.

Passive Eruption

Some scientists believe that occlusal wear of the crown is compensated for by apical cementogenesis. This elongation of the root is known as passive eruption. Although there is little doubt that cementogenesis continues after the occlusal plane has been reached, it is questionable whether coronal attrition is compensated for totally by root growth. Rather, it is likely that bone apposition also occurs in appropriate locations in the socket, so that the entire attachment mechanism responds as a functional unit.

The concept of passive eruption is based mostly on studies of histologic preparations. It was once believed to be a normal physiologic process involving the shifting of the gingival sulcus and epithelium, resulting in obliteration of the cervical collagen fiber groups. This concept is not generally accepted. Every attempt is made by the periodontist to prevent shifting of the attachment epithelium and its associated structures, because such shifting is considered to indicate pathology.

Physiologic (Mesial) Drift

Two movements are associated with physiologic drift. The first is occlusal, resulting from normal wear of tooth surfaces, and the second is toward the midline of the dental arch. This movement is known as physiologic or mesial drift, and it is associated with the wear of the contact points of teeth. These movements affect the tissues of the mesial and distal aspects of the periodontal ligament (Fig. 8–30). The mesial aspect exhibits round interstitial spaces, relaxed fibers, and bone resorption (Fig. 8–30A). The distal aspect exhibits oval interstitial areas, taut prinicipal fibers, and formation of bundle bone. The bundle bone exhibits parallel, longitudinally directed lamellae, and no haversian systems.

Changes Within Physiologic Limit

There are many clinical events that occur naturally, or that can be induced, which lie within tolerable physiologic limits, including those stimulated by orthodontic movement of teeth or by restorative and other procedures. The harsher provedures in tooth restoration, such as gold foil condensation or mechanical separation, may cause resorption of cementum or bone, breaks in the tissues, tears in the periodontal ligament, hemorrhage, and necrosis.

Intense or improperly balanced pressures and tensions produced during orthodontic movement of teeth may also produce these conditions. Generally, the side to which pressure is applied will show signs of necrosis and of bone and cementum resorption. The side subjected to tensional forces, on the other hand, might exhibit blood vessel damage and bleeding, as well as tears in the principal fiber groups.

Changes With Minimal Function

The periodontal ligament narrows with a decrease in function, and exhibits different features. That is, the organization, number, and arrangement of the definite fiber groups are loose, so that a group assumes the character of an irregularly arranged dense connective tissue or membrane. The alveolar bone possesses fewer Sharpey's fibers, and may become aplastic. Similar conditions are noted for cementum.

Once the tissues of the attachment apparatus have experienced atrophic or degenerative changes due to disease, such as those described, they must reorganize if they are once again to function as a ligament. Newly restored teeth, antagonists of bridges or dentures, and bridge abutments that have been in reduced function for some time may therefore not be totally efficient immediately on insertion. The time required for them to attain full function depends on the period necessary for ligament reconstruction. This will, of course, be governed by the extent of atrophy.

CHAPTER 9

Paraoral Tissues and Organs

This chapter is concerned with tissues and organs associated with the oral vestibule and cavity, including the lips, oral mucosa, salivary glands, tongue, tonsils, nasal passage, and sinuses.

LIP

The core of the lip is formed of striated muscle. The composition and structure of its overlaying connective tissue and its surfacing epithelium differ according to location. These differences account for the designation of the external, transitional, and vestibular (mucosal) regions of the lip (Fig. 9–1).

External Aspect

The external aspect of the lip is comprised of keratinous stratified squamous epithelium (epidermis)—more specifically, thin skin. It consists of basal, spinous, granular, transparent, and keratinous (cornified) layers. Components of the stratum basale are juxtaposed to the basement membrane. The cells are cuboidal or low columnar, rich in organelles and mitotically active. In the spinous layer the cells are polyhedral and prickled in appearance. Their mitotic activity is decreased and they contain fewer organelles, but the tonofibrils tend to increase. Flattened cells of the stratum granulosum are identified by dense accumulation of keratohyaline granules, progressively fewer organelles, and no mitotic activity. The stratum lucidum generally does not consist of a definite layer; rather, its cells progress toward keratinization, intermingling with the adjacent cell layers. Organelles are absent and their nuclei have disappeared by karyorrhexis and karyolysis. The cornified layer is thin, comprised of flattened desquamating scales. Below the basement membrane is the

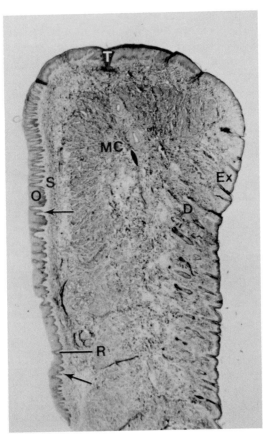

Fig. 9–1. Longitudinal section of lip, showing the external aspect (Ex) or skin, oral aspect (O), transitional (vermilion) zone (T), and muscular core (MC). The oral aspect bears rete ridges *(arrows)*. The dermis (D) contains epidermal derivatives, hair follicles, and shafts. The connective tissue of the submucosa (S) is more diffuse than either the reticular (R) or papillary layer. (Hematoxylin and eosin stain; ×8)

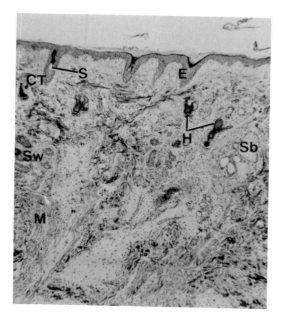

Fig. 9–2. Epidermis (E) of the external aspect of lip penetrated by hair shafts (S). The dermal connective tissue contains hair follicles (H) and sweat (Sw) and sebaceous (Sb) glands. Deep to the connective tissue (CT) is found striated muscle (M). (Hematoxylin and eosin stain; ×32)

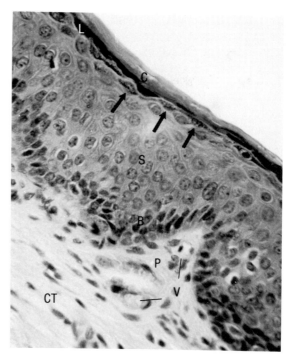

Fig. 9–3. Photomicrograph of vermilion border (transitional zone) of the lip. Illustrated are surface epithelium with rete ridges, with papillary connective tissue (P) containing blood vessels (V) and reticular connective tissue (CT) intervening. The epidermis of the transitional zone consists of basal (B), spinous (S), granular *(arrows)*, lucidum (L), and cornified (C) layers. (Hematoxylin and eosin stain; ×830)

fibrous connective tissue of the dermis (corium), which contains the papillary and reticular layers.

The dermis or connective tissue underlying the epidermis contains hair follicles and sweat and sebaceous glands. Hair shafts and ducts of the glands penetrate the epidermis to gain access to the external environment. The presence of these structures distinguishes the external aspect of the lip from the other regions (Figs. 9–1 and 9–2).

The core of the lip is composed of striated muscle bundles, the orbicularis oris. Strands of muscles of facial expression are intercalated among the fibrous elements of the dermis, providing the skin with mobility.

Transitional Zone

This region of the lip is also referred to as the vermilion border, red zone, and marginal zone. It is present only in humans. As in the external aspect, the epidermal layers are the strata basale, spinosum, and granulosum, a well-developed prekeratinous stratum lucidum, and a very thin, poorly developed stratum corneum (Fig. 9–3).

The connective tissue papillae (papillary ridges) that intervene between the epithelial ridges are numerous, long, and slender, and are well supplied with small blood vessels to accommodate the metabolic needs of the avascular epithelial layer. The relatively thick transparent lucidum and thin corneum layers, together with the tall well-vascularized papillae, account for the intense red (pink) color of the transitional zone. These features and the absence of hair and glands distinguish the transitional portion of the lip.

Development

The gross development of the lip has been discussed in Chapter 4 with the formation of the vestibular lamina (lip furrow band). The external covering of the face and transitional zone of the lip are comprised of keratinous stratified squamous epithelium (epidermis) and an underlying dermis of irregularly arranged fibrous connective tissue.

During the first month in utero the external surface of the embryo is covered by a single layer of cuboidal epithelium. Within a few weeks these cells divide, and the overlying squamoid daughter cells are referred to as the periderm. This protects the cuboidal epithelium. Toward the end of the second month the periderm exhibits increased mitotic activity and stratification to form a multilayer structure. With progressive development the cells located between the basal and superficial cells increase in size and become vacuolated. In the third month the basalmost cells acquire features of the germinating layer. Continued development results in an increase in both the number of layers and the number of cells in each layer. By the fifth month, the epidermis has acquired definitive features, including the formation of epithelial projections into the connective tissue that are known as the papillary ridges or papillae.

Epidermal derivatives (e.g., hair, sebaceous and sweat glands) as well as the epidermis originate from the ectoderm. The former develop concomitantly with the epidermis, as epithelial downgrowths of the layer, completing their development in the connective tissues. The distal terminals of the epidermal derivatives remain continuous with the epithelium. Most sebaceous glands develop and remain associated with hair. They may also be found in areas in which hair does not normally occur—for example, in the deeper recesses of the nasal cavity, in the oral mucosa as Fordyce granules, and on eyelids. Where sebaceous glands occur independently, it appears that development of hair germ ceases while the lateral bud continues to grow.

ORAL MUCOSA

The oral cavity contains two chambers: the anterior, known as the oral vestibule; and the posterior or principal chamber, known as the oral cavity proper.

The vestibule is limited anteriorly by the lips and cheeks and posteriorly by the gingiva and teeth. The oral cavity proper is bound anteriorly and laterally by the gingiva and teeth, superiorly by the palates, and basally by the tongue and sublingual sulcus. Because of functional differences, there are variations in the character and composition of the tissues in the different regions of the mouth.

The oral cavity is the first segment of the digestive system. The architectural pattern of the digestive tube usually includes the surface epithelium, basement membrane, lamina propria, muscularis mucosa, and submucosa. The epithelial layer, basement membrane, lamina propria, and muscularis mucosa constitute the mucosa. The epithelial layer is separated from the lamina propria by the basement membrane, and the lamina propria is separated from the submucosa by the muscularis mucosa. The architectural pattern of the oral cavity is atypical in that not only is the muscularis mucosa totally absent but even the submucosa in such areas as the gingiva, regions of the hard palate, and dorsum of the tongue. When present the submucosa is indistinct and poorly defined, so that the connective tissues of the lamina propria and submucosa merge imperceptibly.

Located in the submucosa are salivary glands. Their secretions may be mucous, serous, or seromucous. Also, a ring of lymphoid tissue is found at the entrance of the pharynx, the tonsils. These are housed in the lamina propria.

Based on anatomic variations in both the epithelium and its underlying connective tissues, the following terms may be applied: lining, masticatory, or specialized oral mucosae. The lining mucosa is found in the more protected areas of the oral cavity, such as the inner lining of the lips and cheeks, soft palate, floor of the mouth, underside of the tongue, and alveolar segment of the dental arch. The masticatory mucosa occupies sites more exposed to chewing forces—that is, the gingiva and hard palate. The specialized mucosa is located exclusively on the dorsum of the tongue.

Keratinous and Nonkeratinous Epithelium

The epithelium associated with lining mucosa is nonkeratinous, that of masticatory mucosa is designated as orthokeratinous, and that of the specialized mucosa can be either. All are comprised of keratinocytes and all advance toward keratinization. Although their features in the stratum basale are similar, increasingly greater differences begin and progress in the suprabasal layers.

Keratinocytes

The keratinocytes of the basal layer of keratinous and nonkeratinous epithelia are similar. They are cuboidal or low columnar and are anchored to the basement membrane by hemidesmosomes.

Their organelle populations are maximal, and their mitotic activity is greatest. Basal layer keratinocytes are dissimilar in that the "tono" elements in nonkeratinous epithelium are filamentous, and in keratinous epithelium are fibrillar.

Keratinocytes of the stratum spinosum in both keratinous and nonkeratinous epithelia are likewise similar in that the cells are rounder, membrane-coating granules are associated with tonofilaments and tonofibrils, and mitotic activity and organelle populations are decreased. The keratinocytes are dissimilar in that the density and size of the tonofibril bundles increase in keratinous epithelium. These course through the cells, with many terminating in the desmosomes; in histologic section these appear prickled. The tonofilaments of keratinocytes of nonkeratinous epithelium are not appreciatively increased, and the spinous appearance is not prominent.

Keratinocytes of keratinous epithelium form a definite granular layer. The cells are flattened, the organelle populations are greatly reduced, the nuclei exhibit degeneration, the internal leaflet of the plasma membrane is thickened, and the tonofibril bundles and their associated keratohyalin granules are increased. Additionally, at the ultrastructural level, other granules are observed, called membrane-coating granules. In contrast, keratinocytes of nonkeratinous epithelium never develop a granular layer. The cells of this intermediate layer are more flattened, the tonofilaments are less numerous, and the organelles are fewer but the nuclei are intact. Glycogen accumulations are prominent.

In the superficial desquamating layer, the components of keratinous epithelium are reduced to keratin-packed discs devoid of both organelles and nuclei. In parakeratinous cells, the nuclei are present as degenerated fragments. On the other hand, nonkeratinous superficial cells are squamoid and contain denser tonofilament aggregations, decreased organelle populations, and well-defined nuclei.

Nonkeratinocytes

Other than keratinocytes, the oral epithelium contains Langerhans' cells, lymphocytes, melanocytes, and Merkel's cells. Langerhans' cells are formed in the marrow and migrate to stratified squamous epithelium. In hematoxylin and eosin stained sections the cells display a clear cytoplasm with intense-staining nuclei. Gold chloride stained sections reveal that they are irregular in shape, with dendritic processes extending throughout the stratum spinosum. Electron microscopic examination indicates the absence of desmosomes, tonofibrils, and melanosomes, and greatly reduced organelle populations. Distinctive keyhole-shaped membrane-bound granules measure 15 to 50 nm long and 4 nm wide. The nucleus is multiply notched, thus presenting an irregular contour. Functionally, the Langerhans' cells, with the cooperation of lymphocytes, are implicated in immune responses.

Merkel's cells are nondendritic, and are located in the basal epidermal layer joined to keratinocytes by desmosomes. These cells are few, except in connective tissue, especially at sites at which the nerve and blood supplies are abundant. The nucleus is notched and the cytoplasm contains dense core vesicles, two features that distinguish them from keratinocytes. On the other hand, the presence of perinuclear filamentous aggregations and occasional melanosomes are features that reveal a keratinocyte-like appearance on cursory examination. The dense core vesicles of Merkel's cells and their close relationship to neighboring nerve endings have led many to believe that these are paraneurons implicated in sensory reception.

Lymphocytes are normally a minor constituent of oral epithelium. They are found in all layers of the oral epithelium en route to the oral cavity, where they are called salivary corpuscles. In the subepithelial connective tissue they aggregate in response to inflammatory processes. With electron microscopy they are identified as large rounded cells having sparse organelle populations and large spherical nuclei with compact chromatin. They do not contain tonoelements or desmosomes.

Melanocytes have been discussed earlier. They originate at the neural crest, and have migrated to the dermoepithelial junction. They are astral-shaped, with dendritic processes extending among the components of the stratum spinosum (see Fig. 2–35). They are neither joined to neighboring epithelial cells nor do they contain tonofilaments. They react positively to silver stains and to dopa-oxidase-tyrosinase techniques. These cells synthesize pigment granules, melanosomes (see Fig. 2–35), which may be taken up by epithelial cells.

Functions

Because the oral mucosa is inextricably bound to the organs with which it is associated and/or contained, its functions must subserve those of the oral and paraoral organs. Accordingly, the functions of the oral mucosa are those generally performed in the oral cavity; these include mastication, speech, respiration, ingestion, digestion, absorption, taste, deglutition, and protection.

Mastication. Biting, chewing, grinding, moving, and mixing the ingested food, although principally functions of teeth, are activities assisted by the lips, cheeks, tongue, and palates. The functional interrelationships existing between these organs and the mucosa are essential in maintaining an effective and efficient masticatory apparatus.

Speech. Similarly, the lips, cheeks, tongue, and palates contribute to speech production. In fact, voice quality depends largely on the size, shape, and control exercised over the various cavities (i.e., nasal, oral, pharyngeal, laryngeal, pleural).

Respiration. Although the oral cavity cannot heat, humidify, or filter inspired air as effectively as the nasal passage, it is a necessary adjunct to the respiratory system. This activity of the oral mucosa is particularly appreciated by those possessing obstructed or congested nasal passages.

Ingestion, Digestion, and Absorption. Food ingested by the oral cavity is fragmented and mixed with saliva by muscle action in masticatory activities. Although food is prepared for swallowing—deglutition—by the moistening and lubricating action of mucous and serous secretions, it is chemically digested by the enzyme ptyalin (α amylase), produced by the major and minor salivary glands that empty their secretions into the oral cavity. Substances dissolved in saliva, if retained in an area covered by thin epithelium and by loose, well-vascularized connective tissue, may be absorbed in the oral cavity. Areas with these characteristics include the sublingual sulcus and the retrobuccal fold. Slow absorption recommended for certain medications may be accomplished by inserting tablets (linguetes) under the tongue or mucobuccal fold.

Taste. This function is accomplished by taste receptors located in mucosal projections or evaginations—the lingual papillae—found on the dorsum of the tongue. Taste will be discussed below (Tongue: Taste Buds).

Deglutition. Food prepared for deglutition is moistened and formed into a soft ball, a bolus. It is moved toward the base of the tongue, and swallowing requires three steps. The first involves biting in centric relation. The second involves pressing the apex of the tongue against the lingual surface of the lower anterior teeth, thereby forcing the midsection of the dorsum of the tongue against the hard palate. Finally, with a single forceful contraction of the mylohyoid muscle, the dorsum of the more anterior part of the tongue is pressed against the hard palate while the base of the tongue is thrown posteriorly, propelling the bolus from the oropharynx to the laryngopharynx and subsequently to the esophagus. Simultaneously, the orifice between the oral and nasal pharynx is closed, and the pharynx and larynx are elevated.

Secretion and Excretion. Products synthesized specifically for internal body use or its surfaces are called secretions. These are in contradistinction to body waste discharges, which are called excretions. A substance may function as both a secretion and excretion (e.g., sweat and salivary secretions). The elaborations of the salivary glands are considered secretions because they help to keep the oral epithelium wet, moisten and lubricate ingested food for deglutition, and contain starch-hydrolyzing enzymes. Waste products, especially lipid-soluble ones, are excreted into the oral cavity with salivary gland secretions. Certain ions, such as bicarbonate, are also excreted into the oral cavity via the ducts of the salivary glands.

Protection. Although diffuse accumulations of lymphoid tissues occur throughout the mucosa as required, protection is offered largely by the tonsillar ring of Waldeyer found at the entrance of the pharynx. Lymphoid tissue functions protectively through the phagocytic activity of the macrophages. Their action is directed against pathogenic organisms, foreign bodies, and moribund vascular elements. Additional protection is offered by antibodies produced by lymphocytes and plasma cells.

Saliva contains a bacteriocidal enzyme, lysozyme, synthesized by serous cells. Lysozyme destroys the cell walls of the microorganisms, thus facilitating access of salivary thiocyanate ions and death of the bacteria. Via the action of lysozyme and the flushing actions of the salivary secretions, dental caries and periodontal infections are reduced. A more comprehensive discussion of saliva is provided later in this chapter (see page 254).

Sensation. The mucosa of the oral and nasal cavities contain both free and encapsulated receptor nerve endings (see Chap. 2). These nerve endings react to specific stimuli such as touch, pressure, vibration, fiber tension, heat, cold, taste, and smell. Reflexes associated with swallowing, salivation, retching, and vomiting are also under the control of nerve endings of the oral mucosa. Because of the varied properties of the nerve endings, valuable information is provided about the conditions of the oral and nasal environments.

Vestibular Mucosa

The oral vestibule is made up of labial and buccal segments, which communicate with the roof of the vestibule (fornix) and the alveolar mucosa. These are continuous with the loose tissue covering the alveolar bone. The alveolar mucosa joins the attached gingiva at the mucogingival line (Fig. 9–4). Mucosal folds called la-

bial frenula, which are medially and laterally situated in the vestibule, tend to limit extensive lip movement. Each part of the vestibule (labial and buccal segments, fornix, and alveolar mucosa) is comprised of three layers: epithelium; lamina propria, including the basement membrane; and submucosa.

Labial and Buccal Mucosa

The epithelium lining the lips and cheeks is relatively thick (up to 500 μm), wet stratified squamous—hence, nonkeratinous. The basal germinating layer is the most prominent. Granular, transparent, and cornified layers are absent. The superficial desquamating layer is comprised of flattened cells. Their nuclei are always present, although smaller and often pyknotic. Between the desquamating and basal germinating layers is the intermediate spinous layer. This is the thickest layer, and its components are larger and rounder than spinous cells in keratinous epithelium. However, the tonofilaments and their desmosomal connections are greatly reduced. Furthermore, the intercellular spaces are smaller and fewer. These features account for the nonspinous and more densely packed appearance of this layer (Fig. 9–5).

The lamina propria consists of a continuous basement membrane, following the contours of

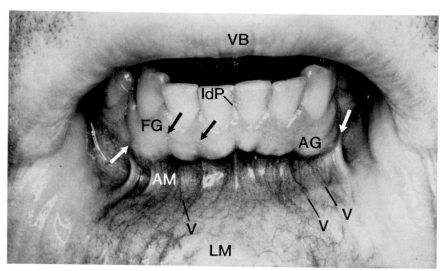

Fig. 9–4. Photograph of reflected lower lip showing oral vestibule and buccal aspect of the incisor segment of the mandibular arch. The upper lip shows the vermilion border (VB) and a section of skin with hair stubble. The rich blood supply (V) of the labial mucosa (LM) is seen because the epithelial covering is very thin and nonkeratinous. Separating the attached gingiva (AG) from the alveolar mucosa (AM) is the mucogingival line *(white arrows)*. Above the stippled area of the attached gingiva *(black arrows)* is the narrow free gingiva (FG), which extends between the teeth as the interdental papilla (IdP).

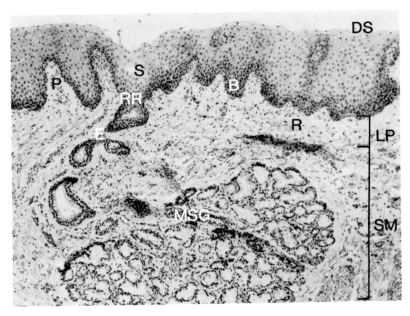

Fig. 9–5. Photomicrograph of tissue taken from the vestibular aspect of the lip. The epithelium is nonkeratinous and consists of the basal (B), intermediate spinous (S), and desquamating surface (DS) layers. The lamina propria (LP) is comprised of the reticular (R) and papillary (P) layers. Note that the epithelial rete ridges (RR) are short and blunt, and that excretory ducts (E) of the mucous end-pieces of the minor salivary gland (MSG) in the submucosa (SM) meander through the papillary layer and exit surfaceward through a rete ridge (RR).

the epithelial ridges and interdigitating connective tissue papillae. Although dimensional and morphologic variations in epithelial ridges and papillae occur, depending on location and masticatory forces, they are generally fewer, broader, more shallow, and often of more irregular profile than those that are covered by orthokeratinous masticatory epithelium. The papillary and reticular connective tissue layers are moderately dense and contain collagenous, reticular, and elastic fibers. These form the meshwork for the rich capillary network and nerves arriving from the underlying submucosa. Occasionally they are collected into larger bundles to provide for greater tissue flexibility and movement. Thus, the possibility of injury to the mucosa by chewing is decreased. The fibrous bundles located deep in the submucosa also anchor the soft tissue to the underlying striated muscles—buccinator (cheek) and orbicularis oris (lips). The cell types present are not unusual, except for mast cell aggregates in the perivascular tissue. Coursing irregularly through the lamina propria are the ducts of the minor salivary glands, which exit via the epithelial ridges ("pegs").

The submucosa is distinguished from the overlying connective tissue by its looser texture and by the presence of secretory end-pieces of the minor salivary glands (Fig. 9–5). Occasional patches of fat and, more rarely, sebaceous glands as Fordyce spots, may be observed. The latter, when present, are generally located close to the lips, especially the corners. The larger arteries and draining veins and lymphatic vessels are similarly found in the submucosa.

Frenula, Vestibular Fornix, and Alveolar Mucosa

The epithelium is nonkeratinous stratified squamous, and structurally similar to that of the other segments of the oral vestibule. The connective tissues of both the lamina propria and the submucosa are especially loose, providing for its accommodating ability. This is a valuable feature in the administration of local anesthetics, and is also significant in the vestibular or peripheral form (border molding) of complete dentures, because these areas require careful and precise fitting to prevent the ingress of air or fluids under the denture. Except for the alveolar mucosa, which is secured to the cortical plates of the alveolar process by a periosteum, the mucosa of the fornix is attached to underlying connective tissue. The mucosal folds of the median and lateral frenula are similarly comprised of loose connective tissue; however, muscular elements may be present.

Oral Cavity

The oral cavity proper is lingual to the dental arches, and includes the gingivae, palates, tongue, and sublingual sulcus. The gingiva on both the vestibular and oral or lingual aspects is covered by masticatory mucosa, as are the anterior segments of the hard palate. The dorsum of the tongue is covered by a specialized mucosa. On the other hand, the more sheltered areas of the oral cavity are composed of lining mucosa.

Lining Mucosa

Parts of the oral cavity composed of a lining mucosa extend from the mucogingival line of the lingual aspect of the dental arch onto the ventral surface of the tongue, up to the specialized mucosa of the dorsum of the tongue. Included are the lingual alveolar mucosa, the sulcus (floor) of the cavity, the underside (ventral surface) of the tongue, and its frenulum. Lining mucosa is also located in the soft palate and its terminal, the uvula. Except for minor differences, the lining mucosa of the structures in the oral cavity proper is structurally similar to that of the vestibular segments.

Lingual Alveolar Mucosa. The epithelium is similar to its vestibular counterpart in that it is nonkeratinous stratified squamous. Variations are slight, involving mostly the dimensional aspects of the tissue. That is, the overall thickness is somewhat decreased and the epithelial ridges are fewer, shorter, and blunter. These differences are also reflected in the underlying connective tissue, especially the papillae, which are fewer and shallower. The fibrous elements of the lamina propria form a delicate meshwork that houses the rich vascular network, which accounts for the intense pink color of the tissue. Except for the secretory units of the minor salivary glands in the submucosa and the periosteal collagenous fibers that attach the connective tissue to bone, the lamina propria and submucosa are substantively so similar that they merge imperceptibly. The looseness of the connective tissue and the presence of elastic fibers provide for tissue mobility and flexibility.

Sublingual Sulcus. The floor of the mouth, or

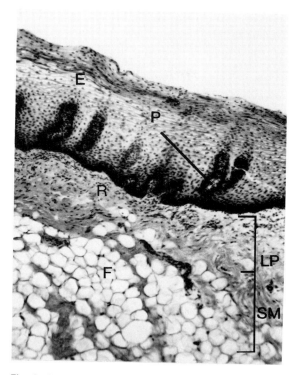

Fig. 9–6. Sublingual sulcus. The nonkeratinous epithelium (E) is thin and the rete ridges and complementary papillae (P) are short and blunt. The reticular layer (R) of the lamina propria (LP) is denser than the submucosa (SM). The latter contains fat accumulations (F) and minor salivary glands. (Hematoxylin and eosin stain; ×50)

sublingual sulcus, is continuous with the lingual alveolar mucosa on the one aspect and the ventral mucosa of the tongue on the other. Because of its sheltered location its epithelium is extremely thin (less than 150 μm) and transparent, resulting in the deep pink color of the tissue. It is nonkeratinous and there are few ridges which are generally very short and blunt (Fig. 9–6).

The basement membrane is very thin, possibly because a firm attachment of the lamina propria to the epithelium is not required. The papillary layer is thin, diffuse, and indistinct, especially in areas in which the papillae are shortest. The reticular layer is denser than either of its adjoining layers. The submucosa is similar in structure and density to that of the alveolar mucosa, except that the secretory units of the minor salivary glands are more numerous and the tissue contains numerous patches of fat (Fig. 9–6). The presence of secretory units and fat cells renders this tissue more diffuse than other lining mucosae. The epithelium and connective tissue core of the lingual frenulum is similar in structure and composition to that of the sulcus, which is to be expected because the frenulum is only reflected mucosal fold.

Ventral Lingual Mucosa. The mucosa of the inferior surface of the tongue is firmly attached to its muscular core. Thus, although thin and nonkeratinous, the epithelium has many short and sometimes branching ridges (Fig. 9–7). Limited mobility of the mucosa is a result of collagen fibers coursing from the reticular layer to those of the muscle sheaths. The connective tissue is thin and diffuse, accommodating the blood and lymph vessels, nerves, secretory units, and patches of fat cells. The submucosa, if present, is very indistinct.

Soft Palatal and Uvular Mucosa. The soft palate and uvula are distal continuations of the roof of the oral cavity (Fig. 9–8). On its nasal aspect the palate forms the floor of the nasal cavity. The mucosae of the oral and nasal aspects of the soft palate are distinctly different.

Oral Aspect. The color of the soft palate and its distalmost segment, the uvula, is deep pink. The soft palate is softer to the touch than the hard palate because its core is comprised of striated muscle rather than of bone. Further-

more, although the mucosa of the soft palate contains an abundance of elastic fibers, its movement is limited.

Food that has been prepared for swallowing in the anterior part of the mouth forms a moist round bolus. In swallowing, the bolus or fluid brushes against the uvula and stimulates the soft palate, with the uvula, to rise and seal the pharyngeal isthmus so that food or drink is prevented from entering the nasal cavity. The lining epithelium is nonkeratinous and thin (less than 200 µm) because the bolus does not abrade the tissue harshly. Unlike the lining epithelium elsewhere in the oral and vestibular cavities, but similar to the pharyngeal epithelium, the palatal epithelium may contain occasional taste buds. Consistent with these conditions, the epithelial ridges and their intervening papillae are broad, shallow, and even absent, especially in the uvular segment.

The connective tissue is thin and diffuse, containing fine collagen and elastic fibers. In the lamina propria the collagen fibers rarey collect into dense bundles. An increase in the cell population and a rich vascular network are consistent with the decreased fiber density. A similar situation is observed in the submucosa. Re-

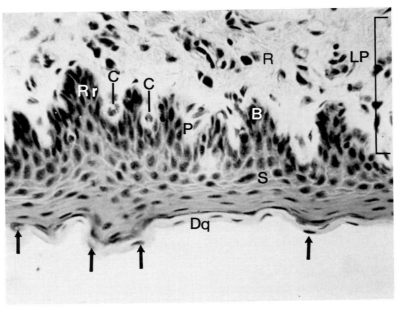

Fig. 9–7.   Section of ventral lingual mucosa showing nonkeratinous epithelium containing basal (B), intermediate spinous (S), and surface desquamating (Dq) layers. Note that the desquamating cells *(arrows)* contain prominent nuclei. The rete apparatus consists of short branched ridges (Rr) and diffuse papillae (P). The latter contain numerous blood vessels (C). The reticular (R) layer of the lamina propria (LP) is similar to that of the papillary connective tissue. (Hematoxylin and eosin stain; × 100)

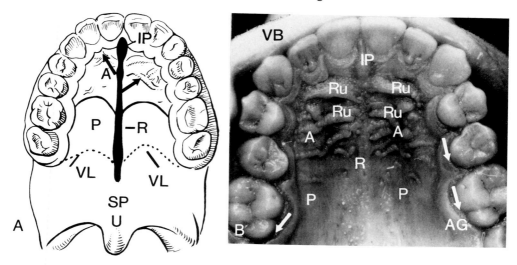

Fig. 9–8. Palate. *A,* Diagram showing premaxillary (incisive) papilla (IP) and anterolateral (A) and posterolateral (P) regions of the hard palate. These are separated by the raphe (R). The relationship of the vibrating line (VL) to the soft palate (SP) is shown (U, uvula; platine rugae, *(arrows).* *B,* Photograph of oral cavity showing vermilion border (VB) of the lip. The relationship of the free *(arrows)* and attached (AG) gingivae is demonstrated. The rugae (Ru) of the anterolateral (A) segments of the hard palate are prominent. The incisive papilla (IP), raphe (R), and posterolateral segments (P) of the palate are demonstrated.

search involving various microtechnical procedures for identifying collagen and elastin has indicated that elastic fibers are usually dispersed among the collagenous elements in both the lamina propria and the submucosa. Therefore, the presence of a definitive elastic fibrous layer is unlikely. The deeper connective tissue, conventionally designated as the submucosa, is packed with secretory units of minor salivary glands. Their secretions are pure mucus, and are expelled on the epithelium via ducts that often are greatly distended (Fig. 9–9).

Striated muscles form the core of the soft palate. These elevate and tense the soft palate and uvular segment, thereby sealing the pharyngeal isthmus. Movements of the soft palate extend posteriorly from a point known as the vibrating line (Fig. 9–8A). It is important in denture construction that the appliance not extend beyond the vibrating line if the maximum postpalatal seal is to be obtained.

Nasal Aspect. The mucosa of the nasal aspect of the soft palate constitutes the floor of the nasopharynx. It is attached to muscle sheaths of the central core. The epithelium is of the respiratory type—pseudostratified ciliated columnar, with goblet cells (Fig. 9–10A). In areas of friction, however, respiratory epithelium is replaced by nonkeratinous stratified squamous epithelium (Fig. 9–10B). The lamina propria is composed of loose connective tissue, with many blood vessels. The collagen fibers are especially abundant, and adjoin the striated muscular core. They secure the mucosa to the muscle sheaths. Elastic fibers that are prominent in the uvular region provide resiliency. Numerous tubuloalveolar glands, the secretions of which are mixed, are located deep in the connective tissue.

Pharyngeal Mucosa

The pharynx is funnel-shaped. It consists of a superior segment, the nasopharynx, a middle segment, the oropharynx, and an inferior segment, the laryngopharynx. The entrance of the pharynx is protected by the tonsillar ring of Waldeyer.

The oropharynx is composed of a thin layer of nonkeratinous stratified squamous epithelium, which may contain taste buds. Accordingly, it is structurally similar to the soft palate. The basement membrane is likewise similar in thickness to that of the soft palate and uvula, but it is thinner than that of the nasal aspect of the soft palate. The epithelial ridges and adjacent papillae are shallow and the connective tissue of the reticular layer tends to be fibroelastin-like housing mucous glands. Unlike the palatal and uvular mucosae, the elastic fibers deep in the

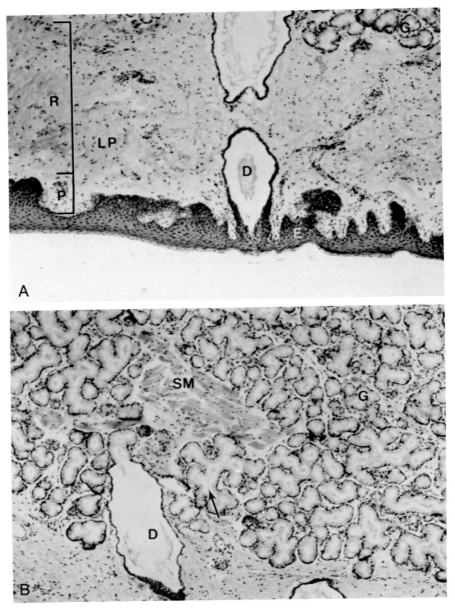

Fig. 9–9. Soft palate. *A,* Mucosa of soft palate covered by nonkeratinous epithelium (E). The morphologic features of the rete ridges and their relationship to the sinusoidal excretory ducts (D) are shown. The papillae (P) are shallow and the connective tissue, as in the reticular (R) layer of the lamina propria (LP), is diffuse (G, mucous end-pieces). *B,* Submucosa contains glands (G) with mucous end-pieces, which may be branched tubuloalveolar *(arrow).* Striated muscle (SM) fibers may intermingle with the secretory units. The excretory duct (D) entering the lamina propria is sinusoidal. (Hematoxylin and eosin stain; ×50)

lamina propria are collected into a definite elastic lamina. This occupies the position of and substitutes for the muscularis mucosa. Because a submucosa is absent in the oropharynx, the elastic lamina adjoins muscle.

The nasopharynx is similar in structure to the nasal aspect of the soft palate and uvula. Hence, it is composed of respiratory epithelium, whose cells rest on a well-developed basement membrane. The lamina propia is comprised of loose connective tissue rich in elastic fibers and housing seromucous glands. A prominent elastic lamina separates the lamina propria from the submucosa. Only in the lateral aspects of the

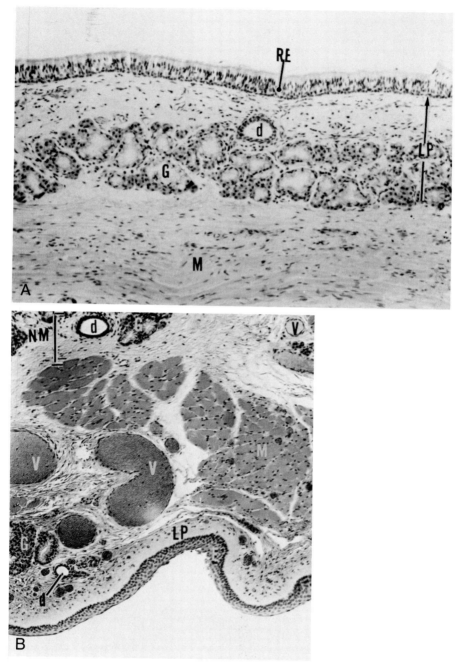

Fig. 9–10. Soft palate. *A,* Nasal aspect of the soft palate covered by respiratory epithelium (RE). The deeper layer of the lamina propria (LP) is a diffuse connective tissue containing tubuloalveolar secretory units of the glands (G). Excretory ducts (d) pass through the lamina propria to empty their contents onto the epithelial surface. The core of the soft palate is composed of striated muscle (M). (Hematoxylin and eosin stain; ×80) *B,* Oral aspect of uvula showing nonkeratinous epithelium (E), lamina propria (LP), and submucosa containing secretory units of palatal glands (G), ducts (d), vascular elements (V), and striated muscle core (M). Note a deeper segment of the nasal mucosa (NM), with secretory units (G) and duct (d) *(upper left).* (Hematoxylin and eosin stain; ×100)

nasopharynx and the pharyngoesophageal junction is a definable submucosa present. Here mucous glands are found, which may extend toward the external striated muscle layers.

The laryngopharynx is located below the superior border of the epiglottis and posterior to the larynx. It opens into the esophagus inferiorly and the laryngeal aperature anteriorly. The mucous membrane is similar to that of the oropharynx in that it is lined by nonkeratinous stratified squamous epithelium and in that the lamina propria is immediately adjacent to the underlying muscular wall. The region of the laryngopharynx located between the aryepiglottic fold and the lateral pharyngeal wall is known as the piriform recess. It is of clinical significance because the sensory nerve to the larynx, the internal laryngeal nerve and its branches lie below the thin mucous membrane. To prevent damage to this nerve, extreme care must be exercised in dislodging foreign objects and bones caught in this recess.

Masticatory Mucosa

Three structures of the oral cavity may be composed of masticatory mucosa—the gingiva, interdental papillae, and hard palate.

Gingiva. The gingiva is commonly known as the gum. It extends from its free terminal (dentogingival junction) near the neck of the tooth to the its junction with the alveolar mucosa, the mucogingival line (Fig. 9–4). The lingual portion of the maxillary gingiva merges with the palatal mucosa.

Parts. The gingiva is divided into three regions: free gingiva, attached gingiva, and interdental papilla (see below and Fig. 8–25). The boundary separating the free and attached gingivae is marked by the shallow gingival groove (see Fig. 8–1). The edge of the free gingiva is known as the marginal gingiva. The epithelium is reflected over the marginal edge to face the crown. The enamel-facing surface of the marginal gingiva, which constitutes the dentogingival epithelial junction, is comprised of the crevicular (sulcular) epithelium and the junctional (attachment) epithelium (see Fig. 8–25). Because of its shape, the segment of the free gingiva located between the teeth (interproximal or interdental space) is called the papillary gingiva or interdental papilla.

Macroscopic Features. The gingiva is generally pale pink. In some people, when thick, it is grayish and, with melanin deposits, it is brown. The free gingiva has a wavy contour and smooth surface, while the surface of the attached gingiva may be stippled (Fig. 9–4).

Microscopic Features. Depending on the severity of functional forces, gingival epithelium ranges from nonkeratinous (10% occurrence) to parakeratinous (75%) to orthokeratinous (15%) (Fig. 9–11). Masticatory epithelium never achieves the degree of keratinization of skin. Accordingly, maximum cornification in the oral cavity is designated as orthokeratinous. In the oral cavity, where the term "keratinous" is used, orthokeratinization is implied, and the terms may be interchanged. The term "parakeratinous" designates epithelium that is less cornified than orthokeratinous.

Orthokeratinous epithelium is generally thick (up to 200 μm). There are many epithelial ridges that are long, slender, and often branched, features that indicate increased functional activity. Because epithelium is cornified, four definitive layers are present, the strata basale (germinativum), spinosum, granulosum, and corneum (Fig. 9–11A). The characteristics of these layers have been described previously. In parakeratosis the superficial cornifying layer is thickened and cornification is incomplete, so that the flattened cell may contain nuclear fragments and a few isolated organelles (Fig. 9–11B). Although a definitive granular layer may be absent, granular cells are intermingled among the underlying spinous and/or overlying cornifying cells. Cells of the germinating layer may contain melanin (Fig. 9–11B).

Replacement of gingival epithelium has been determined by using the mitotic frequency as an index. In general, the turnover rate of gingival keratinocytes tends to increase with age, and the thickness of the stratum corneum tends to decrease with age. Although the precise stimulus for the change in thickness is undetermined, it is most likely diet-related.

The lamina propria of the gingiva is an irregularly arranged dense fibrous connective tissue. Its turnover rate in the gingiva is higher than that elsewhere in the body. There are many long and narrow papillae, especially in areas subjected to harsher stimuli. The interdigitation of the epithelial ridges and papillae provide for tight interlocking. The extreme length of the

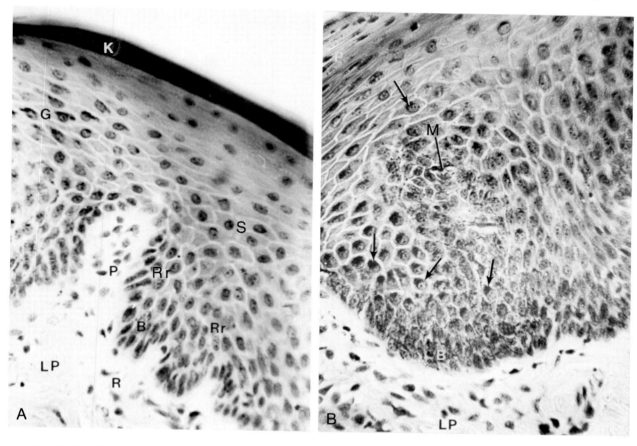

Fig. 9–11. Oral epithelium. *A,* Orthokeratinous epithelium of the gingiva showing the basal (B), spinous (S), granular (G), and keratinous (K) layers. The rete ridges (Rr) are long and often branched, and are complemented by similar papillae (P) of dense connective tissue. The connective tissue of the reticular (R) layer contains more collagen, thus increasing its density (LP, lamina propria). (Hematoxylin and eosin stain; × 430) *B,* Parakeratinous epithelium showing basal (B), spinous (S), and parakeratinous (P) surface cells with pyknotic or karyorrhexic nuclei. Note the melanin accumulations *(arrows)* in the epithelial cells and in the processes of melanocytes (M) (LP, lamina propria). (Hematoxylin and eosin stain; × 430) *C,* Nonkeratinous (NK) epithelium showing basal (B), intermediate spinous (S), and surface desquamating layers. The cells of the latter show well-defined nuclei. Note the shallow rete ridges (Rr). The papillary (P) connective tissue is less dense than the reticular (R) layer of the lamina propria (LP). (Hematoxylin and eosin stain; × 100)

papillae, coupled with the firm attachment of the epithelial ridges to the deeper connective tissue, cause the epithelium capping the papillae to bulge, accounting for the "pitted" or "stippled" appearance of the gingiva (Fig. 9–4). The connective tissue of the reticular layer is even denser than that of the papillae. The increased collagen content is responsible for its increased density, reduced cell population, and decreased vascular supply. Mast cells are relatively numerous.

Interdental Papillae. These are cone-shaped in longitudinal sections. Those of distally located teeth are papillae-shaped. The papillae occupy interdental spaces. Except for minor structural modifications, the papillary tissues are essen-

tially the same as those of the free gingiva (Fig. 9–12). Accordingly, the epithelium varies from orthokeratinous, exposed to masticatory forces, to nonkeratinous, in the more protected areas, or col sites. These are vallate troughs of the interdental papillae that join the vestibular and lingual segments of the papillae, adapting to the outline of the interproximal contact points. The epithelial ridges are long and slender, and are often branched (Fig. 9–12). These accommodate the papillary connective tissue. Although the latter is densely fibrous, the underlying reticular layer of the lamina propria is much denser. This condition is reflected in the decreased number of cells and in the paucity of vascular channels. The base of the reticular layer of the lamina

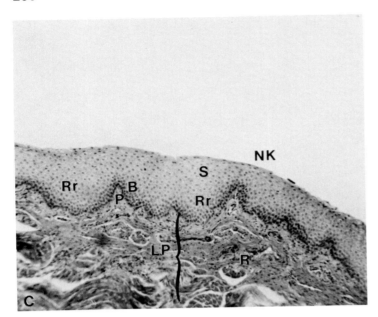

Fig. 9–11 (continued).

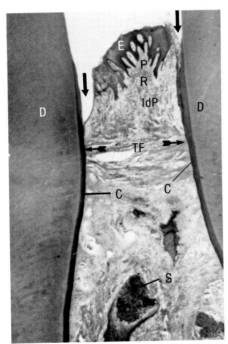

Fig. 9–12. Interdental tissue. The epithelial (E) surface of the interdental papilla (IdP) is nonkeratinous. Note the rete apparatus with long and slender (ridges and papillae), and the moderately dense reticular (R) layer and papillary layer (P). Also note the transseptal fibers (TF) an their relationships to the cementum (C) and crest of the bony septum (S). An artifact, the space created by dissolved enamel *(arrows)*, ovelies the dentin (D). (Hematoxylin and eosin stain; ×8)

propria rests on the transseptal principal fibers of the interdental ligament (Fig. 9–12).

Blood, Lymph, and Nerve Supplies. The arterial supply to the gingiva of the mandibular and maxillary arches differs. The gingiva of the mandibular arch associated with the anterior buccal regions, is supplied by the mental artery and that of the anterior lingual regions is supplied by the incisive and sublingual arteries. The gingiva of the posterior lingual arch segments is served by the inferior alveolar and sublingual arteries, and that of the posterior buccal regions is supplied by the inferior alveolar and buccal arteries.

The gingival tissues of the maxillary arch anteriorly are supplied by the anterior superior alveolar artery, as well as lingually by the greater palatine artery and buccally by the buccal artery. The posterior segments of the maxillary arch are vascularized by the posterior superior alveolar artery.

Lymph from various regions flows from smaller to larger vessels with the paths generally following those of the arteries and companion veins. Lymph nodes drain specific areas: the submandibular node drains the lateral vestibular gingiva; the submandibular and submental nodes drain the anteior mandibular gingiva; the cervical nodes drain the posterior mandibular gingiva; and the superior cervical nodes drain the maxillary gingiva and palate. The buccal and mandibular nodes (parts of the facial node) only rarely drain the gingivae.

Nerves follow paths of the vascular and lymph channels. The vestibular and lingual segments of the mandibular gingiva are innervated by the inferior alveolar and buccal branches of the mandibular divisions of the fifth cranial nerve (V) and the lingual nerves, respectively. The nerve supply of the maxillary gingivae originates from branches of the maxillary nerve (mostly anterior, middle, and posterior superior alveolar nerves). The palatal gingiva receives innervation from the nasopalatine and greater palatine nerves. The buccal nerve also supplies a portion of the posterior vestibular gingiva.

Aging Effects. Aging tends to affect all the tissues associated with the gingiva: epithelium, connective tissue, and bone. Especially significant are the effects on collagen. The fibers tend to become thicker and aggregated into dense parallel bundles, a pattern resembling that of ligaments and tendons. Furthermore, some of the thicker fibers acquire the staining properties of elastin. The precise significance of this property has yet to be determined. At the crest of the alveolus the bone may acquire a cartilaginous appearance and adjacent fibrocytes tend to change morphologically, becoming encapsulated in lacunae.

Hard Palate. The palate forms the ceiling of the mouth and floor of the nasal cavity. It is divided into two regions, the anterior hard and posterior soft palates (see above, Soft Palatal and Uvular Mucosa). The hard palate is further divided into three regions: incisive papillae; median raphe; and fatty anterior laterals and glandular posterior laterals (Fig. 9–8A and B). Color and touch differences are observed between the palates. The hard palate is less pink and less movable than the soft palate. The core tissue of the hard palate is bone, while that of the soft palate muscle is muscle. These structural differences account for those of color and touch.

The epithelial covering of the hard palate is stratified squamous. The degree of keratinization varies from orthokeratinous through parakeratinous to nonkeratinous, depending on the intensity of the abrasive forces. The epithelial ridges or rete apparatuses correlate with the extent of keratinization. They are more numerous, taller, and more slender with increased cornification. The lamina propria, except in areas containing fat cells and glands, consists mostly of a dense fibrous tissue similar to that of the gingiva. Specific regional differences in the structural composition of the mucosa will be described as appropriate.

Premaxillary Palate. This structure is anterior to the palatine or incisive papilla and occupies the anteriormost segment of the hard palate. It is derived from the premaxilla. It therefore includes that segment of the arch bearing the incisor teeth (see Fig. 3–6B). The incisive papilla is an oval mass of tissue that caps the oral opening of the incisive canal. Because of its position the epithelium is generally orthokeratinous, and its underlying connective tissue is densely fibrous. The connective tissue in the suture area often contains epithelial pearls, which are vestigeal remains of the epithelium that covered the bones of the palatal process prior to fusion. Thus, they are not confined to the incisive region but may extend along the length of the raphe region.

The incisive canal contains the nasopalatine nerve and arteries, as well as vestigeal remnants of the nasopalatine ducts. The blind nasopalatine ducts are lined by respiratory epithelium. The underlying soft tissue is compact, and contains irregular patches of hyaline cartilage (remnants of paraseptal cartilage) and mucous secretory units. The excretory ducts of the latter empty into the lumen of the nasopalatine duct. The suggested function of the nasopalatine duct and of Jacobson's organ is that of an extra olfactory mechanism. Jacobson's (vomeronasal) organ, a small channel connecting the nasal and oral cavities, achieves maximum development in humans toward the end of the fifth month, thereafter involuting and generally disappearing before birth. Vestigeal remains may be present in adults, although rare.

Raphe Region. This region of the hard palate is a fusiform mucosal mass that forms the midline of the palate (Fig. 9–8). It extends from the incisive papilla toward the soft palate for a varying distance. The sides of the raphe are formed of lateral segments of the hard palate. The median raphe represents the fusion site of the palatine processes. The surface epithelium is orthokeratinous or parakeratinous, with many long, slender, and often branching epithelial ridges (Fig. 9–13). The connective tissue of the reticular layer is thin and densely fibrous. The cells are fewer, and the vascular and nerve sup-

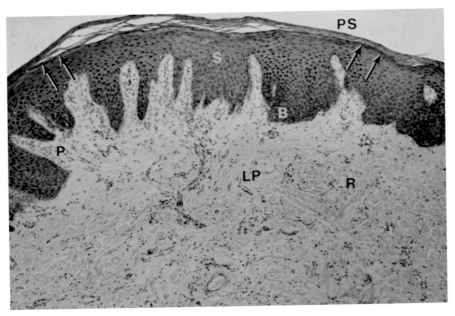

Fig. 9–13. Palatine raphe showing the thick layer of parakeratinous epithelium (PS). This is evidenced by the presence of nuclei in the desquamating cell layers. Note that in this photomicrograph granular epithelial components *(arrows)* are present, which accounts for the possibility of orthokeratinization. The spinous (S) layer is quite thick, and the basal (B) layer follows the contours of the tall and often branching ridges. The papillary (P) and reticular (R) layers of the lamina propria (LP) are densely fibrous. (Hematoxylin and eosin stain; ×50)

plies are correspondingly diminished. The palatine raphe, unlike its lateral territories, has no submucosa, and lacks both fatty and glandular components. The collagen fiber bundles of the reticular layer veer toward the bone to join with those more deeply situated, forming Sharpey's fiber bundles. Thus, the mucosa of the raphe is in fact a mucoperiosteum. The blood, lymph, and nerve supplies of the raphe and incisive papillae, although decreased, are clinically significant—that is, prolonged or excessive pressure may injure the nerves and occlude the vessels with serious consequences, such as inflammation and possibly necrosis. Therefore, particular care must be taken to provide "relief" for these tissues in denture construction.

Palatine Rugae and Fatty Region. The fatty region of the palate is bordered on one side by the raphe and on the other by the lingual aspect of the gingiva. The anterior lateral mucosa of the hard palate distal to the incisive papilla is thrown into irregular folds, forming the transverse palatal ridges known as the palatine rugae (Fig. 9–8). The number, depth, and arrangement of the folds vary, and they tend to become less conspicuous with age. Because of their exposed

location, rugae are covered mostly by orthokeratinous epithelium and their rete apparatuses are long and slender. The papillary connective tissue is more diffuse and has a greater cell population than that of the reticular layer, which is made more compact by dense irregularly arranged bundles of collagen fibers.

The region interjacent between the periosteum and the lamina propria, designated as a submucosa, is packed with aggregations of fat (Fig. 9–14). The subepithelial connective tissue is tightly bound to the bone by guy bundles of collagen originating in the lamina propria. These penetrate and compartmentalize the fat into discrete territories. More deeply, they become intercalated with those of the periosteum.

Glandular Region. The posterior lateral (glandular) regions of the hard palate are located distal to the fatty region and merge with the soft palate. The mucosa is not gathered into folds (rugae). The submucosa contains the secretory units of the minor salivary glands (Fig. 9–15). Except for these differences, the glandular region of the hard palate is not substantially different from the fatty region.

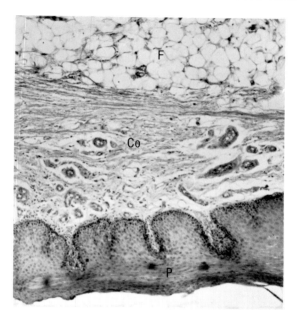

Fig. 9–14. Anterolateral (fatty) segment of the hard palate, showing a thick parakeratinous (P) epithelium. The subepithelial connective tissue (Co) is densely fibrous, and contains few cells. Accumulations of fat cells (F) occur more deeply. (Hematoxylin and eosin stain; ×50)

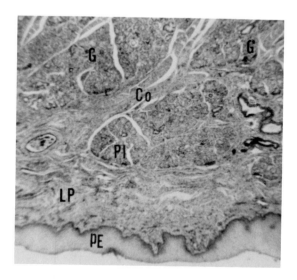

Fig. 9–15. Posterolateral (glandular) segment of the hard palate, which the surface epithelium is parakeratinous (PE). The lamina popria (LP) is densely fibrous, with bundles of collagen (Co) partitioning the deeper glandular elements (G) into pseudolobules (Pl). (Hematoxylin and eosin stain; ×50)

Specialized Mucosa

This type of mucosa is limited to the dorsal surface of the tongue. The epithelium is stratified squamous and is generally orthokeratinous. The rete apparatus is comprised of long, slender ridges. The intermediate and desquamating layers are quite thick, and are formed into epithelial outgrowths known as lingual papillae. Most are filiform in shape, while others may be fungiform, foliate, or vallate. All except the filiform bear taste bunds. On the basis of these features the mucosa, and especially the epithelial covering, is designated as specialized. In the absence of a submucosa, the lamina propria is tightly bound to the muscles, which form the core tissue of the tongue (see below: Tongue).

SALIVARY GLANDS

Glands have been included in the discussions of the various sections of the oral and nasal cavities, as well as those of skin. These include the sweat and sebaceous glands of the skin, the minor salivary glands of the vestibular and oral mucosa, and the glands of the nasal mucosa. The general features of glands have been discussed. The salivary glands, which are located in the paraoral tissues, secrete their products into the oral cavity. They are of the merocrine variety, and their secretions may be classified as serous, mucous, or seromucous. Relative to size, they are designated as major or minor glands (see Tables 9–1 and 9–2).

Functions

There are many and varied substances contained in saliva. Whole saliva is a mixture of glandular secretions, excretions, and gingival sulcus exudates, including leukocytes and bacteria. Certain proteins and cells impart a murky appearance and viscosity to the mixture. Saliva is about 99% water, with the remainder composed of inorganic and organic materials. Its pH varies from 6.7 to 7.4, and its specific gravity is 1.003. Up to 1500 ml of saliva are produced daily.

The organic constituents of saliva include various glycoproteins and proteins, including enzymes (e.g., amylase, ribonuclease, lysozyme, lactoperoxidase, kallikrein, esterase, and acid phosphatase) and mucin. Also found are amino acids, blood clotting factors, hormones, lipids, secretory immunoglobulins, vitamins, urea, and uric acid, among others. The principal inorganic ions include those of sodium, calcium, chloride, bicarbonate, and phosphate. Other electrolytes present in amounts less than 1 mM include F^-, I^-, SCN^-, Mg^{2+} and SO_4^{2-} Bicarbonates primarily, and also various salivary proteins, act as buffers contributing to certain functions. Functions that are a result of the chemical composition of saliva include the following: induction of calcification; limitation of acid action and enamel demineralization; destruction or restriction of growth of bacteria; reduction in blood clotting time; and digestion of starches.

Except for starch digestion and taste, most functions of saliva are associated directly or indirectly with protection. Digestive activities are relatively insignificant. As for taste, it has been reported that a substance called gustin participates in the growth and acquisition of functional competency of the gustatory apparatus. Additionally, in order for taste to occur, the material must be soluble in saliva.

There are a number of protective functions of saliva:

1. Lubrication and moistening of oral tissues
2. Moistening and lubrication of segments of the respiratory mucosa, as required
3. Moistening and preparation of ingested food for deglutition
4. Flushing of debris from tooth crevices and interdental spaces
5. Flushing and/or dilution of sugars, thereby mediating the action of cariogenic bacteria
6. Buffering action to help render the oral environment less susceptible to caries attack and less suitable for growth or certain bacteria
7. Destruction or prevention of growth of microorganisms by immunoglobulins
8. Induction of enamel pellicle formation by calcium binding proteins
9. Aggregation of bacteria by salivary immunoglobulin IgA
10. Enamel recalcification because of the presence of high concentrations of various ions

Furthermore, calcium and phosphate ions in saliva are found in saturated concentrations. Their settling as crystals is believed to be intercepted by a protein group known as statherin.

Structural Features

Irrespective of size or structural complexity, glands consist of secretory end-pieces (units) or parenchyma called acini, tubules, and ducts. Connective tissue binds, partitions, and encapsulates the gland, and also carries and protects the vascular, lymphatic, and nerve elements for the secretory units and their ducts. Secretory units are formed of a single layer of columnar or pyramidal epithelial cells that rest on a basement membrane. The cells are joined laterally by junctional complexes. From their distalmost connections toward the cell base they are typically arranged as tight junction (zonula occludens), intermediate junction (zonula adherens), and a varying number of desmosomes (maculae adherens; see Fig. 2–5). Cells of an end-piece may be arranged so that they appear spherical or pear-shaped (alveolar) or cylindric (tubular). A given gland may contain end-pieces of one or more shapes, which may be organized in the gland in varying degrees of complexity (simple to compound) as described in Chapter 2 (see Fig. 2–18). Glands with a more complex architecture are tubuloalveolar.

The distal surfaces of end-piece parenchyma form the lumen, or secretion-receiving chamber. The lumina of serous units receiving thin watery secretions are smaller than those that are thicker and receive mucus. The lumina of serous units project intercellularly for varying distances; these are called intercellular canaliculi (Fig. 9–16).

Thus, secretions released into the intercellular canaliculi pass into the lumen of the end-piece and then into the duct system proper. Each segment of the system empties into a larger duct until it arrives at the largest portion, the excretory duct, which releases the secretion into the oral cavity. Duct cells rest on a basement membrane, which is a continuation of that of the secretory units. Secretory units and their ducts are housed in connective tissue containing nerves, blood and lymph vessels, fibers, and various cells, inluding fibroblasts, leukocytes, and macrophages, and fat, mast, and plasma cells. In addition to serving the glands individually, as appropriate, the connective tissue components in the larger glands collectively form a capsule of varying density. Branches of the capsule known as septa divide the substance of the

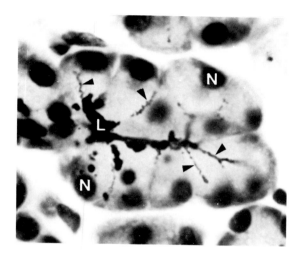

Fig. 9–16. Photomicrograph of serous secretory unit from a parotid gland, showing intercellular canaliculi *(arrow heads)* projecting basalward between the cells from the lumen (L). Note the basal position of the nuclei (N). (Perfused with India ink; ×1470)

gland into territories called lobes, and these are further partitioned into smaller units known as lobules (Fig. 9–17).

Secretory Units

Serous Parenchyma. Except for the minor lingual glands (von Ebner), serous cells are orga-nized into functional units, acini or tubules, located at the duct terminals. Serous parenchyma are located exclusively in acini of the parotid and submandibular glands. They may also occur as serous caps—demilunes—at the tip of mucous secretory units in mixed major salivary glands (Fig. 9–18). Because the carbohydrate moiety in some cells is substantial, some investigators have designated them as *seromucous.* Such cells react weakly to mucicarmine stain. All cells of a given secretory unit may not be engaged simultaneously in producing secretory product(s); rather, they can vary from synthesis-inactive to synthesis-active and secreting. The cytomorphologic features of serous cells are quite constant for any given stage. The cells may be pear-shaped or pyramidal. Distally they do not fit snugly because the intercellular spaces may be modified as intercellular canaliculi. The distal cell portion houses the synthesized product, while the basal segment is occupied by the spherical nucleus (Figs. 9–19 and 9–20). On the other hand, during synthesis, the perinuclear region is densely packed with parallel arrays of endoplasmic reticulum. With routine hematoxylin and eosin staining a blue color is imparted to the area.

Electron microscopic studies reveal the gen-

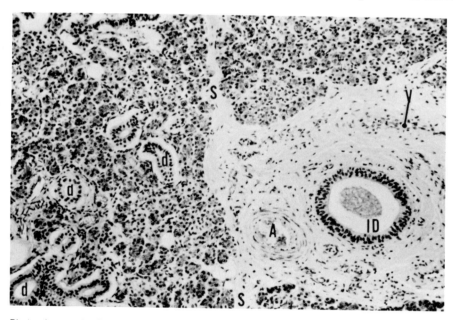

Fig. 9–17. Photomicrograph of secretory parenchyma partitioned into lobules by connective tissue septa (S). The connective tissue mass at which the septae meet contains interlobular ducts (ID), arterioles (A), and venules (V). The composition of the septae is mostly fibrous (collagen), with few connective tissue cells to maintain and protect the glandular constituents, (d, intralobular ducts). (Hematoxylin and eosin stain; ×80)

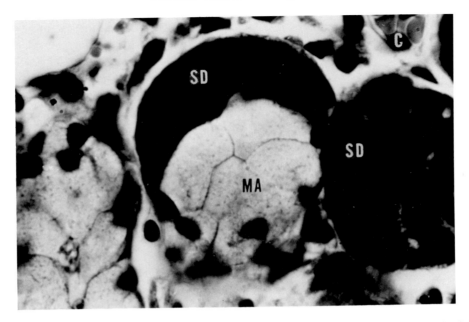

Fig. 9–18. Mucous end-piece (MA) and serous demilune (SD). Capillaries (C) are contained in the glandular stroma. (Hematoxylin and eosin stain; ×800)

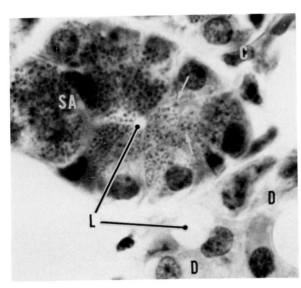

Fig. 9–19. Serous acinus (SA) and intralobular ducts (D) showing differences in cell shape and lumen (L) diameters. The serous cells are pyramidal and contain round basal nuclei and numerous supranuclear granules *(arrows)*. The duct cells are more cuboidal or flattened, and the lumina are large (C, capillary). (Hematoxylin and eosin stain; ×800)

eral cytomorphologic features to be similar to those of other protein-synthesizing cells—for example, the pancreas. Except for the perinuclear arrays of endoplasmic reticulum and the supranuclear dense accumulation of vesicles and stacked membranous saccules of the Golgi complex, other organelles such as mitochondria, lysosomes, peroxisomes, and microtubules are not present in impressive numbers (Fig. 9–21). Secretion granules of different densities that represent varying stages of maturity are modified, concentrated, and packaged en route through the GERL (Golgi-Endoplasmic-Reticulum-Lysosome). The granules accumulate in the apical cell segment (Fig. 9–22). When stimulated, they are released into the intercellular canaliculi or lumen by the fusion of the membranes of the granules and by that of the cell (Fig. 9–23); also see Fig. 1–10). This process, *exocytosis*, provides for the release of the serous product, leaving the cell membrane intact. During the period of secretion the surface area of the luminal wall increases because of additional membranes derived from liberated secretion granules (Fig. 9–23). The luminal diameter is restored to its normal size by the production of pinocytotic (endocytic) vesicles, which may join lysosomes for degradation or may be directed back to the Golgi complex for recycling (see Fig. 1–10). The wa-

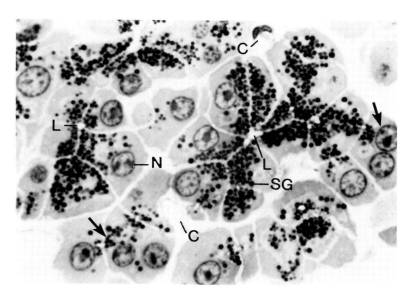

Fig. 9–20. Serous cells with round, basally located nuclei (N), supranuclear secretory granules (SG), and binucleate cells *(arrows)* (C, capillary; L, lumen). (Toluidine blue stain; ×1250)

tery serous gland secretions are proteinaceous digestive enzymes and substances of protective function.

Mucous Parenchyma. These cells are found in both major (submandibular and sublingual) and minor salivary glands. They produce viscid proteinaceous secretions that mostly act to lubricate and protect the oral tissues; they serve digestive functions only minimally. Mucous cells are similar to serous cells in regard to their basement membrane relationship and to certain cytomorphologic features. Mucous cells differ from serous cells in the following ways: the lateral plasmalemma are better defined; the elaboration is stored as droplets; the secretions are thick and viscous; intercellular canaliculi are usually absent (except for cells juxtaposed to serous cells of demilunes and in human labial glands); the lumen is larger; the nuclear configuration changes with accumulation of secretion droplets; and liberation of the secretory product and its effect on the cell membrane at the secreting surface may vary.

Light microscopy of tissue stained with hematoxylin and eosin reveals that mucous cells are pyramidal and larger than serous cells, their nucleus is basal in location, and their apical cell segment appears frothy because the mucigen droplets are washed out in tissue preparation (Figs. 9–18 and 9–24). Special techniques using mucicarmine stain color the cytoplasm intensely

pink. As the mucigen droplets are synthesized and accumulate toward the surface, the nucleus and its surrounding cytoplasm are displaced and compressed toward the cell base.

Electron microscopic examination reveals that the organelles are mainly limited to the perinuclear cytoplasm, and consist of mitochondria, rough endoplasmic reticulum, and other elements. These are not as numerous as in serous cells. The Golgi complex, however, is more extensive, reflecting increased carbohydrate production. It is located interjacent to the nucleus (or endoplasmic reticulum) and mucigen droplets, and is comprised of up to 12 flattened saccules (Fig. 9–25).

Mucigen droplets tend to coalesce en route to the surface. The methods by which the droplets are expelled into the lumen are characteristic for merocrine glands, as described for serous cells (exocytosis), or may involve slight discontinuities created in the surface membrane, as for apocrine glands. Mucigen is converted into mucus.

Serous Demilune. In mixed glands the tubular portions are comprised mostly of mucous cells, and the blind ends may be composed of serous cells. The serous cells may also form a cap over a mucous tubule, a configuration accounting for the name demilune (half-moon) or crescent (Figs. 9–18 and 9–26). Electron microscopic examination of these cells reveals that

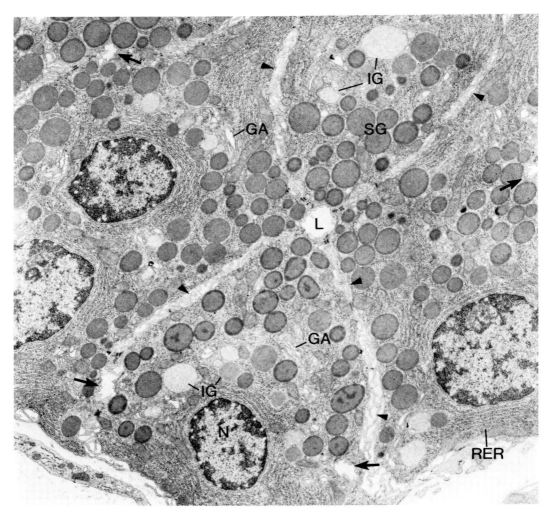

Fig. 9–21. Electron micrograph of serous cells taken from rat parotid gland, showing pyramidal shape of cells, small lumen (L), round, basally located nuclei (N) with perinuclear endoplasmic reticulum (RER) and supranuclear Golgi complex (GA). The immature secretion granules (IG) and mature secretory granules (SG) are more apically distributed. Note the intercellular spaces *(arrow heads)* and the associated intercellular canaliculi *(arrows)*. (×8200) (Hand, A.R., and Ho, B: Liquid diet-induced alterations of rat parotid acinar cells studied by electron microscopy and enzyme cytochemistry. Arch. Oral Biol. 26:369, 1981; reprinted by permission of Pergamon Press, Ltd.)

they contain a centrally located nucleus, well-developed endoplasmic reticulum, and Golgi apparatus. Other organelles are not as abundant. Myoepithelial cells envelop the demilune-mucous secretory unit (Fig. 9–26).

Myoepithelium. Although these cells do not engage in the synthesis of salivary secretions, they are intimately associated with the secretory parenchyma. Secretory units and adjacent intercalated ducts may be embraced by long branching processes of flat stellate cells that are variously designated as myoepithelial, basal, or basket cells (Fig. 9–24). These terms are used

because they contract as muscle cells, they are located between the basement membrane and the bases of the secretory parenchyma, or their processes appear to carry or support the secretory unit (Fig. 9–24). Myoepithelial and secretory cells are joined by desmosomes. As many as eight processes extend from the myoepithelial cell body, which may branch longitudinally to invest the secretory cells and the draining duct. Generally, only one cell envelops an end-piece. The nucleus located in the central cell mass is flattened, reflecting the cell's profile.

Electron microscopic study reveals these cells

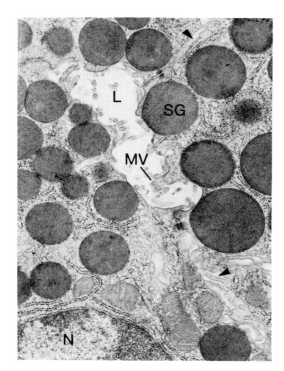

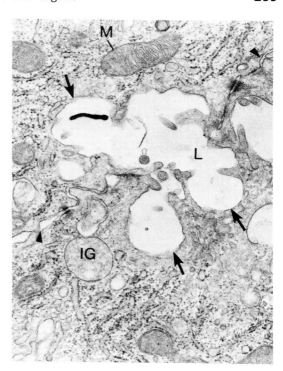

Fig. 9–22. Electron micrograph of resting (unstimulated) serous apical cell segments taken from rat parotid, showing secretory granules (SG), nucleus (N), microvilli (MV), lumen (L), and intercellular space *(arrowhead).* (×19,400)

Fig. 9–23. Electron micrograph of actively secreting cell (stimulated by the β-adrenergic drug isoproterenol) taken from a rat parotid gland, showing immature granule (IG), mitochondrion (M), and intercellular space *(arrowheads).* Note that the lumen (L) is of increased size due to the fusion of the membranes of the secretory granules with the external limiting membrane during exocytosis *(arrows).* (×23,500)

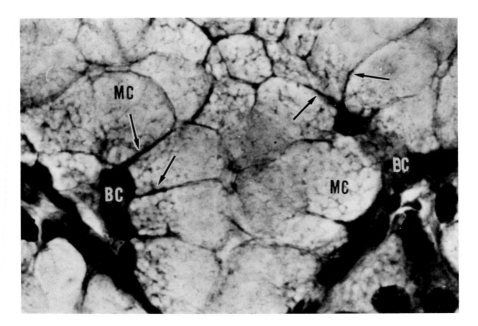

Fig. 9–24. Mucous cells (MC) showing myoepithelial (basket) cells (BC) and their processes *(arrows)* embracing secretory parenchyma. (Hematoxylin and eosin stain; ×800)

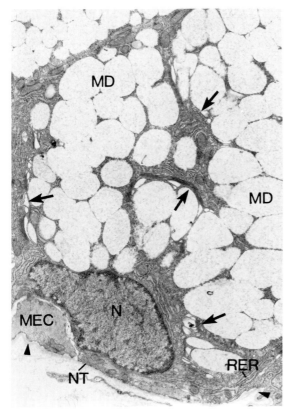

Fig. 9–25. Electron micrograph of mucous cell a from rat sublingual gland showing flocculent contents of secretory droplets (MD), amalgamation of several droplets, basally located squamoid nuclei (N), perinuclear endoplasmic reticulum (RER), and Golgi complex *(arrows)*. The mucous cell bases rest on a basal lamina *(arrowhead)*. Note that the cell process of the myoepithelial cell (MEC) and intraepithelial nerve terminal (NT) lie between the cell base and basal lamina. (×4,280)

to be cytologically similar to smooth muscle in that the cytoplasm is dominated by 6-nm thick actin filaments. These course through the cell body and are longitudinally arranged in the processes. Thicker filaments up to 10 nm occur, but are few in number. Although mitochondria, endoplasmic reticulum, and a Golgi complex are present perinuclearly, their number is not impressive. Many pinocytotic vesicles may be observed along the plasmalemma.

Ducts

From the lumen of an end-piece, secretions flow through a series of ducts of varying sizes. The duct system is generally more complex in the larger (major) salivary glands. Three types of ducts may make up the draining system of a gland. The smallest, the intercalated duct, drains the secretory end-pieces or joins other intercalated ducts. Next in size are the secretory or striated ducts, which drain the intercalated ducts. These are present only as part of the intralobular duct system in major salivary glands. The largest channels, those directing the secretion toward the surface are the excretory ducts.

Intercalated Ducts. These channels drain the secretory end-pieces. Their complexity of arrangement varies; for example, they are simplest in sublingual glands and most developed in the parotids, in which they are long and branching. Optical microscopy reveals them to be simple cuboidal to squamous cells (Fig. 9–27). Nuclei dominate the cell, and their meager cytoplasm stains weakly. Basal cells are frequently squeezed between the bases of the duct epithelium and the basement membrane. The lumen of the duct approximates that of the secretory unit.

Electron microscopic examination shows that the endoplasmic reticulum consists only of a few cisternae in parallel arrangement. The Golgi apparatus, however, is well developed, with dense granules occupying the supranuclear region (Fig. 9–28). Cell membranes of adjacent cells may be plicated and joined by desmosomes. The luminal surface may possess a few short microvilli. The cytologic features of the luminal epithelium, their capacity for mitosis, and the conspicuous presence of basal cells suggest functions other than draining. In this regard it has been proposed that the components of the intercalated ducts also function as replacement sources for spent or moribund cells, as well as adding to or modifying the products of the secretory unit.

Striated Ducts. These channels are the immediate successors of the intercalated ducts. Their location is within the lobules, so they are sometimes called intralobular ducts. Distinct cytomorphologic changes are observed at sites of confluence. Optical microscopy shows that the striated ducts are comprised of simple tall columnar cells surrounding a relatively narrow lumen. In the larger ducts, smaller cells are located between the basement membrane and bases of duct cells. These are probably basal cells. The cytoplasm of duct cells is abundant and stains weakly. Distal terminal bars and basal

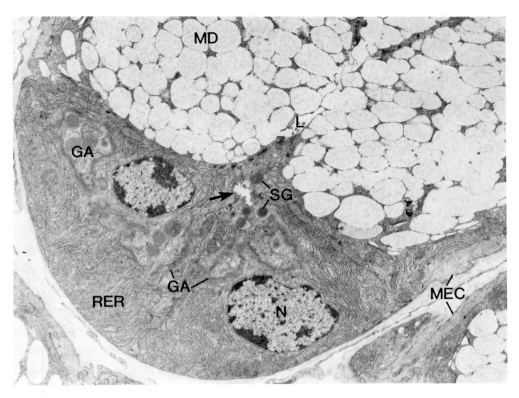

Fig. 9—26. Electron micrograph of myoepithelial cells (MEC) surrounding cells of a serous demilune and its associated mucous cells from a rat sublingual gland. The mucous cells are packed with mucous droplets (MD). Note the relationship of the mucous cells and the lumen (L). The serous demilune cells are abundantly supplied with organelles associated with synthesis—that is, with rough endoplasmic reticulum (RER) and Golgi apparatus (GA) (N, nucleus; intercellular canaliculus *(arrow)*; SG, secretory granules). (×5700)

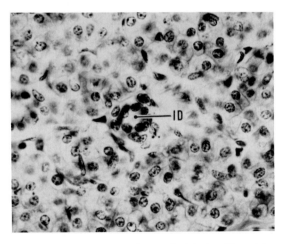

Fig. 9—27. Serous parenchyma of a lobule with an intercalated duct (ID). (Hematoxylin and eosin stain; ×80)

striations are prominent, and the nucleus occupies the central third of the cell (Fig. 9–29).

Electron microscopic observation of the striated ducts reveals the presence of many basal and lateral folds (plications) of the plasmalemma (Fig. 9–30). These divide the cytoplasm of the basal cell segment into processes or folds, which interlock with those of adjacent cells. The folds are responsible for the striated appearance observed in histologic sections. Numerous mitochondria are arranged between the plications (Fig. 9–30). The organelle populations are not numerous. Although the Golgi apparatus mostly occupies a perinuclear position, other organelles, including the rER, peroxisomes, lysosomes, and inclusions, are distributed throughout the cell. The distal cell segment may contain smooth endoplasmic reticulum and secretion granules (Fig. 9–30). Short microvilli are found at the luminal and lateral cell membranes. Interfacing cells bear the characteristic sequences of junctional complexes. Cytomorphologic fea-

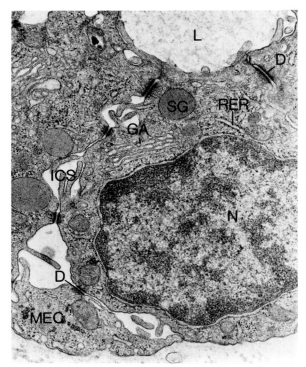

Fig. 9–28. Electron micrograph of an intercalated duct from a rat parotid gland. Note the size of the lumen (L) in relation to that of the low columnar duct cells. Illustrated are the intercellular spaces (ICS) with microvilli, desmosomal junctions (D), connecting duct cells, and myoepithelial cells (MEC). The nucleus (N) occupies a basal position, and the organelles involved in synthesis are not abundant and occupy perinuclear positions (RER, endoplasmic reticulum; GA, Golgi apparatus; SG, apical secretion granule) (×7100)

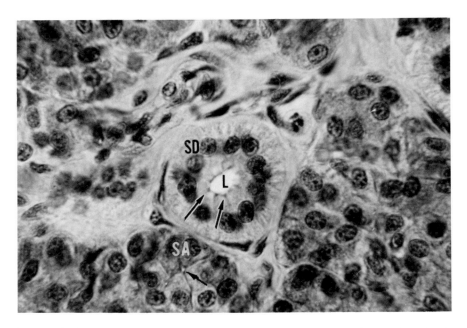

Fig. 9–29.   Serous acini (SA) and an interlobular striated duct (SD). The rounded nuclei and their perinuclear cytoplasmic basophilia are characteristic of serous parenchyma. The striated ducts exhibit well-defined basal striations, rounded nuclei, terminal bars *(arrows)*, and lumen (L). The minute lumen of the serous acinus is indicated by a black arrow. (Hematoxylin and eosin stain; ×430)

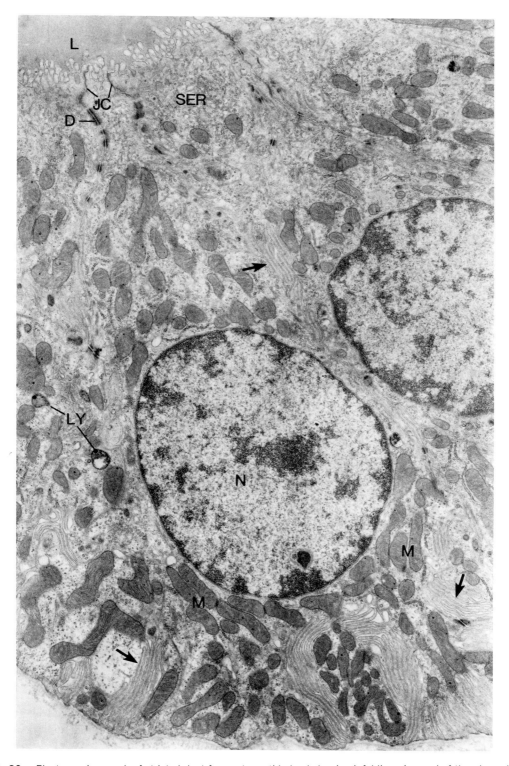

Fig. 9–30. Electron micrograph of striated duct from rat parotid gland showing infoldings *(arrows)* of the plasmalemma basally and laterally. The mitochondria (M) near the infoldings are especially numerous. The apical smooth endoplasmic reticulum (SER) is well developed; and junctional complexes (JC) and desmosomal connections (D) are arranged lumenward (L). The nucleus (N) is basally located (LY, lysosomes). (×10,500)

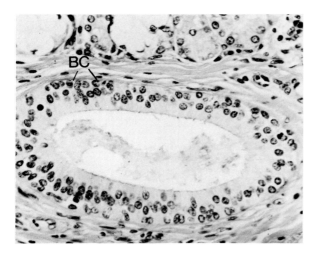

Fig. 9–31. Photomicrograph of human excretory duct from the sublingual gland. The epithelial wall is pseudostratified columnar, with basal cells (BC) packed between the cell bases. The duct is surrounded by interlobular septal connective tissue. (Hematoxylin and eosin stain; ×315)

tures of striated ducts indicate that the cells participate in secretion modification and its conversion into salivary end products, and also suggest that the duct components can recapture valuable electrolytes released by the secretory parenchyma.

Excretory Ducts. These are located in interlobular connective tissue, and are called *interlobular ducts*. They are composed of columnar cells of varying height, which in some glands are pseudostratified (Fig. 9–31). Basal cells lie intercalated between the basement membrane and the duct epithelium. Goblet and ciliated cells may occur as components of excretory ducts. The cells of the excretory duct become stratified on approach to the oral epithelium.

Electron microscopic observation reveals that the cells of the excretory duct exhibit gradual transitional cytomorphologic changes. In general, components of the excretory ducts differ from those of the striated ducts in that the former are taller and often pseudostratified. Furthermore, membrane infoldings are fewer and less complex. Mitochondria, although abundant, do not exhibit preferred orientation, and secretory granules are rare.

Architectural Pattern

The major glands are the most complex. The outer limit of large glands is defined by a dense connective tissue capsule. Branches, or septa,

of the capsular tissue partition the glandular parenchyma into territories called *lobes*. These are divided by septa into smaller parenchymatous divisions known as lobules (Figs. 9–17, 9–32, 9–33, and 9–34). The septal connective tissue becomes increasingly more diffuse as it branches until it forms a delicate network of fibers that binds the constituents of the lobules. This binding tissue is called the stroma. The elements bound by the stroma of a lobule include the basement membrane, basal and secretory parenchyma and their intercalated ducts, myoepithelial cells, smaller striated ducts, and nerves and lymph and blood vessels of microcirculation. These are the intralobular elements.

Septal or interlobular connective tissue contains nerves and blood and lymph vessels, as well as the largest striated ducts. The latter converge and become confluent with the excretory ducts. The denser connective tissue not only carries the largest vessels and nerve bundles but also the excretory ducts. Further, ganglia and pacinian corpuscles are not uncommon in the connective tissue (Figs. 9–17 and 9–32). Because septa are densely fibrous, there are only enough cellular elements present for tissue maintenance.

Major Glands

Three pairs of large (major) glands empty secretions into the oral cavity, the parotid, submandibular, and sublingual glands (Table 9–1). The secretions of the major salivary glands, also known as extrinsic glands, do not flow continuously as do the minor, or intrinsic, salivary glands. They all have branched tubular or tubuloalveolar secretory end-pieces, and all are surrounded by capsules. In the sublingual glands, however, the capsule is poorly developed.

Parotid Glands

These, the largest of the salivary glands, are irregularly shaped and are located in the cheeks immediately in front of and somewhat below the ear, so that their deeper segments extend along the ramus of the mandible (Table 9–1). The secretions gain access to the oral vestibule by way of the main excretory channel, Stensen's duct. The duct opening, as a papilla, is found near the maxillary second molars. The serous acini produce pure serous secretions that drain into long

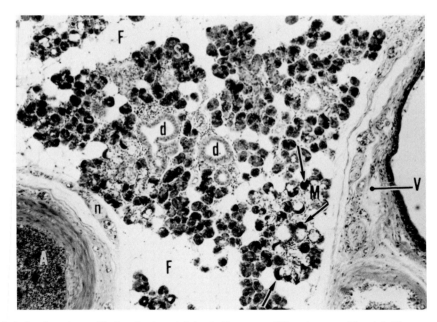

Fig. 9–32. Lobule of a submandibular gland containing both mucous (M) and serous (S) acini. The mucous alveoli may have demilunes (arrows). The septal connective tissue carry artery (A), vein (V), nerve fibers and perikaryons (n), numerous fat cells (F) and other connective tissue cells. Intralobular ducts (d). (Hematoxylin and eosin stain; × 50)

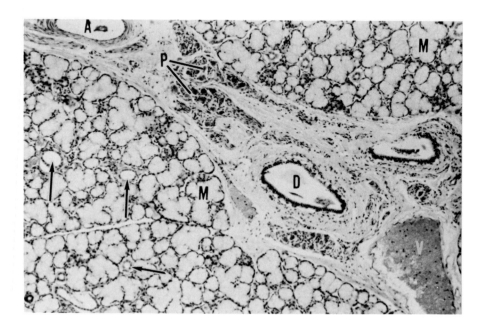

Fig. 9–33. Lobules of sublingual gland separated by septal connective tissue. The latter contains excretory ducts (D), perikaryons (P), artery (A), vein (V), and an assortment of connective tissue cells. The lobules are composed mostly of branched tubular or tubuloalveolar mucous end-pieces (M) and some intralobular ducts *(arrows)*. (Hematoxylin and eosin stain; × 50)

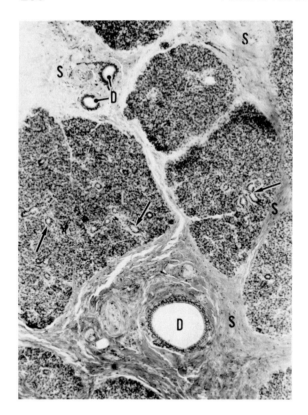

Fig. 9–34. Tissue section of parotid gland showing lobules (L) delineated by septa (S), with interlobular (D) and intralobular *(arrows)* ducts. (Hematoxylin and eosin stain; ×50)

branching intercalated ducts (Fig. 9–34). These in turn lead into an extensive system of striated ducts. The capsule of the gland is well developed, and its septal branches are densely fibrous. Patches of adipose tissue may increase in the septae with age, occurring at the expense of the diminishing secretory units.

Arterial blood is supplied to the parotids through branches of the posterior auricular artery, which originates from the external carotid artery. Blood is drained from the glands by vessels en route to the external jugular vein. Lymph from the gland is received by vessels leading to the superficial and deep cervical nodes.

Nerve supply to the parotid is functionally both sensory and autonomic. Sensory innervation is supplied by elements of the auriculotemporal nerve. Sympathetic innervation of the salivary glands, which accommodates intraglandular vasoconstriction, is supplied by the post-

Table 9–1. Summary of the Structural Features of the Major (Extrinsic) Salivary Glands

Feature	Parotid	Submandibular	Sublingual
Secretory cells	Serous acini	Serous acini; mucous tubules with serous demilunes	Mucous tubules with serous demilunes
Intercalated ducts	Long and branching	Moderate length	Short, poorly developed
Striated ducts	Well developed	Numerous, highly developed	Short, poorly developed
Main excretory duct	Stensen's; opens opposite maxillary second molar	Wharton's; opens at sublingual papilla beside lingual frenulum	Bartholin's; opens with submandibular duct; ducts of Rivinus; open along sublingual fold
Connective tissue capsule	Well defined	Well defined	Poorly developed; prominent septa
Blood supply	Branches of external carotid artery	Facial and lingual arteries	Sublingual and submental arteries
Nerve supply	Ninth cranial nerve through otic ganglion and auriculotemporal nerve; sympathetic plexus	Seventh cranial nerve through lingual nerve and submandibular ganglion; sympathetic plexus	Seventh cranial nerve through lingual nerve and submandibular ganglion; sympathetic plexus
Development	Fourth–sixth fetal week	Sixth fetal week	Seventh–eighth fetal week
Major secretory products	Amylase, proline-rich proteins; statherin	Mucous glycoproteins, amylase, proline-rich proteins	Mucous glycoproteins

ganglionic elements of the superior cervical ganglion that accompanies the arterial vessels. Parasympathetic innervation is from the glossopharyngeal (ninth cranial nerve) via the otic ganglion and auriculotemporal nerve.

Submandibular Glands

These bilateral organs are about half the size of the parotids. They lie primarily in the submandibular triangle, along the medial surface of the mandible posterior and superficial to the mylohyoid muscle. Superior to the muscle, extending anteriorly from its posterior border, is found a deep segment of the submandibular gland. The external opening of the principal excretory duct (Wharton's duct) is situated on the sublingual papilla (caruncle) lateral to the lingual frenulum.

The capsule is well developed (Table 9–1). Its septal branches divide the gland into lobes and lobules. The secretory end-pieces are mostly pure serous (80%) while others are mixed, composed both of mucous tubules and serous demilunes (Figs. 9–32 and 9–35). The serous cells form simple secretory end-pieces. Their components are pyramidal in shape. The lateral and basal plasmalemma are plicated. The lumina of the serous end-pieces are smaller in bore than their mucous counterparts. The ultrastructural features of the mucous parenchyma are similar to those described above. The intercalated ducts are shorter than those associated with the parotid, but longer than those of the sublingual gland. The striated ducts are longer and more branched than those of the parotids.

Arterial supply to the submandibular gland is by way of the facial and lingual arteries. Parasympathetic innervation arrives at the submandibular gland from the facial (seventh cranial nerve) via the lingual nerve and submandibular ganglion. The sympathetic fibers are also associated with the vasculature of the submandibular gland.

Sublingual Glands

As summarized in Table 9–1 these are the smallest of the major salivary glands, roughly the size and shape of an almond. They lie in the central region of the sublingual sulcus covered by its mucosa which forms the sublingual fold. The excretory duct (Bartholin's duct) communicates with that of the submandibular duct, emptying into the oral cavity by way of a sublingual papilla that is located near the frenulum. A few lesser ducts (ducts of Rivinus) gain access to the oral cavity independently along the sublingual fold.

The capsule is so loosely arranged that in most

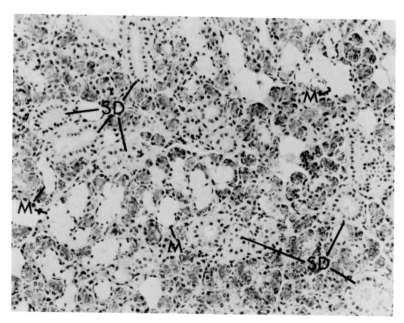

Fig. 9–35. Submandibular gland. Note that most of the secretory units are serous. The mucous secretory units (M) are fewer and are associated with demilunes (SD, striated ducts). (Hematoxylin and eosin stain; ×140)

areas its identity is lost. Only the larger lobules are outlined by septa. The smaller territories, or lobules, on the other hand, are more poorly defined because the connective tissue partitions are thinner and more diffuse (Fig. 9–36). The sublingual gland differs from the other major glands in several other ways: pure serous acini are very rare; the secretory end-pieces are mostly mucous tubules with serous demilunes; intercalated and striated ducts are few and short; and secretory tubules may bypass other ducts, emptying directly into intralobular ducts. The latter are comprised of low columnar or cuboidal cells lacking basal striations.

Arterial supply to the sublingual is via the lingual artery (sublingual branch) and facial artery (submental branch). Autonomic nerve supply to the submandibular and sublingual glands is by way of the secretomotor fibers of the facial nerve, which are directed to the submandibular ganglion through the chorda tympani nerve and lingual branch of the mandibular division of the trigeminal (fifth cranial nerve). Postganglionic parasympathetic fibers arising in the submandibular ganglion reach the gland via the lingual nerve. Nerve fibers of the carotid plexus and facial artery accommodate sympatheitc innervation.

Minor Glands

The minor salivary glands are located in the oral mucosa, and are also known as the intrinsic glands. All regions of the oral cavity contain minor salivary glands except the gingivae, dorsum of the tongue (anterior two-thirds), and anterior segments of the hard palate, including the raphe. With exception of those associated with the tongue, the von Ebner's glands, all are pure or predominantly mucous secretions (Table 9–2). They include those of the lip (labial), cheeks (buccal and retromolar), tongue (lingual), and palate (palatine and glossopalatine). They are small accumulations of mucous units continuously synthesizing and secreting their elaborations. For the most part, intercalated ducts are not well developed, striated ducts are lacking, and their excretory ducts empty directly into the oral cavity. Because of the small size of these glands, capsules are absent and septal compartmentalizations of end-piece accumulations are poorly defined and sometimes absent. This is particularly true for the vestibular and palatal glands.

Vestibular Glands

This is the collective designation of three continuous groups—labials (lip) (Fig. 9–5), buccals

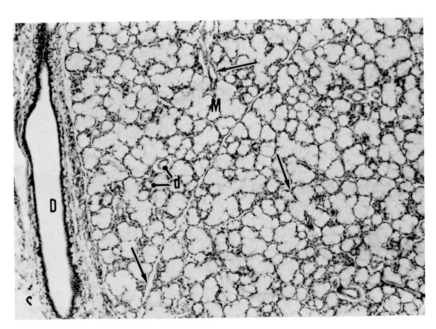

Fig. 9–36. Sublingual gland showing thicker septum (S) containing interlobular or excretory duct (D). The lobules are poorly defined by smaller septal branches *(arrows)* (M, mucous tubules; d, intralobular ducts). (Hematoxylin and eosin stain; ×50)

Table 9–2. Minor Salivary Glands

Location	Name	Type of Secretion
Lip (labial)	Superior labial	Mixed (predominantly mucous)
	Inferior labial	Mixed (predominantly mucous)
Cheek (vestibular)	Buccal	Mixed (predominantly mucous)
	Retromolar	Mixed (predominantly mucous)
Palate		
Hard	Posterolateral palatine	Pure mucous
Soft	Palatine	Pure mucous
Glossopalatine folds	Glossopalatine	Pure mucous
Tongue (lingual)		
Body	Blandin-Nuhn (anterior lingual)	Mixed (predominantly mucous)
Root	Von Ebner (posterior lingual)	Pure serous
	Tonsil, lingual (posterior lingual)	Pure mucous
Floor of mouth	Sublingual (intrinsic)	Mixed (predominantly mucous)

(cheeks), and retromolars (behind and below the molar teeth). All possess spherical masses, the secretory end-pieces, which are comprised of mucous cells. The excretory ducts pass through nonkeratinous oral epithelium, entering by way of the epithelial ridges. The vestibular glands are characterized by their number and location. They are numerous near the midline of the lip but become sparse near the cheek. In the cheek, the number and size of the glands increase more posteriorly. In the lip, the glandular masses may be so shallow that they may be seen or felt. In the cheek, particularly in the molar region, the secretory units are located more deeply in the submucosa. Some may even mingle with the muscle fibers.

Palatine Glands

These also consist of three continuous groups—hard palatine (posterior lateral regions of the hard palate, soft palatine (entire soft palate and uvular process), and glossopalatine (principally near the fauces or the palatoglossal folds). Their secretions are entirely mucous. As in the vestibular glands, intercalated ducts are short and inconspicuous. The glandular accumulations are most numerous in the hard palate, less in the soft palate, and fewest in the uvula. In the hard palate some of the glandular masses may be isolated by collagen to produce pseudolobules (Fig. 9–15). Here the collagen bundles are directed vertically from the epithelium to the periosteum of the palatal bone.

The principal excretory ducts are of two types. The first may be described as a narrow tube lined with columnar or pseudostratified cells that takes an undulating course through the connec-tive tissues and palatal epithelium. The second is sinusoidal-like, with a straighter path. Its wall is composed of several layers of cuboidal or squamous cells (Fig. 9–9). The sinusoidal excretory ducts may function as reservoirs for the secretions.

Lingual Glands

Some investigators consider these to be the most important of the minor salivary glands. They are distributed over the body of the tongue as the Blandin-Nuhn or anterior lingual glands, over the root as the von Ebner or posterior lingual glands, and over the faucial area as the tonsillar or deep posterior glands. Another group not directly associated with the tongue, but within its sulcus, is the sublingual group. The glands of Blandin-Nuhn are located on the underside of the tongue tip, immediately lateral to the midline. They may extend posteriorly for 25 mm. The secretory units of these and of those in the floor of the mouth may be mixed, but they are predominantly mucous. These glands are unusual for intrinsic glands in that a capsule is present, with septal branches organizing the lobules. The more posterior and marginal groups are mostly mucous, with serous elements arranged as demilunes (Fig. 9–37A).

The secretions of the glands are collected by three to five excretory ducts opening into the sulcus in the vicinity of the frenulum. On the dorsum of the tongue, serous secretions in the excretory ducts of the glands of von Ebner flow into and flush the troughs around the vallate papillae (Fig. 9–37B) and lateral tongue folds in which the foliate papillae are located. Also, at these sites, the secretions provide the liquid me-

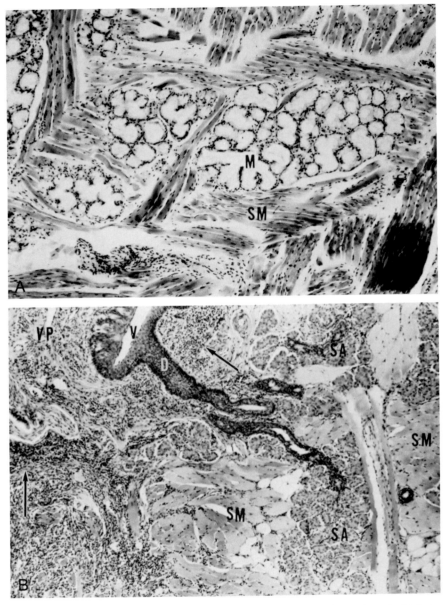

Fig. 9–37. Lingual glands. *A,* Blandin-Nuhn glands. The mucous (M) secretory units are branched tubular elements that intermingle with striated muscle (SM). (Hematoxylin and eosin stain; ×50) *B,* Von Ebner glands. The section contains an excretory duct (D) communicating with the trench or valley (V) surrounding a vallate papilla (VP). Serous acini (SA) are located around the excretory ducts, and some are found among the striated muscle fibers (SM). Note protective lymphocytic infiltrations *(arrows).* (Hematoxylin and eosin stain; ×50)

dium for taste perception and contain digestive enzymes and protective substances (e.g., amylase, lactoperoxidase, lipase, lysozyme). The tonsillar glands are pure mucous, with secretory units that are more deeply situated in the root of the tongue than those of the von Ebner glands. Their excretory ducts empty into the crypts of the lingual tonsils, thus flushing the area of debris.

Clinical Considerations

The intermittent and continuous flow of saliva controls and/or influences the activities in which the structures of the oral cavity are engaged. A review of the functions attributed to whole saliva merely emphasizes the clinical importance of the salivary glands to the oral and paraoral environments. For example, a moist environment significantly reduces the abrasive forces of mastication and deglutition, which in turn influences the character of the oral epithelium. An oral environment well supplied with saliva is less vulnerable to caries arrack. Furthermore, the likelihood of infection and inflammation is increased with a decrease in saliva flow. Certainly, swallowing, speech, and taste are more difficult with reduced saliva output.

TONGUE

The tongue is a large muscular organ occupying the floor of the oral cavity. Because of its specialized mucosal covering and its extremely varied movements, which are executed by the actions of various muscle groups, this organ is well suited to the functions attributed to it.

Functions

Two principal groups of functions are served by the tongue, muscular and sensory. As a muscle, the tongue assists in activities such as ingestion, channeling food toward teeth, mashing and mixing food and saliva, swallowing, and speech. Sensory functions include perceptions of heat, cold, and taste (chemical discrimination).

General Features

The tongue is enveloped by a mucous membrane. Its dorsal surface consists of three regions—the apex (tip), corpus (body), and root (base) (Fig. 9–38). The apex and corpus, which originate from the mandibular swelling of the first pharyngeal arch, comprise the anterior two-thirds of the tongue and are located entirely within the oral cavity. The base or posterior one-third, which arises from the second, third, and fourth arches, lies in the oral pharynx. The body of the tongue is divided longitudinally by the median sulcus. Its bilaterally partitioning connective tissue is the lingual septum. The corpus and root of the tongue are separated by circumvallate papillae arranged as an inverted V. Immediately posterior to the papillae is the sulcus terminalis, and at its apex is the foramen cecum. The dorsal surface bears numerous mucosal outpocketings and folds called lingual papillae. The

base of the tongue shows many irregular surface bulges, which are the external aspect of the lingual tonsils (Fig. 9–38).

Histologic Features

Except for a thin mucosal covering, the tongue is composed of striated muscle and its accompanying connective tissue sheaths. The histologic features of the surface mucosa, both the epithelium and lamina propria, are quite different for the dorsum and underside (ventral surface) of the tongue (see Fig. 9–7).

Mucosa

The dorsum of the tongue is covered by thick orthokeratinous epithelium, which is firmly attached to the lamina propria. The lamina propria of the dorsal surface is compact and securely joined to the peripheral muscle sheaths. The connective tissue is organized into mounds, or lingual papillae. The principal connective tissue mass occupying the lingual papilla is known as the primary papilla and is capped with orthokeratinous epithelium. The latter exhibits prominent rete apparatuses. Their interdigitating connective tissue ridges are designated as secondary papillae (Fig. 9–39). These are numerous, long, and slender, providing a firm attachment of the overlying epithelium. Lingual papillae are of four shapes—filiform, fungiform, foliate, and vallate (circumvallate). A submucosa is lacking in the dorsal tongue surface.

The underside of the tongue, because of its sheltered location and wet environment, is covered by a very thin layer of nonkeratinous epithelium. Its underlying lamina propria is thin and loosely arranged, and contains many elastic fibers. These facilitate changes in tongue shape, thickness, and movement. The connective tissue papillae and epithelial ridges are shallow. In addition, islands of fat, blood and lymph vessels, ducts, and the more shallowly located secretory units of the minor salivary glands are housed in the tissue. The submucosa is thin and extremely diffuse, and similarly contain patches of fat and secretory units of the minor salivary glands.

Filiform papillae provide the dorsal surface of the tongue with its velvety appearance. They may be as long as 2 mm, and are arranged in parallel rows conforming with the V-shaped sulcus terminalis (Fig. 9–38). They are flame-shaped, often with branching terminals (Fig.

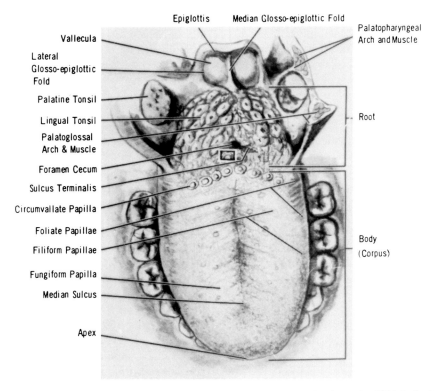

Fig. 9–38. Dorsum of tongue and associated structures. (After F.H. Netter, M.D., from the CIBA Collection of Medical Illustrations, Vol. 3, 1959. Copyright by CIBA Pharmaceutical Company, Division of CIBA-Geigy Corporation.)

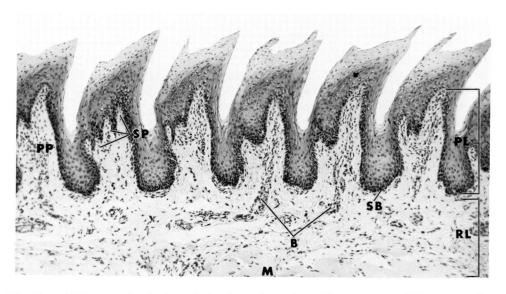

Fig. 9–39. Lingual filiform papilla showing orthokeratinous flame-shaped tips and primary (PP) and secondary papillae (SP) of the papillary layer (PL) of the lamina propria. The reticular layer (RL) overlies the peripheral muscle fibers (M) of the core, (SB, stratum basale; B, blood vessel). (Hematoxylin and eosin stain; ×50)

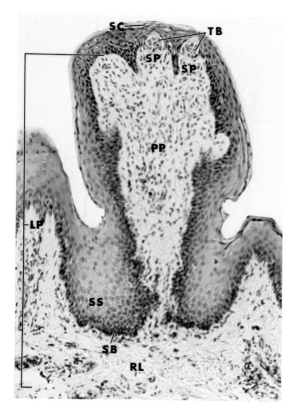

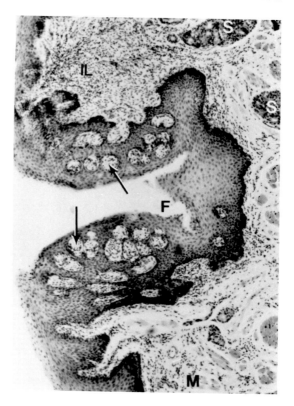

Fig. 9–40. Fungiform papilla showing taste buds (TB) in orthokeratinous epitheium. The primary (PP) and secondary papillae (SP) are formed by the connective tissue of the lamina propria (LP), (RL, reticular layer; SB, stratum basale; SS stratum spinosum (SS); stratum corneum, SC). (Hematoxylin and eosin stain; ×100)

Fig. 9–41. Photomicrograph of foliate papilla. Note the fold (F) and the surfacing nonkeratinous epithelium containing taste buds *(arrows).* Contained in the reticular layer of the lamina propria are the secretory units of minor salivary glands (S) and infiltrations of lymphocytes (IL) (M, striated muscle). (Hematoxylin and eosin stain; ×50)

9–39). A filiform papilla may contain from 7 to 30 secondary papillae. The orthokeratinous epithelium is thick and sometimes appears glassy, with a superficial desquamating layer.

Fungiform papillae are mushroom-shaped and are scattered among the filiform papillae (Figs. 9–38 and 9–40). Although not as abundant as the filiform papillae, they are larger. They may attain diameters of 1 mm and lengths up to 2 mm. The nonkeratinous (sometimes orthokeratinous) epithelium is thin and transparent, so that the rich vascular supply of the underlying connective tissue of the primary papillae renders them intensely pink. Their surface epithelium may contain a few (up to three) taste buds.

Foliate papillae form three to eight parallel mucosal folds on the sides of the tongue at the juncture of the corpus and root (Figs. 9–38 and 9–41). They are most developed at birth and tend to retrogress thereafter, so that by adult-

hood they are inconspicuous or have disappeared. The epithelium is thin and nonkeratinous, and the rete apparatus is poorly developed. Associated with the epithelium are taste buds and excretory ducts of the von Ebner glands. The subepithelial connective tissue is similar in structure and composition to that of the ventral surface of the tongue.

Vallate (circumvallate) papillae number from 7 to 11, and delineate the corpus from the root (Fig. 9–38). They may be over 1 mm in height and up to 3 mm in diameter. As the name indicates, these barrel-shaped structures are countersunk. Unlike the other papillae they do not project above the tongue surface, A furrow surrounds each papilla, so that a space intervenes between the epithelium of the tongue and that of the papilla (Fig. 9–42). As in the foliate papillae, secretions of the minor salivary glands empty into the bottom of the furrows (troughs) to flush and clean them of debris. Up to 35 ducts

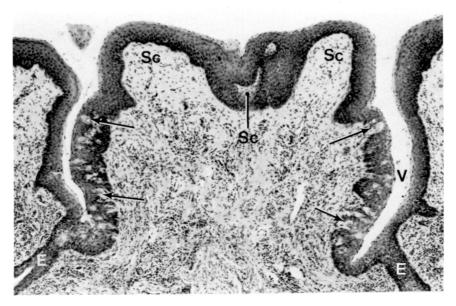

Fig. 9–42.   Vallate papilla showing surface parakeratinous epithelium, which dips into the furrow (V), becoming nonkeratinous and housing taste buds *(arrows)*. The secondary papillae (Sc) are connective tissue extensions of the primary papilla. The primary papilla is rich in cellular elements and blood and nerve supply. Note that the excretory ducts (E) of the von Ebner glands communicate with the valley. (Hematoxylin and eosin stain; ×50)

may open into a single trough. The slope of the papilla may contain several hundred taste buds. The nonkeratinous epithelial wall of the furrow (facing the papilla) contains few taste buds. The orthokeratinous and sometimes parakeratinous surface epithelium contains numerous secondary papillae, and the connective tissue core has many blood vessels and nerves. The deeper connective tissue is less compact; it houses the secretory units of the von Ebner glands among groups of fat cells. These contribute to the decreased tissue density.

Taste Buds

Other than those on lingual papillae, taste buds, in reduced number, are also found in the palatal and pharyngeal epithelia. Vallate papillae discriminate among salty, sour, sweet, and bitter substances that are dissolved in the saliva. Staining reaction of the taste bud components is much lighter than the epithelial cells in which they are located (Fig. 9–43). Optical microscopic examination of taste buds reveals them to be spherical, about 70 μm in height. Each is comprised of neuroepithelial (taste receptor) cells bearing apical microvilli ("hairs"), sustentacular (supporting) cells with apical microvilli, and basal cells. Microvilli of the taste receptor and sus-

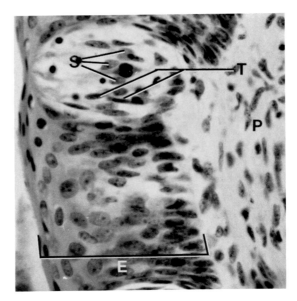

Fig. 9–43.   Photomicrograph of a taste organ contained in the epithelial (E) surface of a vallate papilla. Within the taste bud are slender dark nuclei of the sensory taste cells (T) and lighter rounder nuclei of the supporting (sustentacular) cells (S) (P, papillary connective tissue). (Hematoxylin and eosin stain; ×320)

tentacular cells extend into a corridor, the taste canal. This communicates with the oral cavity through an orifice called the taste pore (Fig. 9–44). The taste receptor and supporting cells are oriented vertically, similar to sections of an orange. Up to 20 neuroepithelial cells occupy the core of the taste apparatus. They are fusiform, with dark-staining centrally located oval nuclei. Only the canal-facing surface of the neuroepithelial cells possesses microvilli, which project into the lumen of the canal. Fluids from the oral cavity may pass through the pore into the canal, coming into contact with the microvilli. The sustentacular cells surround the core of neuroepithelial cells. Although the shape and arrangement of the apical microvilli of sustentacular cells are similar to those of the taste cells, their nuclei are rounder and paler staining.

Basal cells are packed between the bases of the neuroepithelial (taste receptor cells) and supporting cells. These are a replacement source for spent and injured cells of the taste bud. Taste nerve fibers from the lamina propria enter the taste bud, branch, and form knobby endings on both taste and supporting cell surfaces.

Electron microscopic observation shows four cell types, designated as I, II, III, and IV. Cell type IV has been identified as the basal cell. Type I differs from type II in that its rER is less developed. Type III cell, on the other hand, possesses less rER but more sER. Furthermore, this type of cell has "hairs" (microvilli) projecting into the taste pore. It has been suggested that these cell types represent varying stages of differentiation and cell competency.

Lingual Muscles

The musculature of the tongue is comprised of two groups, intrinsic and extrinsic. The former are classified according to their fiber orientation, so they are designated as vertical, transverse, or longitudinal. The vertical muscles are located at the lateral margins of the body. The transverse muscles spread out from the lingual septum to form a network. The longitudinal muscles are located above and below the transverse muscles as two sheets. The one located on the dorsal aspect is known as the superior longitudinal, while that found on the ventral aspect is the inferior longitudinal. The intrinsic muscles serve mostly to perform the intricate movements of the tongue and are organized into discrete groups, which interlace perpendicularly. The extrinsic muscles originate outside the tongue and consist of the genioglossus, hyoglossus, palatoglossus, and styloglossus. These serve primarily to secure the position of the tongue in the oral cavity. Additionally, they help execute the activities associated with tongue movement. The various muscle groups are surrounded by a definitive perimysium of varying densities. Perimysial connective tissue ramifications carrying the vascular, lymphatic, and nerve elements become increasingly more diffuse as a result of fat cells and glandular elements interspersed in the tissue. The binding tissue eventually invests the muscle fiber as endomysium.

Vascular and Nerve Supply. Excluding the palatoglossus muscle, blood supplied to the lingual muscles occurs mostly via the deep lingual artery. Similarly the nerve supply to the lingual muscles (excluding the palatoglossus muscle) is via the hypoglossal nerve. The pharyngeal plexus innervates the palatoglossus muscle.

TONSILS

The opening to the throat (oropharynx) is protected by lymphoid tissue and by the palatine, lingual, and pharyngeal tonsils, collectively identified as the ring of Waldeyer (Fig. 9–38).

Functions

As in the case of other lymphoid tissues, the tonsils function primarily in the production of lymphocytes, which are for local (pharynx) and general use. The latter are distributed via circulation. Lymphoid tissue (nodules) also has a role in antibody production.

Components

Palatine, lingual, and pharyngeal tonsils share four characteristics: they are covered by epithelium; their epithelium is thrown into folds or is invaginated, forming trenches; their internal boundaries are delineated by a capsule; and they are contained in the lamina propria.

Palatine (Faucial) Tonsils

These are about the size and shape of an almond; they are located between the anterior and posterior pillars (palatoglossal and palatopharyngeal folds). Their broad surface, which projects slightly, is covered with 35 or more craters.

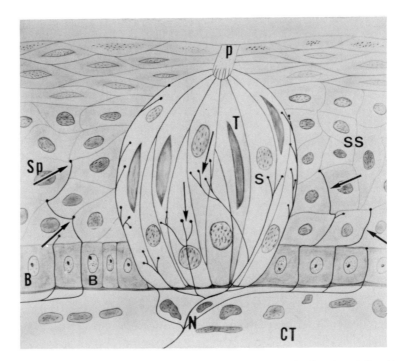

Fig. 9—44. Diagrammatic representation of taste organ (gustatory) apparatus showing the relationship of the cells of the epithelium and taste organ cells to each other and to the underlying connective tissue (CT) and nerve supply. The nerves (N) branch as they leave the connective tissue. Some fibers innervate the basal (B) and spinous (SS) layers of epithelium (white-outlined arrows); others go to the taste bud. Of the latter, some remain outside to innervate the outer cells *(white arrows)*, while others enter the organ to innervate the inner cells *(black arrows)*, or supporting (S) and sensory taste (T) cells. Note the "hairs" (microvilli) of the cells in the pore (P).

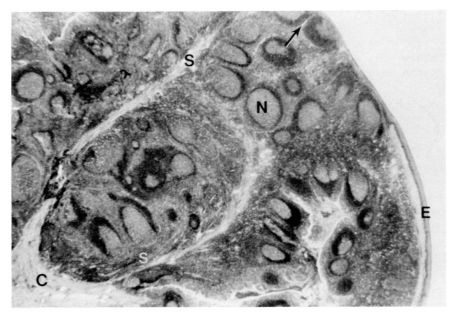

Fig. 9—45. Palatine tonsil demonstrating epithelial surface (E), section of crypt *(arrow),* nodules with germinating centers (N), capsule (C), and partitioning septa (S). (Hematoxylin and eosin stain; ×36)

These are the external openings of the epithelial invaginations, or crypts (Fig. 9–38).

The epithelium covering the tonsils is non-keratinous and continuous with that of the cheeks, soft palate, and oropharynx. At intervals, the epithelium dips into the lymphoid mass to form a network of crypts (Fig. 9–45). The epithelium may be interrupted by the invasion (infiltration) of older (spent) lymphocytes and plasma cells en route to the gastrointestinal tract, where they are eliminated. These cells and the desquamated epithelium may plug the crypts because they are not flushed by the secretions of the glossopalatine glands. The excretory ducts of these glands do not empty their contents into the bases of the crypts, but onto the tonsil surface near the crest of the crypt. Lymphocytes are often present in saliva as salivary corpuscles.

The internal border of the tonsil is composed of a connective tissue capsule that invades the substance of the lymphoid mass as septa, which partition the tissue into lobules. The subcapsular connective tissue contains striated muscle fibers, secretory end-pieces of the glossopalatine glands, and fat.

A layer of lymphoid tissue, over 1 mm thick, is covered externally by epithelium, but is bordered by septa elsewhere. These layers, especially along the crypts, are composed of oval lymphoid units, the nodules, which occur in the lymphoid tissue at varying intervals (Fig. 9–45).

Lingual Tonsil

The lingual tonsil is located at the base or root of the tongue. Raised craters on the tongue surface form the orifices of the crypts (Fig. 9–38). An orifice, its crypt, the subepithelial tissue mass, and its associated capsule form a lingual tonsillar unit (Fig. 9–46). The lingual tonsil is made up of as many as 100 such units.

The root of the tongue and its tonsillar crypts are covered with nonkeratinous epithelium. Their crypts are less branched than those of the palatine tonsils. Disruption of the epithelium by lymphocyte invasion is not severe. Accumulations of cellular debris in the crypts is rare, because they are washed by mucous secretions released by excretory ducts that are generally located at the base of the crypts. Some excretory ducts open onto the surface between the crypt orifices.

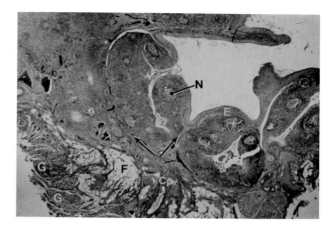

Fig. 9–46. Lingual tonsil showing partitioning of tonsilar tissue into territories by septal branches *(arrows)* of the capsule (Cp). Below the capsule the lingual tissue contains patches of fat cells (F) and glands (G). The latter empty their secretions into the crypts. The tonsilar units are composed of nodules (N), with germinal centers. The epithelial surface (E) is nonkeratinous. (Hematoxylin and eosin; ×8)

Subepithelial lymphoid sheets are separated from the underlying connective tissue by a capsule. Each crypt is surrounded by an inverted pyramid of lymphoid tissue and capsule. The narrow width of the lymphoid sheet generally permits only a single row of nodules.

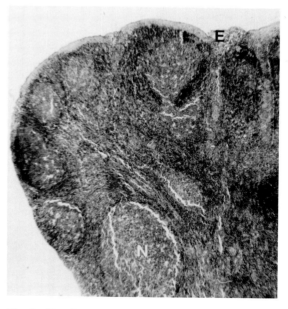

Fig. 9–47. Photomicrograph of pharyngeal tonsil demonstrating surfacing epithelium (E) and nodules with germinal centers (N). (Hematoxylin and eosin stain; ×30)

Pharyngeal Tonsils

These are commonly referred to as the adenoids. They are found on the dorsal wall of the nasopharynx, where they form longitudinal folds about 3 cm long. In histologic section the folds give the appearance of crypts.

The lymphoid tissue is covered by respiratory epithelium, a continuation of the respiratory epithelium of the nasal passage. In areas of friction, however, nonkeratinous epithelium replaces respiratory epithelium. Disruption of the epithelial surface by escaping lymphocytes is more extensive than in the lingual tonsil but not as severe as in the palatine tonsils. The subepithelial lymphoid sheet is up to 2 mm thick, and is intermittently organized into nodules. Septal branches of the capsule project toward the folds to produce lobule-like territories (Fig. 9–47).

The connective tissue of the capsule and septa is not as compact as that of the other tonsils. The subcapsular tissue is diffuse, and contains numerous mixed glands. The openings of the excretory ducts occur randomly in the folds.

Vascular, lymph, and nerve elements enter and leave the lymphoid tissue via the capsules and their septal branches. Lymph vessels surround but do not enter the lymphoid units.

NASAL PASSAGE AND SINUSES

The respiratory system possesses two basic units, conducting and respiratory. The conducting portion directs air to and away from the lungs and is comprised of the nose, nasal cavity, paranasal sinuses, pharynx, trachea, bronchi, and bronchioles. The respiratory portion is composed of alveolar sacs, atria, alveolar ducts, and respiratory bronchioles, which are involved in exchanges between the inspired air and blood capillaries. Because of the importance of the more external segments of the conducting portion to clinical dentistry, the nose, nasal cavity, and paranasal sinuses will be discussed here.

Nasal Cavity

The nasal cavity begins with the anterior nares, which lead into the vestibule. The remainder of the cavity consists of the respiratory and olfactory segments. The paranasal sinuses are pneumatic pockets or chambers of the nasal cavity.

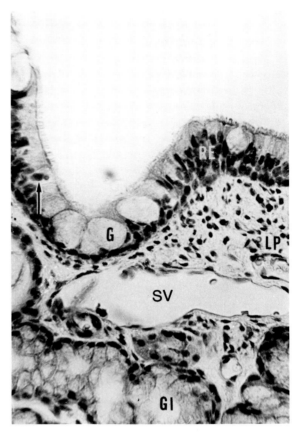

Fig. 9–48. Nasal passage in an area composed of respiratory mucosa. The surface epithelium is pseudostratified, ciliated columnar (RE), with goblet cells (G). The epithelium contains an escaping lymphocyte *(arrow)*. Components of the lamina propria (LP) are sinusoidal vessels (SV), tubuloalveolar glands (GI), and various connective tissue cells. (Hematoxylin and eosin; ×100)

Vestibule

The nasal cavity of the nose is formed of two fossae (chambers), with each separated anteriorly by a cartilaginous septum and posteriorly by a bony septum. The anteriormost enlarged regions are the vestibules. The core of the outer walls of the vestibule, making up the wings or alae of the nose, are composed of cartilage. The external surface of the nose is covered with skin, beneath which is connective tissue and more deeply striated muscle. The connective tissue contains hair follicles and sweat and sebaceous glands. The vestibules of the nose are also lined with skin and connective tissue. The hairs, however, are large, and serve to filter inspired air.

Respiratory Segment

The respiratory portion is composed of ledge-like extensions of the nasal cavity wall called conchae, which tend to increase the surface area of the cavity. The epithelium is typically of the respiratory variety. Goblet cells may be so numerous that they form glandular sheets, pits, or both. The basement membrane is well defined and the lamina propria, especially in the less sheltered areas, is fibrous and contains seromucous glands. Their secretions and those of the mucosal glands maintain a moist sticky surface, so that debris carried by incoming air may be trapped. Foreign particles are then swept toward the nasopharynx by ciliary action and expelled. Although the usual components of connective tissues are present for tissue maintenance, defense cells are especially numerous—lymphocytes, eosinophils, macrophages, and plasma cells. Mast cells occur frequently in the perivascular beds. In this regard, the deeper layer of the lamina propria contains a modified plexus of veins, which can assume sinusoidal features of great distensibility (Fig. 9–48). The large volume of blood passing through these veins assists in heating inspired air.

Several features are shared by various segments of the nasal fossae. These include the following: lymphocytes accumulate in areas of actual or potential need; eosinophils, macrophages, and plasma cell populations are increased; tubuloalveolar glands that produce seromucous secretions are numerous; and the tissue is very vascular. In some regions special (sinusoidal) veins occur that can become greatly dilated and engorged with blood if irritated. Excessive irritation causes the mucosa to expand and congest, or to "stop up" the nasal passage. Insults by allergic reactions and by the common cold may cause the vessels to become engorged, thereby increasing the volume of the tissue. Tissue housing these veins is said to be "erectile." A submucosa is absent in the nasal canal, and the fibers of the deepest layer of the lamina propria serve as a periosteum.

Olfactory (Mucosa) Segment

The roof of the nasal fossae and superior aspect of the nasal septum and superior conchae are covered by olfactory mucosa. The distinguishing

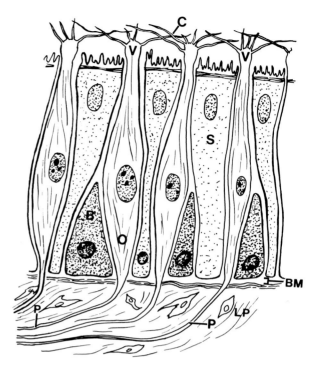

Fig. 9–49. Diagrammatic representation of olfactory organ, showing epithelium and lamina propria (LP). The latter contains collagen, nerve fibers (P), and an assortment of protective and maintenance connective tissue cells. Three cell types make up the olfactory organ supporting or sustentacular (S), sensory olfactory (O), and replacement basal (B). Note the relationship of the three cells regarding location. The terminal of the olfactory cell is bulbous (V) and contains cilia (C); (BM, basement membrane).

feature of this tissue is the presence of organs of smell (olfaction) and serous glands.

Epithelium. The respiratory epithelium associated with the olfactory organ lacks goblet cells. As with the gustatory organs, the olfactory organ is made up of three cell types: supporting (sustentacular), olfactory (sensory—smell), and basal (Figs. 9–49 and 9–50). Because of differences in the shape and location of their nuclei, two zones have been designated: the zone of oval nuclei (supporting cells) and the zone of round nuclei (sensory olfactory cells) (Fig. 9–50).

Supporting (sustentacular) cells are bottle-shaped. They rest, neck inverted, on a poorly developed basement membrane (Figs. 9–49 and 9–50). The nuclei are in the wider portion of the cell, all at the same level. Brown pigment granules are abundant in the cytoplasm and impart

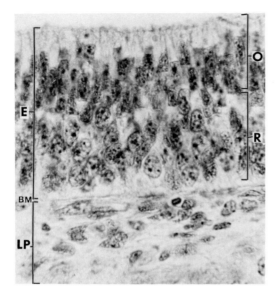

Fig. 9–50. Olfactory mucosa showing epithelium (E) and its zones of oval (O) and round (R) nuclei. (LP, lamina propria). (Hematoxylin and eosin stain; ×960)

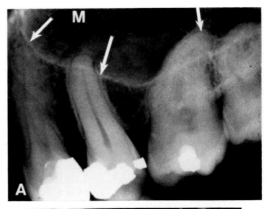

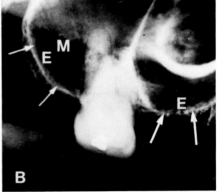

Fig. 9–51. Roentgenograms illustrating the relationship of the maxillary sinus (M) and tooth socket (fundus and lateral walls). *A,* The floor of the sinus *(arrows)* is located close to the fundus of the socket, separated only by a thin sheet of bone. *B,* The extended boundaries of the sinus in the edentulous area (E) are seen to appropriate the space *(arrows)* previously occupied by the tooth and the attachment apparatus.

the tan color to the olfactory mucosa. The free surface of the cell bear microvilli, which give the cell a striated appearance.

Olfactory (sensory) cells are fusiform and are vertically oriented. The cell body is oval, and is located in the middle or basal third of this layer (Figs. 9–49 and 9–50). Olfactory cells are fitted between the necks of the supporting cells. Their nuclei are round and dominate the cytoplasm. The distal tapered process, which extends to the free surface, terminates as a bulbous body that bears up to 12 sensory cilia. The other tapered process, at the proximal end of the cell, is extremely long and slender. It forms an axon process for the cell and, along with those of neighboring sensory cells, forms a nerve bundle in the lamina propria. Here it acquires a neurolemma sheath.

Basal cells are polygonal (mostly pyramidal). They are densely packed in the basal third of the olfactory epithelial layer (Figs. 9–49 and 9–50). Their nuclei are large, round, centrally located, and light-staining. The cytoplasm is pale-staining because the organelle population is reduced, indicating cell immaturity and functional incompetence. It has been suggested that the basal cells are replacements for the injured or destroyed supporting cells. The olfactory sensory cells cannot be replaced. It is estimated that

up to 1% of the sensory cell population is destroyed annually.

Basement Membrane and Lamina Propria. This layer differs from that elsewhere in the nasal cavity in three respects: the basement membrane is so thin that it can only be studied by electron microscopy; the connective tissue is very diffuse; and the tissue is dominated by olfactory nerve fibers and glands. The latter, the olfactory glands of Bowman, possess tubuloacinar end-pieces that continuously elaborate and secrete a seromucous product. The excretory ducts of the glands of Bowman open onto the free surface. Sinusoidal veins are numerous. The connective tissue overlying bone forms the periosteum.

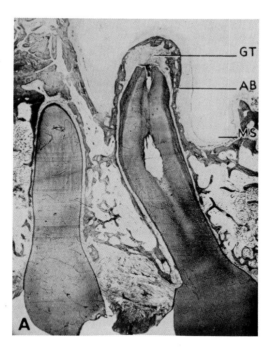

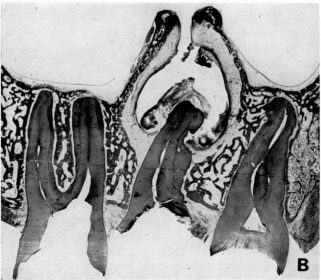

Fig. 9–52. Inflammation at root tip. *A,* Maxillary second molar, in which the pulp is necrotic. Note the relationship of the granulation tissue (GT) to the fundus of the socket and the maxillary sinus (MS) and alveolar bone (AB). The mucosa of the sinus is unaffected at this time. *B,* The infection at the root tip has progressed into the sinus by way of an opening produced in the sinus floor. (Boyle, P.E.: Kronfeld's Histopathology of Teeth and Their Surrounding Structures. 4th Ed. Philadelphia, Lea & Febiger, 1955.)

Paranasal Sinuses

The paranasal sinuses are air cavities located bilaterally in the bones of the face and maxilla. They are named after the bones in which they are located: ethmoid, frontal, maxillary or sphenoid. All communicate with the nasal cavity.

Mucosa

The mucosa of the sinuses, although continuous with that of the nasal cavity, is thinner and less developed. The epithelium of the sinuses is ciliated and pseudostratified. These cells, however, are shorter, and goblet cells are much fewer than in the nasal passage. The cilia direct the secretions toward the minute opening leading to the nasal passage.

The basement membrane is almost imperceptible with light microscopy. The lamina propria is very thin and diffuse. The glands and vascular channels are few, and the specialized veins are absent. As elsewhere, the denser connective tissue is located adjacent to the bone, where it forms a thin delicate periosteum.

Structural Features

A brief structural description of the sinuses is provided here because some sinuses, especially the maxillary, are clinically significant.

The maxillary sinus, or antrum (chamber) of Highmore, is the largest. The maxillary sinuses lie inferior to the floor of the orbits, lateral to the wall of the nasal cavity, and superior to the maxillary alveolar arch. The latter generally extends from the canine through the molar-bearing segment of arches. The sinuses drain superiorly through orifices in the middle meatuses. Enlargement of the sinus chambers, or pneumatization, may continue until the fundus of the tooth socket and the sinus are dangerously close. This is likely to involve the first and second molars. Pneumatization is significant in regard to two clinical conditions, tooth extraction and inflammation.

If the molars displaying the relationship described above require extraction, extreme care must be taken not to puncture the thin bone separating the sinus and the periapical tissue (Fig. 9–51A). In other cases, even where the bony tissue separating the socket and sinus is thick, bone resorption due to disuse atrophy may occur from the maxillary sinus aspect following extraction. Under these circumstances osteoclasia results in an extension of the chamber of the maxillary sinus into the interdental alveolar process, so that only a thin bony plate separates the sinus and the periodontal ligaments of the neighboring teeth (Fig. 9–51B).

Periapical inflammation may be dangerous when the bone separating the sinus and the affected fundal tissue are separated by a thin plate of bone (Fig. 9–52A). The inflammation could produce bone resorption and spread to the sinus (Fig. 9–52B).

Nasopharynx

The nasopharynx is located posterior to the choanae of the nose, and is continued inferiorly with the nasopharynx. Certain aspects of the nasopharynx, such as the pharyngeal tonsils and uvular epithelium, have been discussed previously in this chapter.

The epithelium continuing onto the uvular segment of the nasopharynx from the oral aspect is nonkeratinous. In most other areas, however, it is of the respiratory variety. Two conditions may convert respiratory epithelium to stratified squamous, friction and increased air flow. Mucosal surfaces that rub together will be covered by stratified squamous epithelium. When the air flow is increased, as in the case of one nostril being obstructed by an anatomic defect or by mucosal congestion, the epithelium is stratified. If the friction or air flow is decreased, the epithelium reverts to the respiratory variety.

The basement membrane is usually thick. The connective tissue is compact and contains many elastic fibers, especially in the deeper lamina propria. Great numbers of protective cells, either scattered or aggregated into infiltrations, are present. Typical tubuloalveolar glands with seromucous secretions are abundant. The blood, lymph, and nerve supplies of the nasopharynx are generally richer than those elsewhere in the nasal cavity.

CHAPTER 10

Temporomandibular Joint

The temporomandibular joint, commonly known as the TMJ, is the articulation of the mandible with the opposing surface of the temporal bone. Ends of opposing bones form movable units belonging to a classification of joints called diarthroses. Because of the arched shape of the mandible two articulating units are necessary, and so the term "bilateral diarthrosis" is used. The TMJ is said to be ginglymoid, because the action of the mandible is hinge-like. The ends of the articulating bones are covered by avascular connective tissue, and the joint is stabilized by a collar-like capsule that surrounds the entire movable unit.

ANATOMIC AND FUNCTIONAL FEATURES

The temporomandibular joint differs from other movable joints in that the articulating surfaces are fibrous connective tissue rather than hyaline cartilage, and the body of the articulating bone houses teeth that require specific modes of articulation. Teeth and other components of the attachment apparatus (i.e., periodontal ligament and alveolar process) also form joints, but these are relatively immovable and are designated as synarthroses. The function of the temporomandibular joint is *movement*. Because of the intricate movements that the mandible can make in various planes and directions, activities involved in ingestion, mastication, and speech can be executed.

There are five components to the articulating unit: bone; (temporal and mandibular), articulating disk; synovial membranes; tendons of the lateral (external) pterygoid muscle; and capsule (Fig. 10–1).

The two temporal components are the posterior articular (mandibular or glenoid) fossa and the anterior articular eminence, the lateral aspect of which is called the articular tubercle. The mandibular component is the condyle. The articular disk is positioned between these two bones to form two distinct synovial cavities or joint spaces, superior and inferior. The superior synovial cavity is located between the upper surface of the articular disk and the articulating surface of the temporal bone. The inferior synovial cavity is located on the underside of the articular disk, separating it from the superior surface of the condyle. A capsule envelops the joint and connects the condyle to the inferior surface of the temporal bone. As shown in Figure 10–1, the temporomandibular joint may be conceived as a double joint involving two synovial cavities separated by the articular disk and sharing a common capsule. The superior surface of the articular disk is concavoconvex, and accommodates the depression (fossa) and articular eminence of the squamous temporal component. The disk's inferior surface is contoured to fit the elliptic condyle.

Squamous Temporal

The mandibular (glenoid) fossa is a depression of the squamous portion of the temporal bone. It consists of an anterior and a posterior segment, separated by the petrotympanic fissure. Only the anterior segment of the concavity, into which the condyle fits, is involved in jaw movement.

An elevation known as the articular eminence forms the anterior limit of the temporal component of the TMJ. The anterior extension of the eminence is continuous with the infratemporal segment of the temporal bone. Its lateral extension joins the zygomatic process as the articular tubercle.

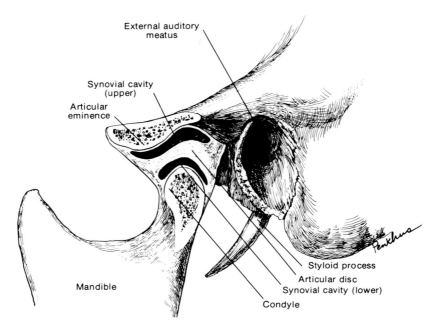

Fig. 10–1. Temporomandibular joint observed in sagittal section. (Clemente, C.D.: Gray's Anatomy of the Human Body. 30th Ed. Philadelphia, Lea & Febiger, 1985.)

Mandibular Condyloid Process

The articulating ends of the mandible are called the condyloid (condylar) processes. They are made up of a head (condyle) and a neck. The long axes of the condyles are oriented in a posteromedial plane, so that lines extended through them will intersect at the anterior margin of the foramen magnum.

The fully formed condyle is from 15 to 20 mm long and from 8 to 10 mm wide. The semi-rounded superior and anterior aspects of the condyle are the articulating surfaces. The anterior slope of the condyle is about half as long (5 mm) as the posterior slope (11 mm). The condyle in the developing mandible is round, and tends to flatten with age. The articulating surfaces merge with the remainder of the condyle—the head with the neck, and the neck with the ramus.

The anterior surface of the neck dips toward the ramus and contains an area (pterygoid fovea) to which the lateral (external) pterygoid muscle attaches. Examination of the condyle from the front reveals that the medial pole projects further medially than the lateral pole does laterally.

Articular Disk

The articular disk separates the condyle from the temporal bone. During movement, the ar-

ticular disk adapts to the shape of the adjoining bones. The disk is divided into three regions or bands—anterior, intermediate, and posterior. The posterior band is thicker than the anterior, and the intermediate band is the thinnest (Fig. 10–2). The thinness of the intermediate band provides for the great adaptability of the disk.

The articular disk communicates directly or indirectly with four structures:

1. The capsule, anteriorly, medially, and laterally
2. The bilaminar area, or the retodiscal pad (retroarticular cushion) posteriorly (Fig. 10–2); the bilaminar area is divided into an upper layer that connects with the temporal bone and a lower layer that connects to the back of the condyloid process
3. The condyle on its lateral and medial poles (Fig. 10–2)
4. Some tendons of the lateral pterygoid muscle anteriorly (Fig. 10–1).

The attachment of the upper layer of the bilaminar area to the temporal bone is loose, thereby providing for the gliding action of the articular disk over the surface of anterior aspect of the glenoid fossa and the articular eminence. The lower layer of the bilaminar area is shorter

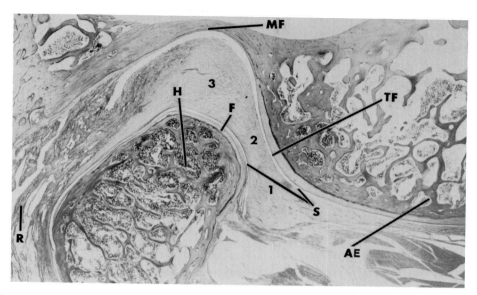

Fig. 10–2. Temporomandibular joint. Photomicrograph of longitudinal section illustrating the mandibular fossa (MF), fibrous lining (F), condyle (H), retrodiscal pad (R), thickened fibrous lining of the posterior slope of the articular eminence (TF), synovial cavities (S), articular eminence (AE), and anterior (1), intermediate (2), and posterior (3) bands of articular disk. (Hematoxylin and eosin stain; × 35) (Courtesy of S.W. Chase, School of Dentistry, Case Western Reserve University.)

because it moves anteriorly with the condyloid process.

Articular Capsule

The girdling structure covering the components of the temporomandibular joint is called the articular capsule, or capsular ligament. Its superior border is attached to the margins of the mandibular fossa and articular eminence. Its inferior border is connected to the neck of the condyloid process (Fig. 10–3). The articular capsule is also joined to the articular disk anteriorly, laterally, and medially.

The capsule is thin, relatively avascular, and very fibrous, except for that aspect that joins and becomes increasingly thickened by the temporomandibular ligament.

Ligaments

The temporomandibular (lateral) ligament is the most prominent of the TMJ. It is a thick, wedge-shaped fibrous band that provides strength and support to the lateral capsular wall. One end of the temporomandibular ligament is inserted into the posterior and lateral margins of the condylar neck; the other end is inserted into the zygomatic process and articular tubercle of the temporal bone. Its fibers are obliquely oriented, coursing posteriorly and inferiorly from its temporal bone attachment.

Two other ligaments, although not integral components of the temporomandibular joint, do influence movements of the mandible. These accessory ligaments are the stylomandibular and sphenomandibular accessory ligaments (Fig. 10–3). The latter, a derivative of Meckel's cartilage, is a squamoid fibrous band extending from the spine of the sphenoid bone to the lingula adjacent to the mandibular foramen. The stylomandibular ligament, a modification of the superficial layer of the deep cervical fascia, courses as a ribbon-like structure from the ramus and posterior border of the mandible to the tip of the styloid process of the temporal bone. Thus, these ligaments connect the mandible to the base of the skull. It is not unusual for them to be structurally more similar to fascia than to ligaments.

Blood and Nerve Supply

The principal blood supply to the temporomandibular joint is via the superficial temporal and maxillary branches of the external carotid artery. Venous drainage occurs by way of the companion veins, which are tributary to the retromandibular vein. Nerve supply is contributed by branches of the auriculotemporal and mas-

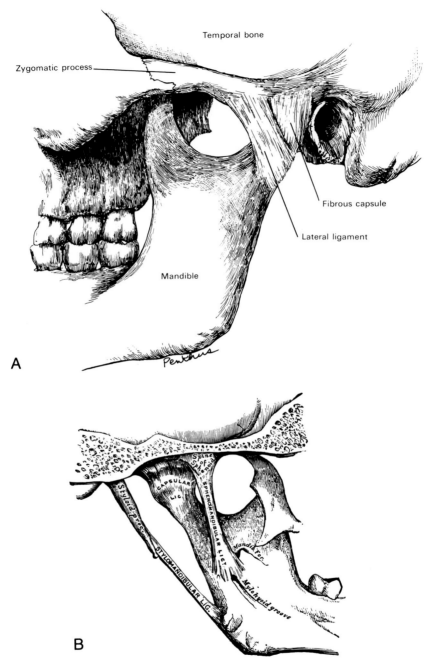

Fig. 10–3. Temporomandibular joint. *A,* Lateral view, showing capsular and lateral ligaments. *B,* Medial anterior view, showing capsular, sphenomandibular, and stylomandibular ligaments.

seteric nerves of the mandibular division of the trigeminal nerve. Vasomotor and sensory endings are present in the connective tissue of the synovial membrane.

HISTOLOGY

The varied composition of the joint components is reflected in the histologic features.

Condyle

The condyle is mainly cancellous bone and, therefore, its central mass is comprised of plates, tubules, and mostly trabeculae separated by spaces occupied by marrow. The spongy core is covered by a thin shell of compact bone. The trabeculae fan out from the more central area and make right angle connections with the endosteal zone of the cortical bone. The marrow spaces, particularly later in development, are filled by yellow or fatty marrow that has replaced the red marrow or myeloid tissue.

In the fully developed condyle, the articulating surface is composed of dense fibrous connective tissue or fibrocartilage. Beneath the fi-brous layer, is found a very thin layer of hyaline cartilage, which communicates with the bone of the condyle (Fig. 10–4).

Articular Fossa and Eminence

The thin bony portion of the articular fossa is composed of compact bone that is continuous with that of the articular eminence. Most of the eminence is composed of bone trabeculae, which tend to thicken with age. Concomitantly, the intertrabecular (marrow) spaces are diminished in size. A thin fibrous tissue layer covers the fossa and eminence uniformly, except on the posterior aspect of the eminence where the tissue forms a much thicker band. The increase in thickness is believed to be a product of increased stress, because it is in this area that the condyle and articular disk slide across the eminence.

Based on the arrangement of the collagen fiber bundles, the thickened fibrous layer over the slope of the eminence can be divided into three layers—inner, outer, and intermediate. In the inner layer the fibers are directed perpendicularly to the cortical surface. The fibers of the outer layer are parallel to the cortical surface,

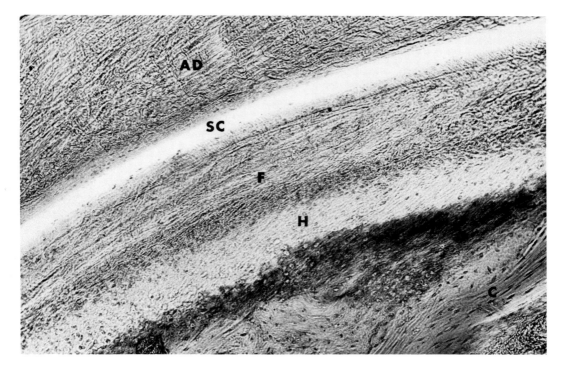

Fig. 10–4. Photomicrograph of section enlarged from Figure 10–2 of the articular surface of the condyle (C). Note the fibrous covering (F), supported by cartilage (H), which rests on bone (AD, articular disk; SC, synovial cavity). (Hematoxylin and eosin stain; × 100) (Courtesy of S.W. Chase, School of Dentistry, Case Western Reserve University.)

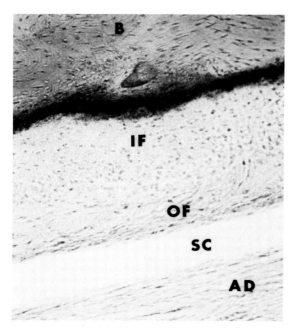

Fig. 10—5. Photomicrograph of section enlarged from Figure 10—2 of the posterior slope of the articular eminence (OF, outer transitional, and IF, inner fibrous, layers; B, bone of articular eminence; SC, synovial cavity; AD, articular disk). (Hematoxylin and eosin stain; ×100) (Courtesy of S.W. Chase, School of Dentistry, Case Western Reserve University.)

and those of the intermediate layer are obliquely oriented (Fig. 10–5). The cellular components of the fibrous bundles are few, consisting mainly of fibroblasts, chondrocytes, and mesenchymal cells.

Articular Disk

Early in the development of the temporomandibular joint there is a period during which the disk (meniscus) is composed of fibrocartilage. As maturation progresses, the fibrocartilage is replaced by dense fibrous connective tissue. The cells of the disk are mostly fibroblasts. The principal fiber is collagen; elastic fibers occur only in small amounts.

The collagen fiber bundles in the intermediate region of the disk are aligned in a parallel fashion in an anteroposterior plane. The fiber bundles of the anterior and posterior segments of the disk, however, do not exhibit a preferred orientation.

Because the fiber population is abundant and the cell population is correspondingly minimal in the articular disk, rich vascular and nerve supplies are not required. These features correlate well with the ability of the disk to withstand great pressures.

Bilaminar Region

The bilaminar region is composed of diffuse connective tissue. Of special interest is the presence of a vascular network, the pseudocavernous plexus or neurovascular zone. With forward movement of the mandible, the vessels of the plexus expand and become filled with blood. The engorged blood vessels assist in filling the space formed by the forward excursion of the mandible. As the mandible moves back or retracts, the blood vessels are compressed and emptied.

The superior layer of the bilaminar region contains numerous elastic fibers that take the form of fenestrated membranes. In the inferior layer the elastic fibers are absent, while collagen components and blood and lymph channels are numerous. The inferior layer of the bilaminar tissue participates in the formation of the lining of the inferior (lower) synovial cavity. Here the inferior stratum may become ruffled, forming villi that project into the synovial cavity.

Synovial Cavities, Membranes, and Fluids

The upper synovial cavity is located between the superior aspect of the articular disk and the fossa of the squamous temporal. The cavity extends from the articular eminence back to the posterior margin of the fossa. The lower synovial cavity is positioned between the condyle and the inferior surface of the disk, and is semicircular in profile (Figs. 10–1, 10–2, and 10–3).

The synovial cavity is lined by a thin tissue sheet, the synovial membrane (synovium). This is bilaminar, consisting of a discontinuous intimal layer of synoviocytes that overlie a loose subintimal layer of areolar tissue rich in vascular supply, containing numerous capillaries (Fig. 10–6). The layer of synoviocytes interfacing the cavity is comprised of two cell types: F (fibroblast-like) cells, also known as A cells, and M (macrophage- or histiocyte-like) cells, also known as B cells. The M cells are astral-shaped and have numerous fine filamentous processes, the filopodia. Electron microscopy shows the cytoplasm to be moderately dense, housing extensive Golgi complexes, mitochondria, and lysosome populations. The F cells do not possess

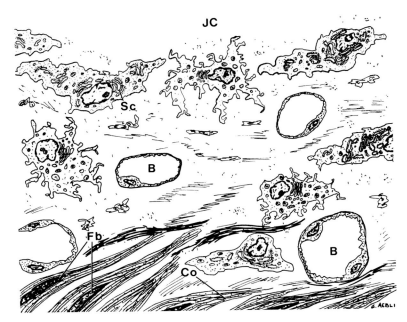

Fig. 10–6. Diagrammatic representation of the synovial (joint) cavity (JC) lined by a layer of synoviocytes (Sc). The loose connective tissue layer immediately subjacent to the synoviocytes is known as the subintimal layer, and it is a highly vascular, loose, areolar connective tissue. The deeper connective tissue is densely fibrous, with fibrocytes (Fb) that maintain the collagen bundles (Co) (B, blood vessels).

filamentous processes. The cell projections, when present, are few, short, and blunt. Additionally, F cells differ from M cells in that the rough endoplasmic reticulum is well developed and the cytoplasm appears to be correspondingly more electron-dense. Because cytologic intermediates are found in the synovium, some investigators believe that A and B cells are functional variants of a single cell. A basement membrane separating the synoviocytes and underlying connective tissue is lacking, and it is not uncommon to find these cells and/or their processes located in the connective tissue.

In areas not exposed to pressure, the synovial membrane is thrown into numerous folds, or villi. The villi are larger and more conspicuous in the inferior synovial cavity.

The joint cavity is filled by a lubricating medium, the synovial fluid, a product of the F type of synoviocytes. This is a viscous, transparent, yellowish liquid. The viscosity of the fluid is believed to be a result of mucin produced by certain cells of the synovial membrane. The fluid is a dialysate of lymph and plasma, consisting of a protein-mucopolysaccharide complex. It is not completely acellular; rather, it contains leukocytes, mostly lymphocytes and monocytes. De-

bris and other foreign materials are removed from the lubricating medium by phagocytic action of cells of the synovium and leukocytes in the fluid.

DEVELOPMENT OF THE MANDIBLE

The development of most oral and paraoral structures has been discussed in Chapter 3. Because of the structural and clinical relationships of the mandible to the temporomandibular joint, the development of the lower jaw has been reserved for this chapter.

Body and Ramus

The intramembranous development of the bony spicules that presage the formation of the mandible is initiated about 6 weeks after fertilization of the egg (see Figs. 4–1 and 4–8). Fusion of the arcs of Meckel's cartilage at the midline of the mandibular arch occurs by the seventh week of development. Although Meckel's cartilage does not participate directly in the development of the body of the mandible, it does dictate the path of growth during development. Meckel's cartilage contributes to the incus and maleus, which are bones of the inner ear. Furthermore, remnants of Meckel's cartilage form the sphenomandibular ligament.

All parts of the mandible (body, ramus, and condyloid and coronoid processes) develop by intramembranous ossification. Only the articulating surface of the condyle and tip of the coronoid process develop by endochondral ossification. The early bone spicules arise spontaneously at multiple sites in a mesenchymal environment. They grow in all directions and fuse with neighboring spicules to form a network (see Fig. 2–45). This osteogenic process succeeds in increasing the length, width, and height of the mandibular arch in its distal growth in a cephalad direction. Elongation, widening, and heightening of the mandible is rapid. By the time the embryo is 50 mm long (tenth week), a triangular cartilage segment called the accessory or secondary cartilage develops immediately posterior to the end of the body of the mandible. A similar cuneiform accessory cartilage is formed for the coronoid process several weeks later. The accessory cartilages are destined to develop endochondrally into the articulating surface of the condyle and tip of the coronoid process (Fig. 10–7). The secondary cartilage of the condyle forms a growth plate for endochondral osteogenesis, which succeeds in lengthening the mandible. The growth plate is spent and replaced by bone by the age of 20 years.

Temporomandibular Articulation

The development of the body of the mandible begins at 6 weeks, and the components of the squamous temporal (articular fossa and eminence) develop a month later.

During the tenth week the accessory cartilage, which is the model for the condylar tip, is produced. As the cartilage model grows and is replaced by bone, the tip of the wedge facing the developing mandible elongates. Within 2 weeks the bone of the model and that of the growing ramus segment of the mandible meet and fuse (Fig. 10–7). Thus, histodifferentiation and organization of the components of the TMJ occur between the tenth and twelfth weeks. By the end of the sixteenth week, the temporomandibular unit acquires definite form. By the eighteenth week the beak-shaped coronoid process, which is the other projection at this end of the mandible, develops intramembranously and fuses with its accessory cartilage (Fig. 10–8A).

Except on the prospective articulating surface, the condyle is composed of cartilage that is eventually replaced by bone. During bone replacement of cartilage, the various zones associated with endochondral osteogenesis are present and active. The connective tissue between the developing condyle and temporal bone becomes organized into primitive synovial membranes, synovial cavities, articular disk, and bilaminar area (Fig. 10–8). The cartilage of the condylar tip never communicates with the inferior synovial cavity; rather, it is separated by fibrous connective tissue (Fig. 10–8B and C).

The articular eminence and fossa begin to as-

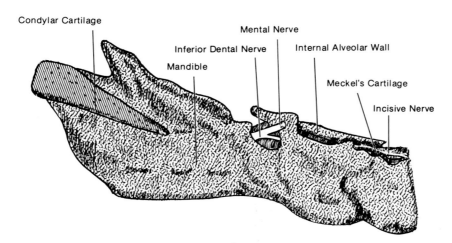

Fig. 10–7. Diagram of lateral aspect of developing mandible of embryo 55 mm long (about 11 weeks). The accessory condylar cartilage has fused with the heightening body of the mandible. Bone is being formed around the nerves and blood vessels so that they come to lie in the mandibular canal. (Modified from Fawcett, E.: The Growth of the Jaws, Normal and Abnormal, in Health and Disease. London, Dental Board of the United Kingdom, 1945.)

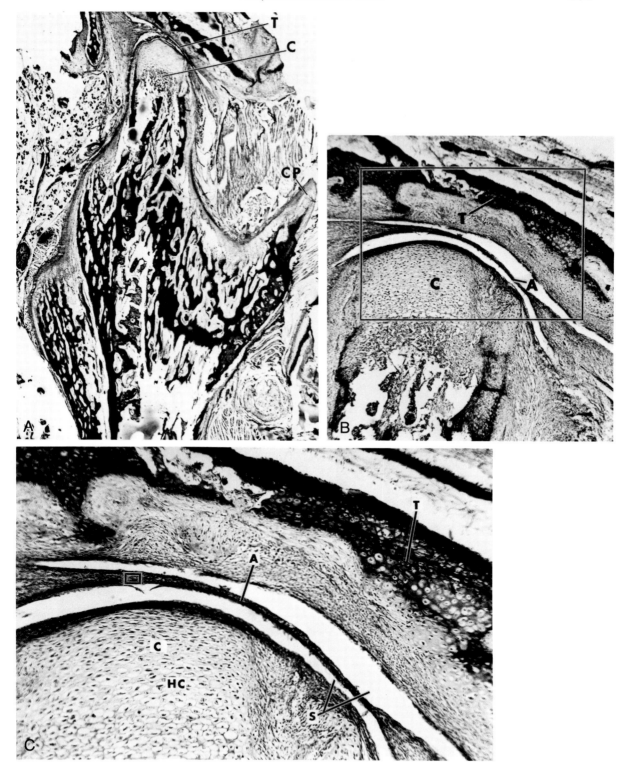

Fig. 10–8. Photomicrographs of temporomandibular articulation in longitudinal section taken from an embryo of 22 weeks. *A,* Union of the condyle (C) and the coronoid process (CP) is shown (T, temporomandibular joint area). *B,* Developing temporomandibular joint of *A* enlarged to illustrate the temporal bone (T), articular disk (A), and condyle (C). *C,* Further enlargement of *B,* showing condyle (C), articular disk (A), temporal bone (T), synovial cavities (S), and hyaline cartilage (HC). (Hematoxylin and eosin stain; *A,* ×35; *B,* ×100; *C,* ×430)

sume definitive form only after birth. In early infancy the fossa is shallow and the eminence is short. During early childhood the fossa deepens and the eminence elongates The period of most rapid growth occurs between the ages of 10 and 11 years. Shortly thereafter the TMJ completes its development. Thus, the bones of the jaws and skull are among the first to begin and among the last to complete development.

Joint Movements

Three movements characterize the temporomandibular articulation: hinge and glide (opening and closing); retrusive and protrusive (sliding back and forth); and lateral (sideward) movements. The hinge movements occur in the inferior cavity and the gliding ones in the superior cavity. These movements may be executed either jointly or independently. For example, in opening the mouth and biting food, both hinge and gliding actions are involved. The movements occur in both cavities, either in or out of phase; however, they must be harmonious (Fig. 10–9). The interrelationships of the components of the temporomandibular apparatus during the open and closed phases of the mouth are shown in Figure 10–10.

The range of movement of the condyle is about 15 mm. Because the distance through which the condyle can travel over the articular disk is only about 7 mm, the movement of the articular disk across the articular fossa is 8 mm.

Muscles Contributing To Mandibular Movement

The movements of the mandible are a result of the action of the jaw and cervical muscles

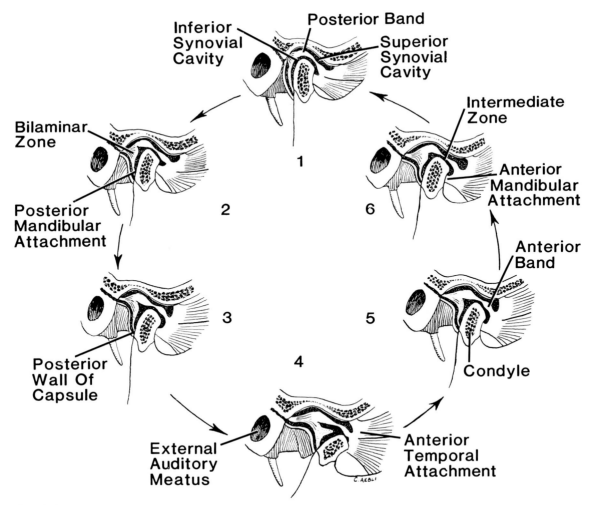

Fig. 10–9. Movement of the condyle and the disk during opening and closing of the mouth. (Redrawn from Rees, L.A.: The structure and function of the temporomandibular joint. Br. Dent. J., *96*:125, 1954.)

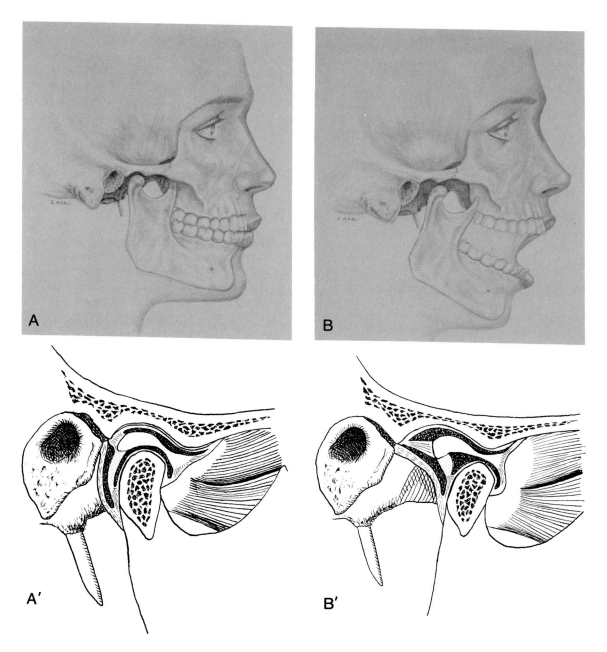

Fig. 10–10. Diagrammatic representation of the interrelationships of the components of the temporomandibular apparatus during the open *(B,B')* and closed *(A,A')* position of the oral cavity. *A'* and *B'* represent sagittal sections through the temporomandibular joint. (Adapted and reproduced by permission from Ermshar, C.B., Jr.: Anatomy and neuroanatomy. *In* Diseases of the Temporomandibular Apparatus. 2nd Ed. Edited by D.H. Morgan, L.R. House, W.P. Hall, and S.J. Vamvas. St. Louis, 1982, The C.V. Mosby Co.)

(Table 10–1). The cervical muscles stabilize the head to increase the efficiency of mandibular movements.

The muscles that assist in executing the movements of the jaw are grouped as elevators, depressors, or protractors. The muscles of mastication that elevate the mandible close the mouth. Other muscles depress the mandible, resulting in opening of the mouth. The action of still other muscles, the protractors, result in forward and backward sliding movements.

The elevator muscles include the masseter, temporalis, and medial (internal) pterygoid muscles. The depressors are the lateral pterygoid, digastric, geniohyoid, and mylohyoid muscles. The lateral pterygoid muscle exerts a protracting action on the mandible. Normal movements of the mandible require harmonious or coordinated interactions of various muscle groups. Lack of coordination results in temporomandibular dysfunction (see below).

CLINICAL CONSIDERATIONS

Temporomandibular disorders can be classified into those that involve pathology of the articulating structures (TMJ problems), and those that involve the muscles of mastication. Masticatory muscle spasm is a major cause of mandibular dysfunction. This condition, known as myofascial pain-dysfunction (MPD) syndrome, is generally considered to be a stress-related psychophysiologic disease, although muscle incoordination and spasm can occasionally be induced by occlusal dysharmonies, mandibular overclosure, or overextension of the masticatory muscles. In many patients persistent clenching or grinding of the teeth (bruxism) may be a contributing factor. Patients with MPD syndrome will complain of dull, aching, radiating pain on the side of the face in the temporal and occipital regions and sometimes extending into the neck. They may also complain of earache. Mandibular movement is limited, the pain is exacerbated by function, and there may be clicking or popping joint sounds. Long-standing MPD syndrome may also lead to secondary degenerative changes in the temporomandibular joint.

The temporomandibular joint is subject to the same pathologic conditions as those involving other joints of the body. The most common pathologic condition is degenerative arthritis (osteoarthritis). This is related to trauma or excessive stress on the articular system. The condition is characterized by degenerative changes in the articular surfaces and, in late stages, by involvement of the articular disk. It is generally associated with pain and some limitation of joint mobility.

About 17% of patients with rheumatoid arthritis will also have involvement of the temporomandibular joint. The inflammation associated with this disease induces the formation of granulation tissue (pannus), which spreads over the articulating surfaces and causes their destruction. As repair occurs, the granulation tissue is replaced by dense fibrous tissue, and even by bone, resulting in limited mandibular movement, or ankylosis. Infectious (pyogenic) arthritis may also cause degeneration of the articular surfaces and lead to ankylosis.

The temporomandibular joint may also be affected by congenital and developmental anomalies, traumatic injuries (e.g., fractures, dislo-

Table 10–1. Muscles of Mandibular Movements

	Muscles of Mastication				Suprahyoid Muscles*		
Function	Masseter	Temporalis	Medial Pterygoid	Lateral Pterygoid	Digastric	Mylohyoid	Geniohyoid
Opening†	0	0	0	+ + (IH)§	+	+	+
Closing	+ +	+ + (Af, Mf, Pf)‡	+ +	+ + (SH)§	+	+	+
Protrusion	+ +	0	+ +	+ +	+	+	+
Retrusion	+	+ + (Mf, Pf)‡	+	0	+	+	+
Lateral excursion	+	+ + (Pf)‡	+	+ +	+	+	+

*The suprahyoid muscles are the digastric, mylohyoid, geniohyoid, and stylohyoid. The relationship of the stylohyoid to the temporomandibular articulation, however, is exclusively one of stabilizing the hyoid bone.
†Action intensity: 0, none; +, minimum; + +, maximum.
‡Af, anterior fibers; MF, middle fibers; PF, posterior fibers.
§IH, inferior head; SH, superior head.

cation), and various forms of benign and malignant neoplasia. The articulating disk can also become displaced, causing clicking or popping sounds or locking and limitation of mandibular movement. These conditions are referred to as internal derangements of the temporomandibular joint.

The accurate diagnosis of temporomandibular joint disease and mandibular dysfunction requires taking a comprehensive dental and medical history from the patient. Information regarding the onset and duration of the symptoms, as well as the presence of systemic diseases, traumatic injuries, bruxism, and emotional stress, should be elicited. Inquiry about factors that may trigger the symptoms is also helpful in establishing a diagnosis. Differential diagnosis is sometimes difficult because both TMJ disorders and MPD syndrome can produce dull, aching, radiating pain; both can cause difficulty in opening the mouth and in eating; and both can lead to clicking or popping sounds in the joint. However, the degree of masticatory muscle pain and tenderness is usually greater with MPD syndrome.

Following evaluation of the patient's dental and medical histories, an extensive clinical examination should be made. This should include examination for dental pathology and evaluation of the occlusion, as well as careful examination and palpation of the temporomandibular joints and masticatory and cervical muscles. Accurate, reproducible radiographs of the temporomandibular joints should also be taken and, occasionally, computerized tomograms (CT scans) and radionuclide studies are helpful.

Treatment will, of course, depend on the condition being accommodated. Degenerative joint disease is usually managed by the use of nonsteroidal anti-inflammatory drugs, limiting use of the jaw, and reducing any physical stress on the joint by establishing a well-balanced functional occlusion. Rheumatoid arthritis is generally treated medically, unless severe restriction of joint movement necessitates surgery. MPD syndrome is managed with pharmacologic agents to relax the muscles of mastication, mild analgesics, soft diet, application of moist heat, and use of occlusal appliances (bite plates). In some patients, physical therapy and/or techniques such as conditioned relaxation or biofeedback are helpful in reducing muscle tension. The techniques employed to treat MPD syndrome should be accompanied by a thorough explanation of the problem to the patient, specifically dealing with possible relationships to psychological stress and how to deal with these factors.

Index

Page numbers in *italics* refer to illustrations; numbers followed by "t" refer to tables.